Hazards in a Fickle Environment: Bangladesh

Advances in Natural and Technological Hazards Research

VOLUME 10

The titles published in this series are listed at the end of this volume.

Hazards in a Fickle Environment: Bangladesh

by

C. EMDAD HAQUE

Department of Geography,
Brandon University,
Manitoba, Canada

KLUWER ACADEMIC PUBLISHERS

DORDRECHT / BOSTON / LONDON

A C.I.P. Catalogue record for this book is available from the Library of Congress.

ISBN 0-7923-4869-9

Published by Kluwer Academic Publishers,
P.O. Box 17, 3300 AA Dordrecht, The Netherlands.

Sold and distributed in the U.S.A. and Canada
by Kluwer Academic Publishers,
101 Philip Drive, Norwell, MA 02061, U.S.A.

In all other countries, sold and distributed
by Kluwer Academic Publishers Group,
P.O. Box 322, 3300 AH Dordrecht, The Netherlands.

Printed on acid-free paper

To my family
(Molly, Avanti and Neil)
and the people of Kazipur

CONTENTS

PREFACE

This book evolved from a collaborative research project between the University of Manitoba, Canada and Jahangirnagar University, Bangladesh, which commenced in 1984 to study the problems of river channel migration, rural population displacement and land relocation in Bangladesh. The study was sponsored by the International Development Research Center (IDRC), based in Ottawa, Canada. It was through this project that I started my journey into disaster research more than thirteen years ago with basically an applied problem of massive magnitude in Bangladesh. I spent two- and-a half-years, in two stages, in Bangladesh's riparian villages to collect the empirical data for this study. Then the growing disaster discourse throughout the 1980s, especially its conceptual and theoretical areas, drew me in further, gluing my interest to these issues. In the 1990s, during my research and teaching at Brandon University, Canada, I realized that, despite the large body of literature on natural disasters, there was no work that synthesized the approaches to nature-triggered disasters in a comprehensive form, with sufficient empirical substantiation. In addition, despite the great deal of attention given to disasters in Bangladesh, I found no detailed reference book on the topic.

Natural hazards and disasters, in my view, should be studied under a holistic framework encompassing the natural environment, society and individuals. Overreaction to the limitations of technocratic-scientific approaches—the control and prevention of physical events through specialized knowledge and skills—has resulted in a call for "taking the naturalness out of natural disasters." Similarly, the behavioral-ecological stress in social studies on individual characteristics has attempted to redirect attention to social structures. There is little doubt that overspecialization in geophysical fields, the recipe of radical political restructuring, and outright rejection of the individual's power to think and act, have inhibited the needed acceptance of disaster research from an integrative perspective and have restricted the development of theory.

In this volume, I outline a holistic framework to analyze the nature-society relationship and its connection to disasters. I then apply this approach to the detailed examination of case study involving floods and riverbank erosion in the riparian environments of Bangladesh. In the concluding chapter, the possible application of the framework to a wider range of environmental hazards is explored. Primarily, the framework illustrates that the human sphere is an integral part of the natural environment while also a differentiated entity—just as social laws are distinct from natural laws. Natural disasters thus occur as interfaces of both natural and human realms.

The book is a timely contribution, arriving at a time when the world has passed through three-quarters of the International Decade on Natural Disaster Reduction (IDNDR). I trust that it will help establish the critical importance of understanding natural disasters in the context of the complex human relationship with the natural environment. In this way I wish to reassert the significance of a unifying approach to disasters studies, one that stresses multiple traditional disciplinary areas (geophysical, biological, social, economic, political, cultural) be assessed, compared, and integrated. From that we can go on to determine how to improve human coping responses to environmental hazards and disasters.

While this book is truly the culmination of a team effort—supported by the IDRC, the Social Science and Humanities Research Council, and the Brandon University Research Committee—its present form and contents are the responsibility of the author.

ACKNOWLEDGEMENTS

This book has its roots in a research project—the Riverbank Erosion Impact Study (REIS)—that was carried out between 1984 and 1989. Although that team effort ended eight years ago, I have continued to pursue my interest in the topic by following the emerging significance of environmental threats and hazards in the present-day world. The series editor, Mohammed El-Sabh, who envisioned the inclusion in this series of a book on Bangladesh's riverine hazards, encouraged me to modify my doctoral work to fill this gap. Petra van Steenbergen of Kluwer Academic Publishers has patiently guided me through the entire process, in terms of format and substance. I am deeply indebted to both of them.

As this study has evolved over the last fourteen years, it would be impossible to compile an exhaustive list of individuals, agencies, and institutions that have encouraged and assisted me along the way. I am most grateful to the people of Kazipur, who not only responded to the questionnaires, but also contributed in various other capacities. My gratitude is extended especially to the following REIS team members who provided insights on specific issues: John Rogge, Raymond Wiest, Larry Stene, K. Moudood Elahi, Matiur Rahman, Ansari Khan, Mohammed Zaman, Ziarat Hossain and Charles Greenberg. Funding for the 1985 survey was received from the International Development Research Centre (IDRC) and for the 1995 survey from the Social Science and Humanities Research Council (SSHRC) and the Brandon University Research Committee (BURC). I am deeply appreciative of these supports.

Without my release from teaching duties through a sabbatical leave, this book would not have been realized. I am thankful to Roger Smith, the Dean of Sciences, and my colleagues in the Department of Geography for providing me this opportunity. During the writing of this book, the Institute of Development Studies, Refugee Studies Program at the University of Oxford rewarded me with a Visiting Research Fellowship. My special thanks go to David Turton and Barbara Harrell-Bond for their assistance and hospitality in Oxford.

David Jones and Marjorie Halarson prepared the maps and diagrams. Glenn Bergen generously gave his time to read and edit the manuscript. Christine Sharpe prepared the typescript with great patience. To all, I express my sincere appreciation and thanks. My final thanks go to my family—wife Molly, daughter Avanti and son Neil—for quietly enduring my detachment from them during the progress of the project.

The publisher and I thank the following organizations and individuals for permission to reproduce or redraw tables and figures used in this book:

Blackwell Publishers (Figure 7.2; Table 7.4); Elsevier Science (Table 4.6; Figure 4.8); Irrigation Support Project for Asia and the Near East Table 4.10; Figure 4.12); M.A.O.A. Khan (Figure 4.5; Table 4.9; Figure 4.13); Plenum Publishing Corp (Figure 6.1); Society for Sedimentary Geology (Figure 4.11); World Bank (Table 4.3; Figure 10.3).

C. Emdad Haque
Brandon, August 1997

FOREWORD

Natural disasters are catastrophes at the interface between physical events and a vulnerable human population. Ever since the beginnings of recorded history, humankind has been faced with a range of these hazards, slow-acting and infrequent in some areas, intense and catastrophic in others. In today's world, the majority of the enormous economic losses due to natural disasters are incurred in industrially developed parts of the world, but the relative impact is much greater on countries with lower per capita incomes. Here the effects on people and the economy continue to be felt for many years after the event.

Bangladesh, probably the nation most affected by natural catastrophes, is literally trapped by water. Lying just above sea level, the country, with a population of around 120 million people over a surface area of only 144,000 square kilometers, is threatened not only by river flooding, but also by cyclones and storm surges from the Bay of Bengal. With the frequent occurrence of massive floods, the number of fatalities (1970: 300,000; 1991: 140,000) and the number of people affected (many millions in each case) reach catastrophic proportions.

These disasters combine two elements: events and vulnerable people. A disaster occurs when an extreme event exposes the vulnerability of individuals and communities in such a way that either their lives are directly threatened or sufficient harm is done to the economic and social structures of their communities. People therefore suffer as a result of both the disaster and their own unpreparedness; but it is also people who will finally solve their own problems, preparing for and mitigating the impact of disasters. Developing a social perspective on disaster mitigation and preparedness is thus vital. It is to this end that *Hazards in a Fickle Environment* advances the profound significance of the societal and human geographical dimensions of natural disasters. Based on first-hand research on riverine hazards in Bangladesh, this book asserts that disaster impacts in Bangladesh, as in most other developing nations, are caused by social, economic, political, and cultural factors as much as they are by the natural environment.

As the book argues, natural disasters should be studied under a holistic framework encompassing the natural environment, society and individuals. In *Hazards in a Fickle Environment* emphasis is placed on social vulnerability to disasters, and the need for an adaptive, sustainable approach to floodplain development. I am sure that readers will find the publication of this book a timely contribution to the **International Decade for Natural Disaster Reduction**. To connect social and cultural development with ecological and environmental policies could produce incalculable benefits for suffering people. This is an initiative which all of us would do well to support.

Mohammed El-Sabh
Editor-in-Chief

26 September, 1997

Advances in Natural and Technological Hazard Series

Part One

Natural Hazards and Human Perspectives

CHAPTER 1

HAZARDOUS ENVIRONMENT AND DISASTROUS IMPACT

The Challenge of Understanding and Responding

The human struggle to deal with natural events is an enduring phenomenon. The stories of Bangladesh's people, land and water together echo this persistent struggle through the arduous landscapes of plains and rivers changing in every season like the country's tempestuous history. Bangladesh and its floodplains, which nourished India's great civilizations, have endured the shock of frequent floods, cyclones, and famines, withstood hard poverty, Mughol and British conquest, domination by Pakistani military rulers and civil travail to be born as an independent nation. Progress and decline, hope and despair—all play out across centuries here.

Today in Bangladesh more than 120 million people face a continual challenge from composite ecological and socioeconomic predicaments: swollen rivers during the monsoon, frequent tropical high winds and tidal surges, dense rural settlements and crowded cities, limited natural resources and industrialization, inequality and extreme poverty, and unstable political regimes. But rich in cultural heritage and spiritual attributes, blessed with a hardworking populace and social cohesion, Bangladesh's people historically have faced the tests from the environment with resilience, courage and optimism. Often, the nation pays a high toll when struck by powerful environmental events, but the people of this land never give up hope.

I have not set out to just offer an empirical narrative nor to merely compile the dramatic geophysical events relating to Bangladesh's rivers and land. Instead, my intention is to explore how the natural environment through the works of the rivers has molded the livelihood and imposed persistent risks on the floodplain inhabitants of Bangladesh. Also of interest to me is how the Bengali society at large and its structures are related to the prevailing hazardous conditions in the country.

We will approach the nature-society interface from a holistic theoretical perspective, integrating the geophysical and social realms and drawing upon the case of riverine disasters in the Brahmaputra-Jamuna floodplain of Bangladesh. How then, we will ask, are individuals in the floodplains, especially the farmers and peasants, coping with the natural hazards? What is the nature of collective coping responses to address riverine hazards and disasters? What are the factors affecting these responses? What kind of institutional and public policies are needed to ameliorate losses and damages from future riverine disasters? Furthermore, how can a sustainable development approach to floodplain planning and management be formulated so that human and economic development can continue, but without adding any considerable risk and potential damage in the future? It is these questions that this book will address.

The remainder of this introduction is divided into six sections, emphasizing the conceptual problems posed by the established approaches, the nature of environmental threats and disasters in Bangladesh, the features of the 'Nature And People Process' (NAPP) framework to analyze environmental hazards, the conceptualization of the hazards problem of Bangladesh, and the overall scope of the study.

1.1. Explanations in Hazards Research: The Study

The play of *nature*[1] creates both order and disorder which manifest themselves as events. Such events and their parameters are governed by complex geophysical and biological processes. Since the beginning of humanity, certain types of natural events, such as volcanic eruptions, earthquakes, cyclones, tidal surges, floods, and forest fires, have often imposed threats, losses and their consequent sufferings upon us. However, the nature of threats from such natural hazards have never been *constant*, rather they have varied according to changing physical and social determinants. The *natural environment*,[2] what the Greeks once referred to as *physis* or the natural order (Bell, 1975), is "neutral" (Burton *et al.*, 1993); it is its human intersects that create *risks* and *hazards* in our context. Human societal and behavioral characteristics, including location and site for living and work, use of matter and energy, progress in technology and communication, and the type of values and culture, along with many other variables contribute to how people deal with risks and threats in the environment. Based on either experience or ingenuity, human cultural elements adapt to natural environmental *variability*[3] over time and space. These processes generate a *band* or threshold of coping ability with extreme environmental events. As long as the environmental variation remains within or close to this threshold, only insignificant damage will take place. When the variability surpasses the coping threshold, that is, goes "beyond the normal band of tolerance" (Smith, 1992: 10), the event may become hazardous, posing threats to lives, properties, resources, and the prevailing psychological state.

In this book I intend to argue that most of the existing approaches to natural hazards studies, those which commonly tend to artificially separate the geophysical/natural and societal spheres provide inadequate explanations. Despite philosophical and epistemological differences between physical sciences and social studies, there are still common areas where an integrative approach is required. A comprehensive understanding of natural hazards and disasters, for instance, solely from a physical or a social-scientific perspective is not possible; in reality these events are the function of an intertwined system of geophysical and social forces.

The segregation of the natural and social spheres—a practice that has profoundly affected research on various aspects of the nature-society interface, such as resources and hazards—has a historical root. Over the past few centuries western institutions associated with knowledge and education in particular have promoted what Kenneth Hewitt calls "the enclosure system" of studying natural calamities (Hewitt, 1983). Defining disasters by the type of natural event has evolved along the line of academic disciplines in a way that "the problem is made manageable by an extreme narrowing of the range of interpretation and acceptable evidence" (Hewitt, 1983: 10). The convenience of treating disasters as a 'special' problem for advancing research in the sciences, particularly in geophysical fields, has outweighed other considerations.

This challenge of addressing risks in the environment has originated from a moral concern: from a technocratic-scientific standpoint, disasters themselves presuppose clear "limits to knowledge and power." There is a strong belief and conviction that this challenge can be taken up by advancing geophysical, geotechnical and managerial capabilities; as such, the real-world problems associated with calamities can be dealt with effectively by society. These views and associated practices have created a dominant stream of thought concerning natural disasters and human responses that strongly believes in technological solutions. This has resulted in a close alliance between established technocratic notions and the prevailing institutions. Yet as Hewitt notes, it appears that "the natural-science-technological fix approach to hazards is itself, essentially, a *sociocultural construct* reflecting a distinct, institution-centred and ethnocentric view of man and nature" (Hewitt, 1983: 8).

Throughout the past three decades many critics have depicted the failure of established technocratic views—which encourage a progressive reductionism—to consider the social dimensions of disasters (e.g., O'Keefe *et al.*, 1976; Hewitt, 1983; Varley, 1994; Cannon, 1994). These contributors, mostly from an array of social sciences, have pursued a distinction between *hazard* (denoting the potential for damage that exists only in the presence of human elements) and *disaster* (referring to an event associated with the impact of a natural hazard). They have suggested that the latter leads to increased mortality, illness and injury, destroys resources and disrupts livelihood when the geophysical characteristics exceed the coping threshold of a society. The so-called critical, alternative views, mostly from social studies such as human geography, social anthropology, and sociology, have attempted "above all, to show that although floods or earthquakes are natural processes, the disasters are not" (Varley, 1994: 1).

The argument for analyzing natural disasters has shifted the attention from geophysical processes to social processes—to a focus on human *vulnerability* rather than natural events. One core element in the critique of technocratic approaches reminds us that although technological devices have been continuously unravelling natural mysteries, global human and material losses have increased steadily during the 20th century. This remarkable progress in science and technology has affected different spheres of life and many domains of the natural universe, yet with only limited success in preventing natural events from adversely affecting humanity. Recent analyses show that although there has been no major increase in the frequency of extremes of geophysical events in the natural domain (O'Keefe *et al.*, 1976), the human and material losses continue to rise. As such, our increased susceptibility to hazards in recent decades has now emerged as a major academic and policy issue. This concern is reflected in the declaration of the 1990s as the United Nations International Decade for Natural Disaster Reduction.

The *vulnerability analysis* of calamities crystallizes the arguments of those who attempt to shed light only upon the social dimensions, and by this seek to "take the naturalness out of natural disasters" (O'Keefe *et al.*, 1976). The vulnerability approach in general is grounded, first, on the fact that "loss in property from natural hazards [disasters] is rising in most regions of the earth, and high loss of life is continuing or increasing among the poor nations of the world" (Burton *et al.*, 1993: 1). Dworkin (1974) and Swiss Reinsurance (cited in Degg, 1992) observe that, since World War II, while more than three-fourths of the total global economic loss from disasters has occurred in the industrialized-developed world, the ratio of loss to income has been much higher in the least developed or developing nations. In addition, the lion-share of disaster-related deaths has occurred in the latter group. The second basis of the vulnerability approach is that geophysical events of similar magnitude and intensity can have significantly different results in socioeconomic contexts as different as developed-rich and poor, non-Western, regions. The task of vulnerability analysis is therefore to establish that the degree of damage and loss is a function of the human context as much as the geophysical hazard itself (Cuny, 1983).

As some recent literature has pointed out, the social construction of disaster based in political economy, although successful in challenging established views, "has failed to make significant inroads into the dominant paradigm in disaster management" (Varley, 1994: 2). The social-focused approach has had little influence on practitioners, regulators, and educators, since: (1) a number of the contributors were arguing from "an avowedly Marxist perspective" (Copans, 1983; Watts, 1983) and thus often from a dogmatic stance; (2) the suggested interventions to reduce vulnerability were political in nature, and thereby difficult to accept by many quarters; (3) the development planning in question "must be, broadly speaking, socialist" (Susman *et al.*, 1983: 280); and (4) an overemphasis on social spheres undermined the needed balanced treatment of natural processes and causes of hazards. Overall, the lack of a wider acceptance of the vulnerability approach points to the political and ideological differences between its proponents and the established institutions.

In my view, the failure of the so-called alternative social theories of disasters to make an appealing influence are embodied in their conceptual and epistemological shortcomings. By separating nature, implicitly or explicitly, from the social sphere, the social-focused views undermine the relevance of geophysical processes. As such, the vulnerability approach offers a narrow, partial analysis, failing to encompass the entirety of the disaster process. Indeed, the approach must be credited for its persuasive explanation of the social aspects of disasters, but it gives a fragmentary perspective of natural disasters as a whole.

I suggest that the nature and people relationship may not be comprehended under a dichotomized structure since these domains are not so easily reducible or separable in all areas. The issue is a complex one. In some areas, the natural and human domains are distinct, governed by completely different sets of principles and forces. In other areas, however, nature and people belong to an inseparable entity. For instance, *homo sapiens sapiens* cannot be removed from nature in terms of biological laws. Without a holistic framework, which is capable of treating the distinctiveness and the wholeness of natural and social processes concurrently, an acceptable explanation of disasters may not be found.

A careful review of the so-called alternative views of natural calamities reveals that by focusing primarily on macro-level determinants, these approaches in turn reduce human individuals solely to subordinated elements of macro structures. Further, they fail to underscore people's abilities for autonomous decision-making. In a broader sense, the social-focused approaches consider human perceptions, intentions, and ideas—that is, the 'mental processes'—all as dependent variables of socioeconomic structures. As such, a form of 'human reductionism' seems embodied in the vulnerability approach. Undeniably, the influence of social processes upon people and their behavior is significant, particularly in the context of coping with natural hazards. However, a complete removal of proactive, autonomous aspects from individuals' cognitive and decision-making abilities restricts our understanding of the human and natural-event interface. One major objective of this book is to address why an assessment of the relationships between social-structural variables and personal perceptual and behavioral features is important for explaining human coping process in a hazardous environment.

The context in which human decisions are made and the way the environment is perceived are valuable aspects of understanding people's behavior in a hazardous environment. Today, ethnocentrism and determinism in many analyses limit the acceptance of different interpretations and understandings of unfamiliar contexts. For social scientists, the sense in which established views are "unashamedly indifferent to history and to human and environmental diversity" is of paramount significance (Hewitt, 1983: 14). "When looking at hazards in a cross-cultural context, and disasters in non-Western, non-industrial contexts," writes Hewitt, "one begins to have the suspicion that the authority of the dominant view derives from much the same source as [Edward] Said [the author of *Orientalism*] . . . sees the European view of 'The Orient'" (Hewitt, 1983: 28).

Despite the fact that a considerable volume of literature on natural calamities, their causes, patterns, and consequences, has thus far been accumulated in the industrialized-developed world,[4] a relatively far smaller effort has been made to understand the problems stemming from natural disasters in non-industrialized nations. Such a paucity of natural hazards research, in particularly in relation to disastrous human impacts, has jeopardized any effective global and regional disaster prevention and mitigation endeavors. This gap is more acute than it first appears since, globally, the poor nations suffer disproportionately from natural disasters. By concentrating on the issues of the least developed or developing nations in general, and of Bangladesh in particular—one of the poorest and most hazard-prone nations in the world—this study attempts to fill in this lack of non-Western perspectives on the nature-society interface.

My objective then is to emphasize that neither the natural domain nor the social spheres should be claimed as the more necessary area of study than the other for natural hazards and disasters research. There are two reasons for this assertion: (1) the relative roles of natural (geophysical and biological) and social, economic, and political processes differ from one event to another; and (2) the physical, biological, and human spheres function through some very dynamic processes to modify each other. "The 'natural' amd human'," Blaikie *et al.*, write, "are so inextricably bound together in almost all disaster situations, especially when viewed in an enlarged time and space framework, that disasters cannot be understood to be 'natural' in any straightforward way" (Blaikie *et al.*, 1994: 5-6). Likewise, disasters cannot be explained only from a societal perspective.

In sum, the broad focus of this book is the nature-society interface. With specific consideration given to *riverine hazards*, namely riverbank erosion and flooding, we will explore the natural environment and the particular social formation of Bangladesh, and the interrelationships and interfaces between geophysical processes of the Brahmaputra-Jamuna[5]—one of the mighty rivers of the world—and the inhabitants of its banks. The book attempts to understand the complex, intertwined relationships between rivers and floodplain occupants, to explain the processes that generate hazards and often lead to disasters, and to finally develop planning tools for ameliorating human suffering through disaster prevention and mitigation.

1.2. Environmental Threats and Disasters in Bangladesh

The large floodplains of Bangladesh, namely the Ganges, Brahmaputra-Jamuna, and Meghna, along with the coastal plains have always been attractive lands for settlement. The opportunities for agricultural practices and an expanding sedentary population are supported by an annual inundation of these low-lying alluvial plains. The regular flooding by the great rivers commences at the appropriate time for cultivation and lasts for a desired duration for paddy rice and other major crops. Further, the gentle relief, a comfortable tropical climate with a mild winter, and a regular water and moisture supply have facilitated human occupance of these floodplains over the last three millennia.

The benefits from the use of the riparian environment, however, are obtained at a cost of devastating losses from several related natural hazards. The inhabitants of the plains cannot avoid environmental risks; several types of riverine hazards continually pose threats to their livelihood, crops, houses, and other established infrastructures.

Most of the land here is characterized by the deltaic formation of Quarternary sediments deposited by the mighty river systems. The combination of the deltaic sediments with a high water content from the annual monsoon, a low degree of compaction, and a large amount of runoff materials from watersheds causes the principal rivers to change their bed configurations. During any season the river channels may shift laterally, encroaching onto the land by up to a kilometer. The dynamic nature of the young delta and the close association between riverbank erosion and flood hazards are well recorded by Arthur and McNicoll (1978). Alluvial sediments, they observe, have

> both short-run and long-run consequences. In each monsoon some banks are eaten away, while new land is built up elsewhere—sometimes only to disappear again in a future season. Quite extensive land areas may be involved in these year-to-year shifts, with a large impact on the local economy. Over the centuries there has been a steady eastward shift in the active delta region, leaving major parts of the western districts (the so-called moribund delta area) no longer flushed and renewed by the annual flooding. This trend has been marked by spectacular shifts in some of the main channels—most recently in the late 1770s and early 1800s, when the Brahmaputra changed course to form the Jamuna River and the junction between the Ganges (Padma) and Meghna Rivers was first formed (Arthur and McNicoll, 1978: 27).

Natural environmental fluctuations in the timing and amount of the monsoon rains not only aggravate lateral shift and bank erosion, but also cause high floods, severe crop damage and resultant famine and starvation. An examination of the historical record of natural calamities, scarcities, and famines in the districts of Bangladesh between 1757 and 1970 reveals that the frequency and spatial variability of these disasters were clustered in both time and space (Currey, 1988). For instance, floods were very frequent in the 1770s, 1780s, and 1790s, but were almost non-existent in the 1920s and 1930s. Although there are no reliable estimates of the socioeconomic damages of these disasters, they created disruptions that had ripple effects throughout the national economy, not just in the flooded areas and not just during the period of actual inundation (Roger *et al.*, 1989a).

The geography of Bangladesh, as a humid tropical region, also makes it susceptible to tropical cyclones and tidal surges. All the physical and meteorological requirements for developing catastrophic tropical cyclones are present in the region. They include: (1) the water temperature of at least 26 degree celsius throughout the year to a depth of 60 meters (Hastenrath and Lamb, 1979), (2) some vertical shear in the horizontal wind near the prospective cyclones (Haque and Blair, 1992), (3) the concavity of the bay and its estuaries (Dube *et al.*, 1982), and (4) the shallowness of the deltaic coastal shelf. These tropical cyclones have accounted for much loss of life over the decades through coastal flooding; two major human catastrophes of the 20th century were caused by the cyclones of November 12, 1970 and April 29, 1991 in Bangladesh, which each caused the deaths of hundreds of thousands of people (see Haque, 1995a).

The above citations confirm that the geophysical characteristics of the Bengal delta are quite variable over time and space. When these variations in nature are analyzed in combination with the socioeconomic and demographic conditions of the country, a complex and extremely vulnerable human ecological perspective becomes apparent. Today, more than 120 million people live in Bangladesh in an area of only 144,500 square kilometers, which is about the size of the state of Wisconsin, USA. Although, the early decades of the 20th century showed a slow demographic growth against ecological and technological constraints, since World War II, population growth has accelerated sharply as a result of decline in mortality—associated with fewer famine deaths, some improvements in public health, and a control over some epidemic diseases. Due to a predominance of the agrarian resource-base and a slow pace of industrialization and urbanization, the vast majority of the population still live in the countryside. They are spread all over the plains: on the natural levees, interior plains, emerging mid-channel islands (locally known as *char*) and reclaimed lands in the estuaries. Consequently, the density of rural population per square kilometer in the floodplains exceeds 1,200—one of the highest in the world.

The economic conditions of the country are also precarious. After experiencing a long colonial rule by the British East India Company and the British Raj, and a semi-colonial relationship with Pakistan (West Pakistan until 1971), the political and economic affairs could not be determined by the local people until 1971 when Bangladesh became an independent nation. Now, with a per capita GNP of only around US $220, Bangladesh ranks as one of the poorest economies in the world. Because of the predominance of subsistence agriculture, the country is one of the most intensively cultivated areas in the world, often supporting in excess of 15 people per hectare. While more than 80 percent of the people live on income from agriculture, more than 50 percent are functionally landless, statistics which reflect a relatively skewed distribution of landownership.

The demographic and socioeconomic settings thus have made the massive population impoverished and marginalized. Along with these factors, the high environmental susceptibility of most of the country to extreme ecological events compounds the insecurity and risks faced by the millions. Table 1.1 lists the major and worst of the environmental disasters over the last two centuries. Historical records show that during 1960-1970, 13 tropical cyclones struck coastal areas. There were at least six major floods in the 19th century and there have been 14 in the 20th century. The Ganges and the Brahmaputra-Jamuna rivers, with their erratic patterns, have shifted and widened their channels throughout the 20th century, displacing millions of people from their usual place of residence. The combined effects of floods and riverbank erosion often create immense havoc for the country at large. For example, several hundred thousand people often lose their homes during a very high flood season. During the 1988 flood, more than 45 million people were displaced from their residence and over 2,000 died (Haque and Zaman, 1993). The disruptions caused by such periodic disasters are lasting, especially for the most vulnerable and poorest disaster victims.

TABLE 1.1. Major Environmental Disasters in Bangladesh in the Last Two Hundred Years

Year	Event	Impact/Human Casualties
1769-76	Great Bengal Famine	"Eliminated almost a third of Bengal's population"
1784-88	Floods and famine; radical shift in course of the Brahmaputra river (commenced in 1787)	Unknown
1873-74	Famine	Unknown
1876	Bakarganj Cyclone and tidal waves	400,000 deaths
1884-85	Famine	Unknown
1897	Chittagong Cyclone	175,000 deaths
1918-19	Influenza epidemic	400,000 deaths
1970	November 12 Cyclone and tidal waves	400,000 - 500,000 deaths
1974	Famine	500,000 deaths (official count 30,000 deaths)
1987[a]	Floods	1,657 deaths
1988[b]	Floods	2,379 deaths
1991[c]	April 29 Cyclone and tidal waves	150,000 - 200,000 deaths

Sources: Arthur and McNicoll (1978: 29)
[a] Brammer (1990b)
[b] International Red Crescent press release, September 26, 1988
[c] Haque (1995a)

Environmental disasters have had so devastating an impact upon Bangladesh's people and economy in recent decades that these issues have already been recognized as a matter of international concern (Roger *et al.*, 1989a; UNDP, 1989; World Bank, 1990). In effect, the composite ecological, demographic and socioeconomic problems have made the country a challenge to human occupance. Today any serious work on global environmental disasters is likely to incorporate the case of Bangladesh, chiefly because of the extreme state of its natural hazards and vulnerability.

As mentioned above, the empirical focus of this book is the riverine hazards and consequent disasters in Bangladesh. Two specific riverine hazards, namely riverbank erosion and flood, are singled out for investigation. The complete destruction of human settlements and arable land by the events of high floods and river channel changes render a large group of people landless and homeless each year in the Brahmaputra-Jamuna floodplain. As stated earlier, the area consists of one of the most densely settled rural habitats in the world. Loss of a small parcel of land due to riverbank erosion may thus be devastating for the rural population, in particular for marginal peasants. As well, the riverine hazards generate a process of involuntary migration among the potential and actual victims of the disaster.

Several postulations guide this study. The coping responses to hazards, and the decision-making process in choosing migration destination are considered to be constrained by macro-level factors. In turn, the coping process of the disasters-induced displacees are postulated to be related to a number of geophysical, socioeconomic, demographic, political and cultural factors. In addition to direct loss, the impact of such displacement may create enormous problems in terms of indirect socioeconomic cost or chain effects. Such socioeconomic costs primarily are linked to economic and resource recovery, adaptation constraints of displacees to new sociocultural settings, and the acceptance and assimilation of the displacees by the natives or original settlers at the place of destination. These processes, consequently, not only create social and economic problems for local communities but also pose serious threats to development efforts of the national policy-makers and planners. By focusing on natural hazards and involuntary migration, this study attempts to assess the floodplain occupants' perceptions of hazards, coping strategies with the flood and erosion problems, and involuntary migration behavior in order to understand the dynamics of disasters and their impacts.

1.3. Approaches to the Nature-Society Relationship: A Conceptual Prologue to Hazard Studies

The study of the relationship between nature and people has been a traditional concern of both anthropology and geography, along with several other conventional disciplines. The theme has received continuous attention, with varying degrees throughout the 19th and 20th centuries and a renewed interest in recent decades. This increasing interest in the environment and hazards has three sides. First, there has been a steady increase in disasters, especially in the developing world since World War II. The trend has been compounded by an increase in *chronic risks* and *hazards* in the industrialized world (Cutter, 1994). Second, our knowledge about hazards has changed considerably in recent years; the simplistic categorization of natural extreme events as the outcome of natural processes or technological failures is no longer effective. In addition, hazards or disasters are now rarely treated as singular phenomenon, but are linked instead to the complex intersections of geophysical, social, technological and cultural systems. Third, the ideas and methods of responding to hazards have also been modified significantly. There is a wider recognition that disasters are largely imbedded in political, social, economic and technological domains, factors often impossible to separate from the impacts of the events (Cutter, 1993; 1994; Cuny, 1983). Consequently, coping responses and the management of risks have also become quite complex, entailing new response and management approaches at the micro and macro levels (see Mitchell, 1990).

Although there is an emerging need to replace the narrow disciplinary orientation in order to deal with newer types of problems and issues, a review of related literature cannot escape traditonal approaches. There are two reasons for this incapacity. First, the disciplinary concepts, methodologies, experiments and observations have each evolved individually. Over time, they have become more incompatible. Second, it is important to understand the historical discourse and approaches to nature-society interfaces within each discipline. During the last half century, significant contributions to natural hazards research have been made by anthropologists, geographers, sociologists, psychologists, engineers, biologists, and philosophers. The concurrent interest and focus in the physical, biological, and human spheres, however, have been mainly confined to the integrative disciplines such as anthropology and geography. In the following section, the nature-people discourse in geography and anthropology are critically reviewed.

1.3.1. THE GEOGRAPHICAL APPROACHES

The Early Approaches: Causal Reasoning

A large body of literature on natural hazards and the nature-society relationship has accumulated in geography. Interpretation of this relationship has historically been framed in a dichotomy—that the domain of nature is extraneous to society and people. The explanations for this view have developed through a whole range of stages and have been based upon a variety of other postulations. Gold (1980: 29), in this respect, points out that "possibilism and determinism were to serve as the end-points of geographical debate about man and environments until after World War II." The concept of *environmental determinism* concedes that human activity is causally linked with the physical environment in which the latter functions as the control for human behavior. *Possibilism*, however, admits the human individual as an active rather than a subordinated agent to his or her environment. This suggests that people act to create an environment, and choose from a wide range of alternatives those characteristics which best fit their cultural context. The common adoption of causal reasoning (Harvey, 1969: 402) and exclusion of explanatory variables other than the physical environment has eventually resulted in the demise of both of the foregoing approaches.

In view of increasing evidence of human capacity to alter and influence environmental conditions, the focus on nature's effects upon people began to shift to identify the human role in changing the natural landscape. Two approaches were prominent here: *landscape geography* and *geography as a human ecology*. In the former approach, landscape was seen as the end product of an interaction between physical landscape and human-made elements (Sauer, 1925; 1959; Freeman, 1961). The criticisms of the landscape approach can be summarized as follows: (a) the preoccupation with visible forms of human occupance of an area ignores the underlying processes (Wagner and Mikesell, 1962: 5; English and Mayfield, 1972: 6); (b) the commitment to macro-scale units as subject provides superficial results of human activity (Brookfield, 1964; 1973) and the failure to account for longitudinal analysis of the history of human progress (Grossman, 1977).

With the latter approach, geography as human ecology, Barrows (1923) attempted to provide another alternative to a determinism. He defined the discipline of geography "as human ecology," in which the scope of geographical study was confined to human biological nature and relationship to the natural environment. Barrows argued that geographers should focus on the mutual relationships between

> man and his environment form the standpoint of man's adjustment to the environment, rather than from that of environmental influence. The former approach is more likely to result in the recognition and proper evaluation of all the factors involved, and especially to minimize the danger to assessing to the environmental factors a deterministic influence which they do not exert (Barrows, 1923: 3).

Nonetheless, due to its narrow definition and a bias towards synthesis of phenomena, the early ecological approach failed to receive widespread acceptance (Ackerman, 1963; Holt-Jensen, 1980). More recently this approach has enjoyed a revived interest with the emerging concern over environmental problems during the last two decades (Mitchell, 1979: 10). In support of the criticisms of the ecological approach, Chorley (1973), and Hewitt and Hare (1973) asserted its serious shortcomings in application to human geographic studies since the concept casts social humans in passive, subordinate and ineffective roles. But unlike the biological models of ecosystem that only produce negative feedbacks, social systems need to incorporate positive feedbacks that people create through conscious action (see Stoddart, 1967).

Positivist Approaches: Human Subservience to Universal Laws

In the face of the progressive development of capitalism and technology and an unprecedented "recession of physical environment" (Craik, 1970), the focus on the 'man-environment relationship' in mere causal and adaptation terms became subservient to the study of spatial forms and processes of phenomena—basically a quest for laws of general validity. The application of *positivism* in human geography soon became the mainstream trend with strong advocacy from Ackerman (1945), Ullman (1953), Schaefer (1953), and Harvey (1969). The positivist approach necessitates a set of primitive postulates relating to people and nature. Following a number of analysts (viz., Holt-Jensen, 1980; Johnstone, 1983a; 1983b; Haque, 1988b), these premises can be identified as: (a) that both nature and people function within a definitive structure of universal laws and are expressions of the formation of two mutually independent entities; (b) that a law is a generalization of unrestricted range of space and time, and thereby possesses a repetitive character; (c) that events involving human decision-making have a determinate cause, identifiable and verifiable from reality on universally agreed methods and criteria; and (d) that natural scientific study of society and its members is feasible.

The preoccupation with laws of general validity and the application of *systematic-nomothetic* methods in the analysis of social elements led positivists to disregard completely the subjective role of the human individual. As Sayer (1979) points out, this circumstance arose from the fact that the "debates about the character of people and Nature were considered to be 'metaphysical' and/or meaningless, or at least irrelevant" to them. With the emergence of the critical approaches in geography since the 1970s, the positivist approaches have been under stern attack among Anglo-American geographers; now almost all of the premises of the approach have been refuted. For example, Guelke (1971) and Sayer (1979) assert that a positivist assumption of the unity of method and explanation for all sciences, whether social or natural, wrongly disclaims the role of *a person as a social being* in nature. Pred (1967) and Olsson (1975) further point out that an a priori postulation that human actions are repetitive and thereby predictive denies the dialectical changes in social and human realms through time and space (also see Guelke, 1971).

Behavioral Approaches: Person as Respondent to Environmental Stimuli

In order to overcome the shortcomings of a positivist approach based on *normative* reasoning, the *behavioral approach* emerged with an emphasis on examining human behavior from positive experience of real-world situations. The epistemology of the behavioral stance concedes the human as a responder to stimuli, and hence seeks to determine how different individuals respond to certain stimuli, and how the same individual responds to stimuli in different environmental conditions (See Wolpert, 1964; Porteous, 1977). There are four basic elements inherent in the behavioral approaches in geography. These are: (1) that the environment which people perceive may vary markedly from the true nature of the real world; (2) that the individual interacts with the environment, responding to it and reshaping it; (3) that the focus is the individual rather than the group; and (4) that the approach requires multidisciplinary interactions (Gold, 1980).

However, in a critical assessment of this approach, Bunting and Guelke (1979) argue that behavioral and perception studies are based on two false premises. Namely, (1) that identifiable environmental images exist which can be measured accurately, and (2) that there are strong ties between such images and behavior. Leach (1978) is also critical of behavioral epistemology. He rejects behavioral concepts by maintaining that any attempt to explain behavior solely in terms of ideas, intentions or preference is very limited. However, Leach also cautions that a disregard of the role of ideas or subjective-cognitive components leads to a mechanical view of people. He argues that it is the underlying economic forces that mould the cultural and intentional appearance of people, and reflect the complex, intertwined relationship between the historical development of society and individual behavior (see Jensen-Butler, 1981; Cox, 1981).

Structuralist Approaches: Human as Part of Nature and as an Agent of Conscious Action

Based upon the criticisms of positivist and behavioral approaches, *structuralism* appeared as an alternative explanation of human behavior in relation to natural and social progress. The prime epistemological characteristic of structuralism is expressed in an axiom that explanations for observed phenomena or *superstructure* are embedded in structures which underlay all these phenomena (i.e., the *infrastructure*), but are not easily identifiable. It also postulates that natural sciences deal with the result of processes (i.e., superstructure) but not the processes themselves (i.e., infrastructure)—that is, results change in a predictive manner identifiable from experiments.[6] In the case of human affairs, the processes are changing continuously, at the same time influencing the contents of the superstructure (Smith, 1981; Peet and Lyons, 1981). As opposed to positivist and behavioral approaches, these assertions imply that the patterns of superstructural appearances cannot be employed to identify the processes although they may offer indications of tendencies.

In the area of the nature-society relationship the structural perspective has not yet been able to offer any comprehensive theory, though it does provide a relatively coherent framework. In refuting a positivist view of nature as external to humanity, several analysts (e.g., Leach, 1978; Burgess, 1978; Sayer, 1979; 1982) insist that although nature and society are separate entities, they are differentiated parts of a united nature. Laws of nature and laws of social action are different and distinctive from each other: the former phenomena are universal but the latter are historically specific and are subject to change through human practice. Therefore, to structuralists, it is fundamental to recognize the distinctiveness between the natural laws and social laws. Sayer's remark is worth citing here: "We can only work as Nature does, by working with its given materials; we can change Nature by obeying Nature's law . . . [whereas] we can change the very terms of social laws, by disobeying them" (Sayer, 1979: 28-30).

Further, the transformation of nature, Burgess (1978) suggests, is linked with the dialectic of the mode or *modes of production*[7] (Burgess, 1978; also see Peet and Lyons, 1981). Thus, in any given historical period, the interaction of people with nature is locked into a determinate structure of social relations. The concept of nature alone is ineffective for understanding such a relationship; this requires instead a focus on the relations of human societies with nature (Smith and O'Keefe, 1980). Glacken (1967), in this regard, argues that the understanding of nature is impossible without integrating it with an understanding of history; separately each remains an empty idea.

Principal criticisms against structuralism center around three issues. First, its materialist epistemology stresses functional economic relations that reduce the role of values, ideas, and individual human action to the status of "epiphenomena" that are derivative of the mode of production. Thus, as a consequence of its bias towards structural relationships, the approach represents an "inauthentic model of man" (Ley, 1978: 46-50). Second, structuralist epistemology, which maintains that knowledge is obtained not by the accumulation of evidence but by the development of theories, eliminates the scope for verification of these theories against evidence (Jones, 1980; Johnstone, 1983b). Finally, structuralism is overly influenced by an "organistic mode of thinking," in that it maintains that parts of a system cannot be explained independently from the whole because the parts are dynamically interrelated and interdependent. Consequently, an element of functional reasoning stands for its logic; the tautology embedded in the approach makes it empirically untestable (Duncan and Ley, 1982).

Despite its shortcomings, structuralism provides the necessary basis for surpassing the frail aspects of positivist and behavioral approaches. The behavioral epistemology disapproves of the laws that operate beyond the realm of human existence. In contrast, positivists disregard the subjective decision-making ability of individuals. The structuralists unify these approaches, as is reflected in their account of the subjective decision-making capacity of people, in the context of opportunities and constraints provided by social structure and economic laws.

In order to overcome the limited capacity of orthodox structuralism, particularly of *idealism* and *structural Marxism*, several scholars advocate the adoption of *realism* as a conceptual framework. As one of the adherents of this philosophy, Bhaskar asserts that "perception gives us an access to things and experimental activity access to structures that exist independent of us" (Bhaskar, 1979: 9). The proponents of realism further maintain that while orthodox structuralism represents the superstructure as the mechanical effects of processes in the infrastructure, a plausible alternative is to view these relationships in terms of a *dialectic* process within themselves and between them (Bhaskar, 1975; 1978; Keat and Urry, 1975; Williams, 1981). Following the realist approach, if infrastructure represents social formation and individuals belong to the superstructure, then society functions as a guide to the actions of individuals through the process of socialization. The actions of individuals are also intrinsically related to the process of reproduction and transformation of society (see Johnstone, 1983b).

Finally, the strength of the structuralist approaches lies in that they provide a scope for analyzing people's dual role—as part of nature and agent to change nature as a conscious subject. The significance of structuralism is undeniable since an understanding of the nature and people relationship in a dialectic perspective does match with the needs of contemporary geography. More so, the "total content of connected natural and social phenomena are again coming up on the agenda, if not for other reasons than the practical ones" (Hagarstrand, 1976: 331).

1.3.2. ANTHROPOLOGICAL PERSPECTIVES

Paucity of Hazard Research

Historically, anthropologists have focused on ethnographic studies. Still, the early ethnographic studies that have dealt explicitly with natural hazards and disasters have been scanty (Belshaw, 1951; Keesing, 1952; Wallace, 1956a). Apparently, the discipline has devoted itself to the study of "normal" patterns of human behavior and has considered hazards "isolated and annoying interruptions of norms" that create "breaks in patterns" (Wallace, 1956b: 15). The tendency has been to treat extreme natural events as ecologically marginal, remote or peripheral issue. Torry (1979a) believes that more plausible reasons for this avoidance tactic "lie at a deeper, epistemological level." He suggests that "there is a long-established penchant in anthropology for conceptualizing a social system interacting with its physical environment in terms of 'resilience'—the ability of a system to resist external perturbations" (Torry, 1979a: 521). This homeostatic view of environmental hazard-losses typically has regarded natural events as constraints on traditional social organization rather than forces that may lead social change.

The development of the cultural ecological approaches in the discipline, however, has provided new impetus to the anthropological study of natural disasters (Oliver-Smith, 1986; Hansen and Oliver-Smith, 1982). For example, the sociocultural adaptation of pastoral nomads to drought-prone environments has been evaluated by several ethnographic studies (e.g., Barth, 1958; Sweet, 1965). Nonetheless, these type of studies are commonly cultural-ecological interpretations of coping strategies to 'marginal' environments, not to environmental 'extremes.' Recent anthropological work in disaster areas have concentrated on the role of social organization, indigenous values and knowledge, and local resources in resettlement of disasters-induced displacees (Hansen and Oliver-Smith, 1982; Oliver-Smith, 1986; 1992; Zaman, 1991a). These studies have provided important inputs to the management of disasters-related displaced populations.

The Debate on the Nature-Society Dichotomy

Despite the historical lack of natural hazards research in anthropology, especially involving extreme natural environmental events, an interesting discourse on the nature-society relationship has reappeared in recent years. In 1996, the arguments of those challenging the nature-society dichotomy—"one of the central dogmas of western scholarship"—came together in Philippe Descola and Gisli Palsson's (1996) edited collection *Nature and Society: Anthropological Perspectives*. This work provides a comprehensive account of the variety of perspectives on natural environment and people relations, and challenges the "traditions which insist upon separating the two on both empirical and theoretical grounds" (Descola and Palsson, 1996: i). The central issue thus is the problem posed by the dualistic concept of "nature-culture" and its epistemological implications.

Furthermore, Descola and Palsson argue that the interpretations of nature and human relations in anthropology in general have reflected a strong ethnocentric view. In many cases the nature-culture dichotomy supported a series of analytical tools for apparently antithetical research programs. For example, "materialists" tried to adopt natural scientific models of causal explanation anticipating a stronger foundation and a wider scope for social studies. The proponents of this view regarded nature as the basic determinant of social action—analogous to natural determinists in geography. In turn, very little attention was given to how non-Western societies conceptualized their 'environment' and their relation to it (see Simmons, 1993). In a similar way, symbolic anthropologists applied the nature-culture opposition as an analytical tool to logically frame myths, rituals, food and body symbolism, and other aspects of social life. Descola and Palsson (1996) assert that while each of the above approaches stress a particular aspect of the polar opposition (i.e, nature shaping culture versus culture imposing meaning on nature) they however take the dichotomy for granted, sharing an identical, universalistic conception of nature. In Descola and Palsson's words:

> Anthropology can no longer restrict itself to the conventional social analysis of its beginnings; it must rethink its domains and its tools to embrace not only the worlds of *anthropos*, but also that part of the world with which humans interact (Descola and Palsson, 1996: 14).

1.3.3. THE NEED FOR A HOLISTIC APPROACH

As it appears from above approaches to the nature-society interface in geography and anthropology, the weaknesses of any particular philosophical position have generally been cited as the cause for its outright rejection by other approaches. More often, with incompatible goals and an inflexible epistemology, the persecutors of certain approaches have deliberately overlooked the possibility of integrating and accommodating each other (see Harvey, 1973). As a result, the appearance of "semantic quibbling" has become common, and is reflected in strong counter attacks between academics (see Ley, 1978; Sayer, 1979; Eyles and Lee, 1982).

Attempts at reconciliation through the adoption of a holistic approach have been growing in recent years. For example, Beaujeur-Garnier (1976) argues that a unity and cohesion of approaches in studies of geophysical and social spheres are essential to fill a gap that has long existed due to narrowness of the traditional disciplines. Harbert (1979) and Johnston (1982) likewise view differential disciplinary routes as complementary, rather than as alternatives, suggesting that they can be used in an eclectic way. This new realization then calls for a change in the fundamental divisions of academic labor—specifically, the removal of the disciplinary boundaries between the natural and social sciences. This requirement is imbedded in the relational understanding of *anthropos* and the environment: that is, "the person is part of the environment and, likewise, the environment is part of the person" (Descola and Palsson, 1996: 18).

1.4. Nature and People Process (NAPP) Framework: A Conceptual Tool for Disaster Analysis

In order to understand riverine hazards and their disastrous impacts upon people's livelihood in Bangladesh, this book uses a framework that focuses on the structures and processes which concurrently unify and differentiate nature and people. The general basis for the Nature and People Process (NAPP) framework is that disasters are intersects of geophysical, biological, social, political, economic, and cultural spheres. The "process" idea is used to conceptualize the dynamic relationships between various subsets and associated variables. It is important to take a broader view since

> [t]he key to a successful didactic approach to natural disasters is to consider them as whole phenomena, in which the demands of the problem [emergency rescue, the risk in unsafe shelter, the management of mass casualties] replace narrow disciplinary specializations. Thus, in the emergent field of disaster studies it is imperative that the approach be fully interdisciplinary (Alexander, 1991: 209).

As such, the NAPP framework advocates a balanced treatment of people's construction of society and the influence of social structures upon them , a balance which questions the established view of outright dualism between nature and people. The following section outlines the conceptual premises of the NAPP framework, followed by a schematic illustration of how disasters occur in the interface of the natural environment and social domain.

1.4.1. THE FEATURES AND THE PREMISES

Human life and affairs evolve through their intersections with the natural universe. They also involve the *phenomenal world* of our mental systems. Interpretations of nature thus can be varied and difficult. Nature is often defined as an *objective* domain which includes all but the human cultural spheres and is governed by universal laws; but it is also interpreted as a socially constructed phenomenon. The latter notion can be found in the arguments put forward by behavioral-ecological approaches which suggest that "it is important to assume some phenomenological notion of *intentional environment*; the 'affordances' of the environment vary from case to case but depending upon its 'meaning' or the way in which it is perceived" (Descola and Palsson, 1996: 18; also see Ingold, 1996; Carello, 1993). In sum, for an understanding of natural and social processes, we must recognize the fact that we are part of the natural environment, and that the natural environment is part of the way we conceptualize nature. Society and nature hence "embrace an irreducible system" (Descola and Palsson, 1996: 18).

The Spheres: The Whole and the Differentiated Parts

Figure 1.1 illustrates the NAPP framework in an attempt to explain schematically how and why a disaster is an outcome of a complex process involving both the natural environment and societal realm. This approach is based on the idea that an understanding of natural disasters requires us to trace a framework under which these so-called spheres (i.e., natural, biological, and human) intersect, and eventually create some impediments against meeting human needs and wants. Universal laws govern the natural spheres, which also include the biological aspects of humans. With a limited capacity to influence nature and its features, humans in the past were predominantly subject to these laws (see Kates, 1971: 438). Prior to the Industrial Revolution, biological imperatives primarily determined human capacity to adapt to the threats that stemmed from geophysical variations.[8]

In defining the human sphere, the NAPP framework posits that the *laws of nature* and *laws of social action* are distinctive from each other. The latter phenomena are historically specific, non-repetitive, and subject to change through human action. As opposed to positivists in geography and materialists in anthropology who consider that human actions can be explained by universal laws of unrestricted range in space and time, treating them as predictable phenomena, the NAPP framework views people as both creators of their own needs and dependent elements of social structures (Figure 1.1). Burgess (1978) clarifies that "the idea of self-creation" of people who attain their goals is essentially the idea of the *historical* rather than universal quality of these needs and aspirations. I argue along with many others (e.g., Hewitt,1983; Mills,1959) that a denial of the differentiated existence of the human cultural sphere from nature leads to

> that dangerous transfer of notion of 'predictability' as feature of scientific method, to the partial and scientistic notion of *predicting society*, which is tantamount to controlling it. This is not the stuff of scientific enquiry but technocratic wishful thinking, a rhetoric that sees society as running smoothly only when fully predicted and managed by its 'experts' (Hewitt, 1983: 23).

Consequently, the NAPP model proposes that the intersections of natural environment and human culture should be analyzed in the context of *social formation* (Figure 1.1). The term *social formation* implies a "specific mix, prevailing within a particular society at a specific time" of social relations, reproductions, technologies and cultural characteristics (Gregory, 1981: 309). Furthermore, this suggests that the human coping needs and responses are historically defined, and may not always exhibit a repetitive behavior. The autonomous decision-making capacity of people should be accommodated to our understanding of social formation, people's livelihood, and hazard coping capability.

The NAPP framework then questions the validity of a universal "package" of social progression and its associated coping modes with environmental threats. That is, a definitive and predictive modelling of society may not always be feasible. Such an attempt to formalize the human "modes of coping" with natural environment in accordance with "techno-social stages," under a universalistic model was initially made by Robert Kates (1971) and later modified and extended by Burton *et al.*, (1993). By determining the modes of coping by development stages of society they attempt to predict perceptual, behavioral, technological, and institutional characteristics in terms of 'adjustments' with the 'natural event systems.' Thse characteristics then correspond to a unilinear path of development stages: pre-industrial (folk), mixed, industrial, and post-industrial.[9]

The issue of unilinear economic stages of social progression bacame prevalent in development studies discourse during the 1970s. Deriving from this period, the *dependency theory* suggests that 'underdevelopment' of postcolonial societies is causally

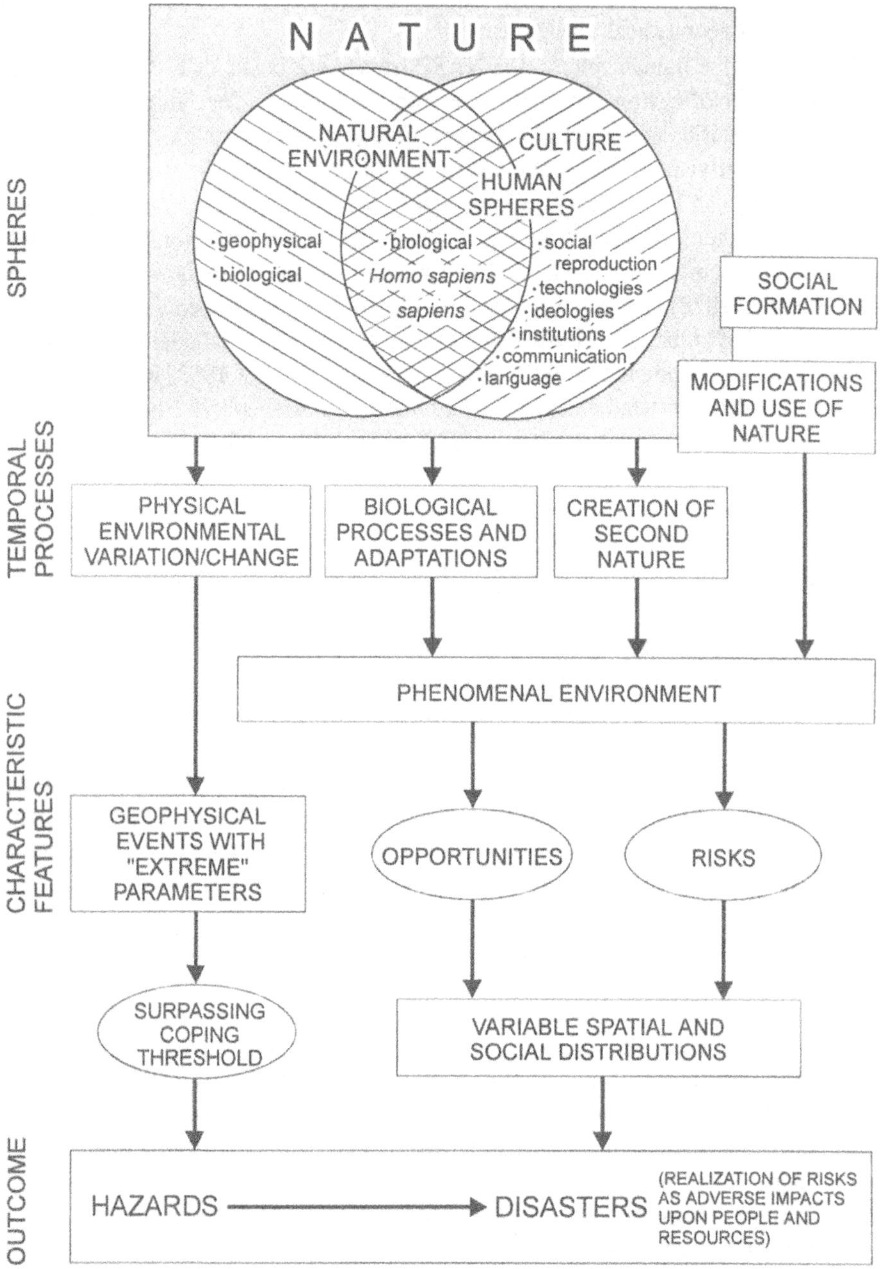

Figure 1.1. The nature and people process (NAPP) framework: relationship and outcome

linked to asymmetric political and trade relationships with the industrialized world (Frank, 1966; Amin, 1974; 1977; Emmanuel, 1972). With convincing arguments, rejections of this attempt at universalizing human culture under the principles of general laws have appeared in several anthropological works (e.g., Ingold, 1996; Hornborg, 1996; Hviding, 1996). Hornborg (1996), for instance, rightly insists that an appropriate epistemology concerning nature and society emphasizes "embeddedness, self-regulation, and local autonomy" while a universalistic conception overemphasizes objectification and decontextualization. Hviding (1996) also criticizes the conventional approaches for their inability to accommodate alternative 'ethnoepistemologies' and their corrective tendency to modify certain domains of indigenous knowledge so as to make them compatible to the dominant modern science. These trends impede any serious effort to understand the role of social formation, belief and practice. In effect, in examining the objective patterns of social changes and subjective worldviews of people, plurality in interpretations seems to be a more viable scheme than a singular, universal model.

Temporal Processes and Their Characteristic Features

Dynamic natural geophysical processes involve both temporal and spatial dimensions. In the human context, both slow and catastrophic natural events are relevant to this type of analysis. For example, statistical *variation* in atmospheric parameters in terms of human time scale is expected, whereas *climate change*, which refers to a major shift in the trend is still a matter of conceptual abstraction rather than experience (see Hare, 1995). People of each social formation, based on their experience, intuition, beliefs, rituals, and symbolic understanding of geophysical processes generate some "expectation of the average conditions" (Smith, 1992: 10). Their coping abilities with such socially constructed "expected" conditions of environmental variation are related to several societal characteristics: (1) symbolic mechanisms, including culture and its constituent norms and roles, (2) technological innovations and their application, (3) social and institutional mechanisms and control, and (4) distributional mechanisms, which specify the movement of people, activities and resources (Micklin, 1973). It appears that both perceptual elements and objective environmental conditions are significant in determining the coping threshold (see Mileti, 1980).

Two dynamic subsets, namely geophysical processes and the human vulnerability characteristics, generate hazards (Figures 1.1 and 1.2). Any variation in either of these two spheres through time and space changes the nature of risk from a specific hazard (de Vries, 1985). Until recent recognition of the adverse effects of human activities on the environment, the general trend in the relationship between people and nature has been viewed as progressing toward harmony and order (Kates, 1971). Contemporary nature and people discourse, however, reveals that the human use and transformation of natural forces have created new sets of problems. Today, floods can originate from a natural variability in meteorological conditions or from human actions such as deforestation or dam-failure. Drawing a boundary between conventionally accepted "natural" (or Acts of God) and 'human-made' hazards and determining them generically have become very difficult.[10]

Similarly, until very recent times, in particular in Western societies, nature was generally considered to be extraneous to humans due to people's limited capacity to *alter* natural conditions. The present trend of new mass-produced life-forms using biotechnological tools (i.e., "non-natural" ways) validates the creation of a "second nature" (see Burgess, 1978; Figure 1.1). The new techniques of human reproduction (Strathern, 1992), transgenic manipulations of animals, and development on xenotransplantation (Papagaroufali, 1996) are likely to further obscure the established boundaries between humans and non-humans and alter their relationship.

Hazard and Disaster Process

In the process of making a livelihood and in meeting the basic needs and cultural wants, people make and intersect with nature; transforming nature into resources by modifying its different facets. Many natural elements are, however, not favorable to human existence and needs, and often create adverse effects. These *potential* threats to human lives and resources, in terms of loss and damage, that emerge from the intersections between extreme geophysical events and a vulnerable human community can be defined as *hazards* (Figure 1.2).

Risk is embodied in the systems of natural environment and people intersections. No person can live in an entirely risk-free environment; this concern for risk can be traced back to the prehistoric time (Covello and Mumpower, 1985). Though the concept of *risk* is sometimes treated as synonymous with *hazard*, *risk* has "the additional implication of the chance of a particular hazard actually occurring" (Smith, 1992: 6). Proceeding from this point, the distinction can be made that hazard is a "potential threat to humans and their welfare" whereas risk is "the probability of hazard occurrence."[11] A common practice in behavioral studies is to classify risks in terms of *voluntary* action on the part of the individual and *involuntary* where extraneous forces, such as geophysical events or socioeconomic structures leave no room for personal choice. But such a generalization is difficult to adopt universally. As Palm (1981; 1990) observes, in California's earthquake-prone zones, despite the general awareness of the presence of hazard, people involved with the real-estate market continue to invest in housing development in the area. The primary motivation here is to maximize profit. Although such a behavior is apparently a kind of voluntary action, one might argue that the market system (i.e., macro structure) actually motivates the behavior of the investors and developers.

Categorization of voluntary risk-taking behavior with distinctive boundaries is thus problematic, although the relative human autonomy in relation to macro structures remains an issue. Heavy rainfall may wash away homes in a wealthy hillside residential area in Oakland-Berkeley Hills in California, USA, as well as those of the poor in Rio de Janeiro, Brazil. For the wealthy Californian, living in the hazardous hillside environment is a relatively *voluntary* act, where it is not to the same degree for the poor

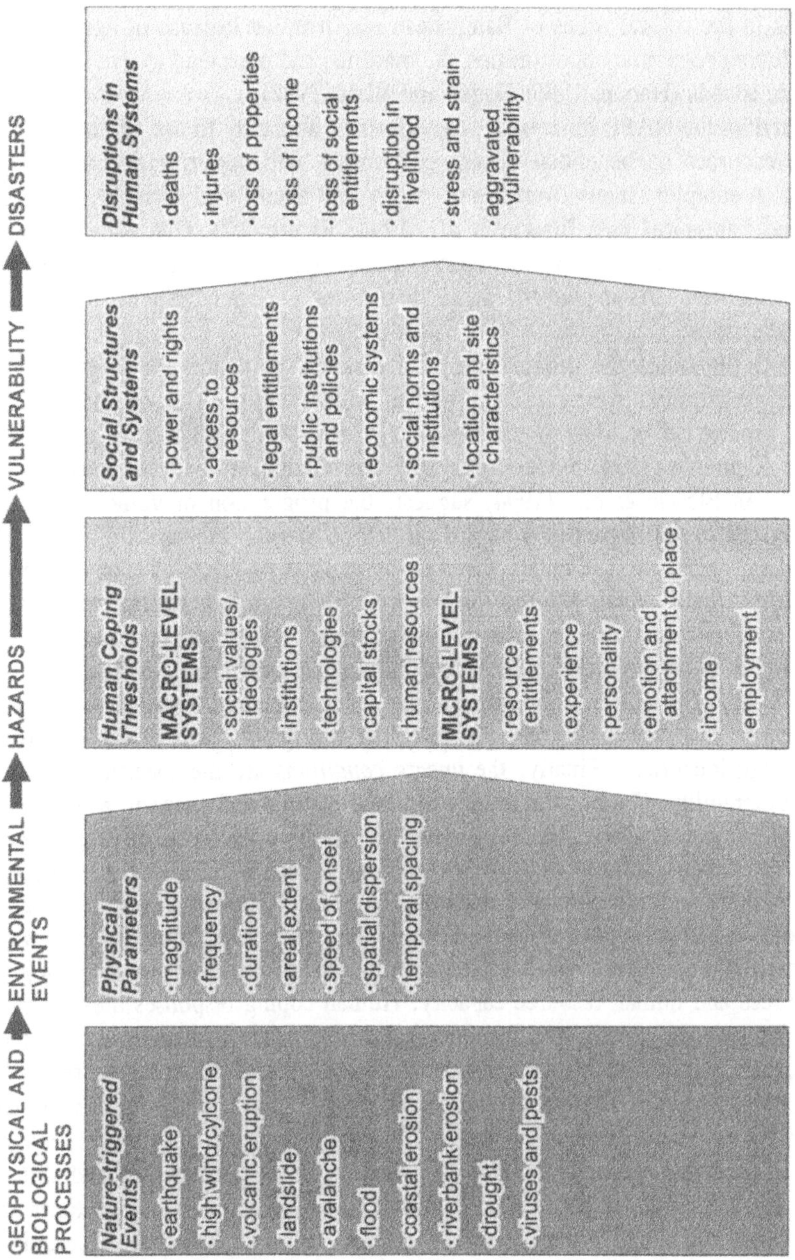

Figure 1.2. The progressive process of disasters

Brazilian job-seeker. Blaikie *et al.*, (1994: 10) argue that if the structure of urban landownership and rent means that the closest they can get to economic opportunities is a hillside slum, people will locate there regardless of the landslide risk. My own investigations in the coastal zones of Bangladesh reaffirm that because of agricultural and salt-producing economic opportunities, the marginal and poor tend to live in highly cyclone-prone islands (Haque, 1995; Haque and Blair, 1992).

As depicted by the NAPP framework, the efforts of humanity to use and transform nature into resources or beneficial phenomena create both *opportunities* and *risks* (Figure 1.1). A complex, interwoven sets of macro- and micro-level variables function as stimuli and "pressure" (see Blaikie *et al.*, 1994; Figure 1.2). Consequently, the realization of *hazard* or *risk* as adverse impacts upon people and their livelihood can be defined as a *disaster*. As explained earlier, a disaster is correlated with the social coping characteristics.

Figure 1.2 illustrates the progression of a disaster, depicting the intersections between the geophysical processes and human vulnerability. The details of the *vulnerability* process can be found in the 'Disaster Process Model' (Blaikie *et al.*, 1994) and in Terry Cannon's (1994) model concerning the relationship between hazard and vulnerability. As Blaikie *et al.*, (1994) suggest, the progression of vulnerability to disasters is linked to the impact of a hazard on people through the series of levels of social factors that generate it. Overall, there are three links or levels of economic and political spheres which explain the disaster process. First, the *root causes*, which are functions of economic structure, legal definitions of rights, and gender relations, reflect the distribution of power in society. Second, *dynamic pressures* translate the effects of underlying factors or causes into the vulnerability of *unsafe conditions*; these include reduced access to resources as a consequence of the way global or regional pressures function through localities. Finally, the *unsafe conditions* are the specific forms in which the vulnerability of a population is expressed in time and space in association with a hazard. As stated above, the human coping threshold, in terms of geophysical variability, is profoundly influenced by the socioeconomic capability, which reflects both macro-level systems and micro-level characteristics of resource users (Figure 1.2). The former factors influence social values and ideologies, which in turn affect the way nature is constructed socially, as well as exering influence upon institutions, technology, capital resources and human resource capacity. Human coping responses thus are not only determined by the macro-level systems, but also by people's own perceptions, subjectivity, and decision-making behavior. From an ecological-behavioral perspective, Kates (1971) suggests that variation in the perception of a specific natural hazard can be accounted for by a combination of factors—the way in which the risks are perceived, the nature of personal encounters with the disaster, and elements of individual personality. In order to underscore the autonomous role of individuals, Kates takes an extreme position, claiming, "such perception appears to be independent of common socioeconomic indicators" (Kates, 1971: 441).

Several analysts (e.g., McCormic, 1989; Haque, 1988a; Haque and Hossain, 1988; Porter and Brown, 1996) indicate further that if the macro systems facilitate a wider range of choice and encourage an autonomous role in decision-making process, individuals tend to become innovative, proactive resource-users. For example, in a free market, capitalistic system with democratic political institutions, a relatively higher degree of political, legal and economic 'freedom' is enjoyed by the citizens; state and other institutions function as supportive means here. In contrast, in totalitarian economic systems, or in limited market-based economies, micro-level systems are deliberately subjugated and controlled. Eventually, the role of the individual becomes a subservient element of the macro structures. However, in reality it is difficult to separate human individuals and the social realm. The prominence of an individual's role, by and large, is relative to the prevailing sociopolitical and economic systems, which in effect create varying degree of vulnerability and coping capacity.

In sum, the NAPP framework suggests that disasters be understood through a holistic conception which recognizes that nature and people are differentiated parts of one entity. Disasters are interfaces of the natural and human subsets; they cannot be regarded as matters of only natural or social affairs. Depending on the type of hazards and vulnerability, the relative emphasis should vary, but a simultaneous coverage of geophysical, biological, social, economic and political perspectives is finally essential to explain disasters adequately.

1.5. Conceptualization of Riverine Hazards in Bangladesh

The central focus of this book is an examination of the characteristics of human coping responses to riverine hazards and the impact of disasters upon people's livelihood and displacement. Specifically, we will look to the impacts of riverbank erosion and flood disasters upon the vulnerable inhabitants in the Brahmaputra-Jamuna floodplain of Bangladesh. The floodplain environment is a composite outcome of the complex interplay between physical environment (which is governed by natural laws and exists independent of the social realm) and social processes by which people objectify their needs from floodplain soil, water, vegetation, and other natural outputs. The flood and bank erosion events of the river are largely an *appearance* of natural laws.[12] Meanwhile, the human use of the floodplain environment is not entirely dependent on the momentary subjective choice or free will of individuals in general, but on historical process and the state of social formation. It is these underlying forces or the nature of social structures that mould significant parts of the social, cultural, and psychological appearances of individuals. These elements operate dynamically, placing individuals within certain social *opportunities* and *constraints,* which themselves function through a variety of economic, political, and legal means and institutions.

Concurrent with this human context, the natural events of riverbank erosion and flood occur with varying magnitude, areal extent and temporal spacing. Given the current state of knowledge of erosion along the banks of large braided rivers and during high floods, their occurrences cannot be precisely theorized in terms of temporal and spatial distributions (see Coleman, 1969; Khan, 1988; 1995). The impacts of these natural phenomena become hazardous mainly in terms of four features: (1) randomness in the spatial and temporal dispersal pattern of the natural events; (2) direct material loss of individually or collectively owned properties; (3) indirect loss through deprivation of disasters victims from social entitlements; and (4) sufferings from psychological strain and stress.

The human coping responses to these riverine hazards are often characterized by a few positive outcomes, usually confined to certain groups or social classes. The positive outcome of river erosion hazard may take place in a number of ways. For instance, it may occur when some individuals gain directly through a change in resources, especially by losing non-cultivable land and subsequently gaining emergent cultivable land. But more commonly, gain takes place through changing social relations and accessibility characteristics; specifically, by the large landowners who operate as a social class, using to their advantage the existing legal, political, and sociocultural apparatuses.

In contrast, for many in the society coping responses are characterized by adverse impacts through which individuals directly lose land and other properties because of a narrow coping threshold. The consequential effects are devastating for the socioeconomically vulnerable groups, particularly for marginalized peasants and the victims of previous disasters. More importantly, the indirect losses of social entitlements for the latter groups have far reaching effects for them. The process of spatial displacement also involves sociocultural deprivation and dislocation, which plays a profound role in determining the intensity of indirect disaster impacts.

In effect, the state of individuals in the social structure becomes the central focus of the analysis of impacts of riverbank erosion hazards. In other words, the nature of the social context within which each individual reacts creates certain constraints; these social relations and institutions deserve prime consideration in assessing the effects of disasters. The response process to natural hazards and response strategies for manipulating them, both represent attempts by resource-users at gaining power and mitigating loss; actions such as these will be assessed in light of the premises of the NAPP framework. From the same perspective we will consider the process of involuntary migration in terms of displacement from the place of origin and coping with the place of destination.

The status of landownership and an individual's relation to it will be given special consideration in the empirical study of Bangladesh. The significance of land as the principal source and means of livelihood, as well as in terms of its simultaneous link with physical and social processes is enormous. Occupation, income, and education are regarded as other pertinent socioeconomic variables in the context of natural disasters impacts on the floodplain inhabitants. We will evaluate the functions of these variables in terms of their appearances regarding the effects of the underlying socioeconomic base and broader social relations. Four explanatory variables will then be used in assessing the coping response behavior in the study area. These include: (1) personal encounter and experience of disaster, (2) attitudes toward nature, (3) attachment to place, and (4) hazardousness of an area. We will examine these variables not only in terms of an individuals' unique attributes—as put forward by the behavioral-ecological approaches to hazard studies—but also in terms of their relation to larger social characteristics; in other words, encompassing how these explanatory variables are socially constructed.

1.6. Scope and Organization of the Study

In general, most disasters studies fall under three orientations: (1) geophysical/technical-emphasis to understand the natural processes and events in terms of geotectonics, climatology, and biology (e.g., Bryant, 1991; Smith, 1992; Kovach, 1995), (2) human adjustment behavior-emphasis to study individuals' response patterns to extraneous natural events that are extreme in relation to coping threshold (e.g., Kates, 1962; 1971; Saarinen, 1966; Burton *et al.*, 1968; 1993; Hewitt and Burton, 1971; White, 1974), and (3) social/vulnerability-emphasis which "puts the main emphasis on the various ways in which social systems operate to generate disasters by making people vulnerable" (Blaikie *et al.*, 1994: 11; also see Cannon, 1994; Varley, 1994). This book makes a considerable departure from these established treatments of disasters, and arises from a unifying, holistic NAPP approach to respond to a call which has appeared in the last decade. The approach integrates the natural and social realms to explain where universal laws function, but simultaneously puts stress on the social laws and systems that largely determine human coping conditions. Following World War II, with increasing success in influencing natural event systems, public policies in the industrialized nations in particular embraced a geophysical-technocratic approach. Such actions reduced extreme natural events but failed to minimize losses and damages. In the last two decades, the vulnerability approach to disasters developed as a rejection of the assumption that disasters are caused by external natural events, instead explaning how society creates conditions in which people face hazards differently. The vulnerability approach, although furthering our understanding of the disaster process in the societal orientation, actually avoided the necessary accommodation of natural factors. In effect, the approach failed to make any inroads in the mainstream disaster management institutions due to its preoccupation with resource-distributional issues (Varley, 1994).

I intend to reintroduce a unified approach into disaster studies, providing greater explanatory capacity, yet avoiding the qualifications of an equally deterministic approach rooted alone in a technocratic or political economy paradigm. I take the view that nature-triggered disasters, the intensity of their impacts, and the effectiveness of coping measures are intrinsically related both to the nature of geophysical events and the social structure. The relationships between human coping behavior in a hazardous environment and the condition of individuals within the social structure are given special emphasis.

The individual's creative role in influencing superstructural or macro-level features is also accommodated in the macro-micro relational analysis. By shifting focus of disaster study to the Nature and People Process signals an advance for both environmental determinism and vulnerability analysis.

The book is presented in four parts:

Part One consists of three chapters, aiming at introducing the issues and problems, and critically reviewing the literature on disasters-induced displacement of population and human coping responses to environmental hazards, risks and disasters. This part thus will set out the conceptual perspective of the book.

Part Two evaluates the geophysical—hydrological and meteorological—characteristics of the Brahmaputra-Jamuna river system to determine the nature of non-human aspects of the riverine hazards. From a structuralist-holistic perspective, we will analyze social class formation, land resource management, population growth and its spatial distribution and the process of vulnerability.

Based on first-hand field surveys conducted in 1985 and 1995, the magnitude, nature, and characteristics of erosion and flood disasters impacts, particularly on land-loss, loss of social entitlements, and coping response types at the micro level are assessed in Part Three. Presented in Chapter 9, the case of the erosion disaster victims in urban squatter settlements is incorporated to provide a comprehensive coverage of the issue.

In Part Four, management, planning, and public policy issues are analyzed in conjunction with riverine hazards and development problems. An attempt is made to leave the reader with conceptual and applied guidelines that indicate how the NAPP approach can be incorporated into efforts of understanding and ameliorating environmental disasters.

CHAPTER 2

NATURAL DISASTERS-INDUCED DISPLACEMENT
An Overview of An Emergent Crisis

2.1. Environment and Population Displacement: A Prelude

In the last chapter I argued that threats in a natural environment and their manifestation as disasters must be analyzed in association with the human dimensions of these phenomena. In order to explain the relationship between nature and the social processes that largely determine the type and extent of risk to individuals or groups, an analytical framework is here formulated that stresses the social construction of nature and risk. This framework is a critique of the false separation of *hazards* from the human domain, a view characterized by a technocratic-systematic approach to the environment. The Nature and People Process (NAPP) model underscores the social-structural processes by which individuals are positioned and conditioned in relation to risk, vulnerability, and access and entitlement to resources. As well, it emphasizes the way humans as individuals and collective entities influence social changes and the use of nature. The model thus integrates the physical sphere with the human domain in the explanation of human manipulation and struggle against risks in a given environment.

Conceptually, the model attempts to provide a foundation for the 'access' model proposed by Blaikie and his colleagues—a demonstration of "how social systems create the conditions in which hazards have a differential impact on various societies and different groups within society" (Blaikie *et al.*, 1994: 46). That is, social processes largely determine the site, location, technological and cultural tools and other tangible and aesthetic resources within which nature is embodied. But this context for nature and human relations changes when the extremity and suddenness of a natural event surpasses the coping threshold of a given society; the event then generates certain 'appearances,' the losses and disruptions of everyday life. Displacement of population due to the loss of homes is one of many significant appearances of the nature-society interface. In this chapter, these issues concerning environment and population displacement are critically reviewed and elaborated in order to lay the necessary base for their empirical examination in the context of Bangladesh.

During the last twenty-five years, particularly since the United Nations Conference on Human Environment in 1972, there has been an increasing interest in the complex relationships between environment, resources, people and economic development.

Despite a great deal of recent interest and an enormous expansion of environmental discourse at different institutional levels, including changes in public attitudes towards ecology, the focus of the debate has remained, by and large, on the impact of industrialization and modernization (i.e., *post-facto* elements). Similarly, as expressed by El-Hinnawi:

> [p]ublic awareness of environmental hazards is initiated mainly by 'trigger events'. Only when a severe earthquake or cyclone causes massive damage in a country, or when drought affects a whole region driving millions of people to flee and causing tremendous human suffering, do the policy-makers and the public become alert to such environmental hazards and their consequences (El-Hinnawi, 1985: 2).

Today, global economic expansion is proceeding on an unprecedented scale while the magnitude of losses of human lives and resources due to environmental events is increasing dramatically. If we are equipped only with a limited awareness of this situation, El-Hinnawi warns that we cannot meet these challenges.

A 25-year average, over the 1969-93 period of 'natural-triggered'[13] disasters exhibits that every year more than 144,300 persons were killed, 46,600 injured, and 120,660,300 victimized (IFRCRCS, 1995: 99). Globally, these events also brought about the homelessness of 4,613,000 persons each year. According to 1994 statistics, some 160 million people were being affected by disasters other than wars. Then in the mid-1990s the volume of environmentally induced displacees (EIDs), what Myers and Kent (1995) call "environmental exodus", was estimated to be around 25 million in the mid-1990s. These EIDs are people who have been forced to leave their traditional habitat, temporarily or permanently, because of noticeable environmental events that jeopardized their existence and seriously affected their usual livelihood (El-Hinnawi, 1985: 4). Currently, the number of EIDs is unexpectedly larger than the count of conventional refugees (around 22 million) and is growing due to the fact that "environmental disasters such as floods, droughts and earthquakes are uprooting more and more people every year" (ICIHI, 1986: 10).

Meanwhile, many of the postcolonial states of Asia, Africa and Latin America are experiencing a rapid population growth. While economic growth is either stagnant in many of these least developed nations or its benefits only further marginalize the impoverized members of society, a rapid population increase is aggravating people's overall vulnerability to extreme natural events. The EIDs are principally concentrated in areas of high population density, leaving them especially prone to natural events. To compound the problem, the most densely populated regions of the world are in the "ring of fire" which is geologically the most earthquake prone area; also, many other densely populated regions are relatively more prone to floods and droughts than sparsely populated regions. The areas where EIDs are concentrated include sub-Saharan Africa, southern Asia, southeast Asia (especially China), a few pockets in Central America (Mexico in particular) and northeastern Latin America (Brazil).

A crude estimate by Myers and Kent (1995) projects that the total number of environmentally displaced persons may even surpass the mark of 50 million by 2010 AD From a neo-Malthusian viewpoint, they rationalize this projection by arguing that such a massive size will emerge as more and more marginalized people will "overload" the already degraded physical environment. Myers and Kent's study, nonetheless, rightly points out that while today many of the EIDs are uprooted by environment-related events, it is these marginalized-displaced populations who cause much of the ongoing deforestation, desertification, soil erosion and then account for the aggregate environmental decline originating in other natural and human-made events. The growing complexities and scale of the problem necessitates that we urgently "widen and deepen our understanding of environmental refugees by establishing the root causes of the problem" (Myers and Kent, 1995: 9).

Because of its formative and exploratory nature, research on environmentally induced migration lacks any distinct and comprehensive frame of reference (Oliver-Smith and Hansen, 1982: 2-3; Lassailly-Jacob and Zmolek, 1992). The studies accomplished in the 1970s and early 1980s are largely characterized by a particularistic approach, in terms of causal-types of hazards as well as cultural regions. Since the mid-1980s there has been a gradual shift in research on environment-related displacements to the broader global and taxonomical issues. This new direction was introduced by El-Hinnawi (1985) in his influential work on 'environmental refugees.' After remaining relatively ignored for decades, both in research and pedagological arenas (Rogge, 1987: 1; Jacobson, 1991: 238), the field of *forced migration* is now emerging as a significant and necessary area. Recently, several international workshops and symposiums have taken place on environmentally displaced populations, chiefly organized by the international donor agencies (such as UNHCR, IOM, RPG, CIDA, IDRC, ODA) to further understand and increase awareness of the phenomenon (IOM, 1996; Terrill, 1984; Varley, 1994). Much of the discussion and debate, however, still is confined to applied policy perspectives.

In this chapter I will review two specific aspects of natural hazards-induced displacement. An overview of the literature on definitional and taxonomical issues is presented as an introduction. Next, Section 2.2.2 narrows the concept and definition of natural disasters-induced displacement (NDID) from the broad category of involuntary or forced migration. Section 2.2.3 then turns to the relocation, resettlement and adaptation problems of the NDIDs during the post-disaster phase.

2.2. Migrants, Refugees, and Displacees: Taxonomical and Definitional Problems

2.2.1. VOLUNTARY AND INVOLUNTARY MIGRATION

The wide range of factors and forces affecting population movements lead to varied migration patterns—voluntary, seasonal or periodic, forced migration due to political or religious persecution, forced migration due to a lack of economic supports, and forced migration due to environmental changes. The phenomenon of *involuntary migration* deserves distinction from that of free or *voluntary migration* in terms of its genesis, process, and consequences (see Oliver-Smith and Hansen, 1982: 2-4). Similarly, migration induced by natural hazards needs to be distinguished not only from voluntary migration but from other involuntary migrations that are causally linked to other ecological, social or political phenomena.

Migration results from some degree of prior deprivation (see Oliver-Smith and Hansen, 1982: 3); people move away from a particular location to satisfy unfulfilled needs or desires which could not be met (see Wolpert, 1964; ICIHI, 1986). Within this general context of migration, several analysts distinguish involuntary migration with differing sets of criteria (e.g., Petersen, 1958, 1975; Eichenbaum, 1975; Oliver-Smith and Hansen, 1982; UNHCR, 1991). For instance, in formulating his classification of migration, Petersen (1975: 317-27) places emphasis on the 'will' of migrants as the decisive element. He suggests that in "forced" migrations the migrants have no choice concerning their move while in 'impelled' migration they retain some degree of power in making the decision to move. But in both cases the migrants play a *passive* role since their 'will' is subservient to the broader causes (i.e., natural phenomena or sociopolitical institutions) that determine and direct the situation.

The concept of *pervasive-intensive* continuum of hazards is used to encompass the composite characteristics of the physical events. For example, droughts are pervasive phenomena, with a slow onset; in contrast, tornadoes are intensive hazards with high energy outputs, highly localized, with a very rapid onset (Burton *et al.*, 1978: 29-31). Although it is normal to view displacees of natural hazards as *involuntary migrants*, it is important to note that in coping with many pervasive type of hazards, the choice of whether or not to move before the onset of the anticipated disaster remains with the migrants. Examples of such population movements include drought or flood-induced migration of individuals, families or other social groups.

Decisions in such cases are influenced by the intensity of the pressure exerted upon the migrants, by their perception of future hazards, and by their perceived ability to cope with the expected hazard loss and damage. There is a need to recognize the difference in the degree of involuntariness among natural hazards-induced migrants, namely those who take *flight* from their place of usual residence in anticipation of hazards, and those who are *displaced* or forced to move by the onset of a hazard. The former scenario relates to the case of a pervasive type of hazard, when people have sufficient time to evaluate their decision to move, while to latter is common in the case of intensive hazards.

Movements resulting from ecological pushes are classified by Petersen (1975) as *primitive* migrations. He, cautions, nonetheless, that this term does not imply "the wandering of primitive peoples as such, but rather a movement related to man's inability to cope with natural forces" (Petersen, 1975: 319). In terms of a general migration framework, Petersen proposes a classification of two major types of migration: (1)

innovating and (2) *conservative*. Persons who migrate because they want to improve their situation participate in the former type of migration; in contrast, *conservative* migrants "move geographically in order to remain where they are in all other respects." Migrations resulting from sudden natural catastrophes can thus be termed *conservative* since the migrants do not move to better themselves, but rather in response to changes in their prevailing conditions. This situation applies to hazards-induced migrants in Bangladesh.

With reference to the general classification of involuntary migration phenomena, it is necessary to distinguish "refugees" from other *involuntary migrants*. In many instances, populations uprooted by a sudden and uncontrollable event have no choice of a particular destination but are freed to seek refuge wherever they can; they then become 'refugees.' Refugee populations thus constitute a distinct category of involuntary migrants. The first international convention referring to refugee protection was adopted in 1933, where humanitarian norms were developed and instruments were approved. The Constitution of the International Refugee Organization was formulated and approved by the international community in 1946. The definition of a refugee then evolved rapidly during the post-World War II period. The United Nations 1951 Convention relating to the Status of Refugees, which was later amended by the 1967 Protocol, attempted to define refugee status. These Protocols, which were adopted into international law, defined 'convention refugee' in the following manner:

> Convention Refugee means any person who, by reason of a well-founded fear of persecution for reasons of race, religion, nationality, membership in a particular group or political opinion, (a) is outside the country of his nationality and is unable or, by reason of such fear, is unwilling to avail himself of the protection of the country or (b) not having a country of nationality, is outside the country of his former habitual residence and is unable to return to that country (Young, 1989: 2).

The concept of refugee movement is extended by Olson (1979) beyond the conventional sources of conflict created by humans, to include people displaced from their residence by natural disasters. He identifies five types of external compulsions that alone or in combination can cause refugee migration: physical threats, such as floods or volcanic eruptions; economic insufficiency, such as drought or famine conditions; religious persecution; ethnic persecution; and ideological persecution (also see Kolenic, 1974). Rogge (1977) points out that because of their non-political characteristics, such *ecological refugees* are seldom eligible for assistance from the international agencies that normally help other refugee groups. He further notes that while the initial impetus for their movement is ecologically induced, in many cases political motivations within the larger spectrum of the migration process may be discernible (such as was the case in Ethiopia in the 1970s and 1980s).

Overall, it is evident that involuntary or forced migration is distinguishable from that of free or voluntary migration—first, in recognizing the *diminution of power* or *lack of power* to decide to undertake the migration, and second, by comprehending the *absence of a motivation* to move from the place of origin. The spatial mobility of persons in such cases either functions as a means of escaping a threatening situation or as a means of survival, often with an orientation toward reestablishing past conditions. In this book, our scope for considering "involuntary migration" is restricted to natural disasters-induced movements; the migrants therefore are viewed both as "flight-takers" who attempt to escape future hazards, and as victims of disasters—"displacees" who, following the loss of homestead, are forced to move to a new place.

2.2.2. 'ENVIRONMENTAL REFUGEES' OR ENVIRONMENTALLY DISPLACED PERSONS (EIDS)

Defining 'Environmental Refugees'

The term *environmental refugee* is not one which is officially accepted by national and state governments, or international agencies, yet it is widely used within both the academic community and development agencies. Despite the increasing significance of environment-related population displacements,[14] analysts are still far away from reaching a common definition of environmental refugees. In fact, many writers have encountered the problematic task of defining the people who are displaced by environmental processes and changes (Cardy, 1994; Kibreab, 1994; Lassailly-Jacob and Zmolek, 1992; McGregor, 1993; 1995; Westing, 1992a; 1992b).

Many analysts use the term 'environmental refugees' with different meanings and boundaries, and many others employ different terms such as "ecological refugees" (Rogge, 1977; Islam, 1992), "environmental exodus" (Myers and Kent, 1995), and "internally displaced persons" (Garcia, 1992) to discuss identical issues (see Mahmud, 1994: 63-65). The ambiguities which exist in such varied usage are obvious. To complicate matters, varying criteria have been considered in defining environmentally displaced persons. A *temporal dimension* (whether one is temporarily or permanently displaced) was advocated by both El-Hinnawi (1985) and Jacobson (1988; 1991); *causal factors* such as nature or human activities were used by Lonergan (1995) and UNHCR, IOM, and RPG (1996), while political or environmental sources were stressed by Kibreab (1994) and McGregor (1993; 1995); furthermore, *legal aspects* were employed by Lassailly-Jacob and Zmoleck (1992). This lack of clarity may impede the real causes and needs of the EIDs, while the urgency of their recognition by the national and international communities arises from the necessity of assistance and protection of their livelihood.

Following the growing number of people displaced through environmental disasters, the term "*environmental refugee*" has met with some degree of wider acceptance (El-Hinnawi, 1985; Jacobson, 1988; 1991; Kibreab, 1994; McGregor, 1993; 1995; Tickell, 1989a; 1989b; Westing, 1992a; 1992b). El-Hinnawi introduced the term to take into account "those people who have been forced to leave their traditional habitat, temporarily or permanently, because of a marked environmental disruption (natural and/or triggered by people) that jeopardized their existence and/or seriously affected the quality of their life" (El-Hinnawi, 1985: 4). He omitted from the definition people displaced for political reasons or by civil strife, and voluntary migrants who look for better jobs purely on economic grounds. This leaves three broad categories of "environmental refugees": i) those *temporarily displaced* because of an environmental stress such as an earthquake, cyclone or industrial accident, after which people return to their usual habitat; ii) those *permanently displaced* and resettled in a new location because of permanent environmental changes that affect their original habitat, including riverbank erosion-induced displacees in Bangladesh or displacees uprooted by the Sardar Sarovar Dam on the Narmada River in India; and lastly, iii) individuals or groups of people who search for a better quality of life owing to a deterioration of the resource base to the point where it no longer supports their basic needs.

By emphasising the *habitability of people* in different places, Jacobson (1988) attempts a further categorization of 'environmental refugees.' Accepting El-Hinnawi's category of temporary displacees due to a localized disruption such as earthquake or flood, she makes a distinction between 'permanent' changes in people's habitat and the processes that undermine their livelihood. Jacobson's approach to the study of EIDs is based on the premise that the degree and nature of damage to people's habitability should be the principal criterion for classifying them. She notes that natural disasters are the second largest cause after land degradation of the growing number of environmental refugees. Statistically, more people are now being killed or displaced by cyclones, avalanches, earthquakes and other natural events than ever before. This trend is indicative of a causal link which suggests that human activity is a major component in determining the severity of these disasters.

The intensity of environmental events is also used to classify EIDs. In Westing's (1992a; 1992b) view there are three categories of environmentally induced displacement. First, he believes that uprootment triggered by acute events such as flooding, cyclone, and locust outbreaks are largely due to inappropriate settlement by the victims in a disaster-prone area. The second type of environment-related population movement is precipitated by a gradual worsening of the environmental quality. Such long-term resource base deteriorations include nutrition loss, soil erosion, deforestation and increased salinity. *Overpopulation*, in terms of the 'carrying capacity'[15] of an area is the main cause of this type of population movement. The third category, according to Westing (1992a), includes the cases which stem from conflicts of culture and systems of production. Such 'clashes' usually occur when land formerly used for subsistence farming, livestock or slash-and-burn cultivation is appropriated from indigenous populations and redistributed for mechanized commercial agriculture, urbanization and forms of industrialization. Westing asserts that in all of these cases the carrying capacity of the region becomes overwhelmed, causing large-scale population displacements.

Lonergan (1995) views population displacement from a perspective of 'environmental change.' Some of these changes are the results of gradual environmental degradation and are thus termed "cumulative changes." Yet such slow geophysical processes are accelerated through the interaction of human activities, including deforestation, land degradation, soil erosion, salinity, siltation, waterlogging, desertification and climate changes. Many abrupt environmental changes intervene with great gravity in the normal way of life and result in either temporary or permanent population displacement. As suggested by Lonergan (1995), the causes of sudden, often catastrophic natural events are in many cases indistinguishable. This issue will be elaborated in the next section.

Definitions and discourse never entirely meet the human needs they are meant to address. The elaborate taxonomies of forced migration and the identification of 'environmental refugees' are a superficial and simplistic attempt to encompass the complexity and multidimensionality of environmentally triggered population movements. A separation of 'environment' as a predominant causal factor from other factors may not always be feasible as the causes are often overlapping and intertwined. Despite the obvious problems, an attempt at simplifying and classifying environment-related population movements is necessary, particularly for reasons of formulating international and national policies and mobilizing resources to assist them. But the conceptual basis of such classification should be clear and capable of logically explaining why and how the forced migration of people is generated.

To sum up, it appears that the predominant paradigm for forced migration and environmentally triggered displacement takes a neo-Malthusian viewpoint (e.g., Cardy, 1994; El-Hinnawi, 1985; Jacobson, 1988; Lonergan, 1995; Westing, 1992a; 1992b). This perspective in essence argues that environmentally induced displacements are a symptom of ecological decline and stress, and often signal the deterioration or change in the habitability of an area (see Jacobson, 1988). The problem of forced population movement is then seen in terms of the resource-population relationship and its balance. That is, in certain regions 'overpopulation' in terms of 'carrying capacity' primarily causes involuntary population migration (see Westing, 1992a; 1992b).

This neo-Malthusian notion has several shortcomings in explaining nature-society interfaces, especially the complex interrelationships between population, resources and nature. The 'carrying capacity' is often wrongly conceptualized in terms of non-renewable and potentially non-renewable resources, a fallacy which gives rise to *a priori* parameters for the carrying capacity of a site, place, or ecosphere. The concept of carrying capacity, used in a narrow and inappropriate manner, hides the fact that although the utility of natural resource capitals are measurable in quantitative terms, knowledge of human ingenuity and future human capacity to manage them is limited (Simon, 1981). A more important application of neo-Malthusian ideas in environmentally induced displacement turns our focus to 'appearances.' Accordingly, the main causes of nature-society disharmony are rooted in inappropriate management strategies and techniques on the part of institutions and in individuals' limitations to cope with a given environment due to personal or group characteristics. For example, floodings which are associated with deforestation in upper catchment areas are portrayed as mainly caused by subsistence peasants' lack of knowledge and consideration of their wood gathering activity. This explanation, however, fails to uncover the underlying societal relationships, production systems, institutional structures and the associated values, belief systems, traditions and culture that affect social and individual behavior. Overall, the gaps in the neo-Malthusian paradigm of forced migration reveal the urgent need for developing an alternate framework capable of explaining and encompassing the underlying forces and processes within the relationship of nature and society.

'Environmental' Category of Displacees

A major debate in the discussion of forced migration or displacement centers around the application of the term 'environment' in referring to displacees uprooted due to ecological changes. One group of analysts applies the term "environment" both to distinguish *natural* elements as a set of distinct causal factors (El-Hinnawi, 1985; Jacobson, 1988; 1991; Myers and Kent, 1995) and to establish it as a domain where a change triggers human migration or causes displacement (Lonergan, 1995; UNHCR, IOM and RPG, 1996). In turn, several researchers who oppose this practice argue that the adoption of the term "environmental" to describe people displaced by natural or human-made disasters and ecological degradation can imply a false separation between the inseparable categories (McGregor, 1993; 1995; Kibreab, 1994; Suhrke, 1991). This debate needs a critical assessment.

As described in the preceding section, a widely used model concerning the forced displacement of population is based on causal variables to group the explanatory reasons or causes into natural and anthropogenic categories (Beyer, 1989; Newland, 1981; Smyser, 1985-86; Westing, 1992b). Lonergan (1995) and UNHCR, IOM and RPG (1996) suggest that these causal factors lead to environmental change and eventually trigger involuntary population migration (Figure 2.1). The *natural causes* identified in this scheme include earthquakes, volcanic eruptions, and flooding, all characterized by short-term, localized and severe events (see Figure 2.1). The *anthropogenic causes* include global warming, acid rain, pollution of riverways, industrial accidents, reservoir construction, nuclear testing, hazardous waste site construction, megaprojects and depletion of ecological resources and consequent environmental degradation. These events may be characterized by short-, medium- and long-term time frames.

The dualistic natural-anthropogenic causal model is based on an understanding of nature-society interrelationships. This approach fails to recognize that the human sphere is an entity inseparable from nature, and thus the model shifts the necessary focus from understanding the mechanisms of 'appearances' or features of nature-society interfaces (e.g., the process of occupance of risky sites, or the process or production of carbon dioxide or chloroflorocarbons) to a determination of causal factors. Alternatively, an understanding of the dialectic relationships between nature and society in which both domains affect each other continuously explains more thoroughly forced population movements—an apparent "event" of nature-society systems.

A bifurcation of causes of forced migration overlooks the complexities and difficulties of separating them, although for some practical purposes, such models may provide some useful simplicity. As emphasized by the NAPP framwork, the natural and human

Causes	Events/features		Time frame	Scale & intensity	Predict-ability	Revers-ibility	Main organizations involved
Natural causes	earthquakes, volcanic eruptions, flooding, etc. necessitating disaster relief		short-term	local and severe	limited or none	yes, in some cases	gov't and int'l relief agencies, self help
Anthro-pogenic causes	global warming, acid rain, pollution of riverways, etc.		long-term and grad-ual	global, regional and inc-remental	to some degree	partial and diff icult	bilateral gov't programs and multilateral schemes (GEF) envir. org's
	industrial accidents		short-term	local and severe	limited or none	some-times	owners of affected facilities & national gov'ts
	predictable env. causes of displace-ment		medi-um to long-term	local and some-times severe	yes	no	national gov'ts, dev-elopment agencies
	depletion of ecolog-ical reso urces and/or environ-mental	localized problems	medi-um to long-term	basically local, & could be serious	to some degree	some-times partial	national gov'ts, de-velopment agencies, private sector, own effort
		irrever-sible problems	long-term	local to regional and serious to severe	to some degree	no	national gov'ts, bi-lateral gov't programs, int'l org's

Figure 2.1. Environmental Change as a cause of migration

Source: UNHCR, IOM, and RPG (1996).

spheres overlap with one another—an entire separation of the causes of ecological and social change is not viable. One problem area, in terms of overlapping natural and anthropogenic factors is global *climate change*. The global climate change modelling predictions indicate many emergent manifestations including sea-level rise and massive population displacement from the coastal regions (IPCC, 1990; Myers, 1993; Myers and Kent, 1995; Tickell, 1989a; 1989b; 1990). Still other manifestations include intensification of rainfall regimes in the tropics and subtropics (Downing, 1991), increased severity of the cyclone and monsoon systems, and an increased incidence of severe and protracted droughts. Until very recently, scientists have been facing enormous difficulty in separating the role of human activities in climate change from that of the normal natural processes.

While there is a general consensus that ecological changes contribute to large-scale population movements, it is difficult to determine their role with *certainty*. Most scholars tend to agree that long-term resource-base deterioration is one of the principal underlying causes of environmentally induced movements, although the chain of events may begin with any number or combination of social, political and economic factors. As such, the environmentally displaced persons only show the surface of underlying, more pervasive processes (see Mahmud, 1994: 70-71).

A separation of environment from daily social, political, and economic life tactically and conceptually harms the goals of academic research and institutional policy. McGregor (1993; 1995), Kibreab (1994), Suhrke (1991) and Suhrke and Visentin (1991) justifiably criticize the use of the term 'environmental' to refer to people displaced by ecological features such as natural and human-made degradation. Recent literature on disasters demonstrates repeatedly the unquestionable role of humans, either in causing sudden environmental events or in making populations more vulnerable (Wijkman and Timberlake, 1984; Cannon, 1994). For instance, some floods form due to natural climatic variations while others are caused by deforestation, which is often caused by development in the upper reaches of the drainage basin.

Evidence is continuously revealing that the economic, social, political and natural causes of population displacement cannot be easily distinguished. As McGregor convincingly illustrates, resource users may become vulnerable when institutional forces (e.g., market, state, NGOs, etc.) directly or indirectly undermine their usual coping methods or adequate relief does not arrive, as occurred in the Dust Bowl disasters in North America in the early 20th century (see Warrick, 1983). Furthermore, McGregor argues that migrants' actual decisions to flee commonly demonstrate much more complexity than a simple "environmental" push as implied in studies of the effects of climate change. Migration is only one of a variety of simultaneous or consecutive survival strategies pursued by families, such as selling assets and wage-labor, eating sparingly or undertaking short-distance moves (also see Haque, 1988a; 1988c; 1993; Haque and Zaman, 1990; 1993).

The broad categorization of environmentally related displacees and the consequent staggering statistical figures of EIDs have been subject to stern criticisms. As mentioned earlier, Myers and Kent (1995) suggest that by 2010 AD the total 'environmental exodus' will reach 50 million. Similarly, the United Nations Environment Program (UNEP) estimates that the number of people forced to leave their homes because of destruction of their means of livelihood will reach one billion by the end of the first decade of the 21st century (cited in RPG, 1992: 2). Understandably, such statements primarily attempt to increase public awareness, however, in doing so they may not constitute a useful point of departure for advancing scientific knowledge. Social scientists, as Suhrke (1991) puts it, are generally reluctant to accept these types of claims although many NGOs and development agencies find them useful.

The forceful arguments put forward by several theorists against the distinct and separable nature of environmental and societal causes of population displacement offer a useful critical perspective. While the specificity of EIDs is often unclear due to the complex and interwoven nature of relevant determining factors, the need for spelling-out the varying kinds of needs of different types of displacees is undeniable. In order to avoid the controversy, the IOM (1996) in its Statement of Principles omits the term 'environmental refugees' and replaces it with the broader term of 'environmentally displaced persons.' This designation refers to people "who are displaced within their country of habitual residence or who have crossed an international border and for whom environmental degradation, deterioration or destruction is a major cause of their displacement, although not necessarily the sole one" (IOM, 1996: 2). Still, the debate over using the term 'environmental refugees' is far from over in the discourse of involuntary migration. The critical literature on the issue indeed sheds light on the implications of overgeneralization and conceptual ambiguity. In order to underscore the predominance of environmental changes as triggering factors of the 'event' of population displacement there is a need for explaining the poverty (economic), environment and population nexus as a prelude to EIDs.

Environmentally Induced Displacees (EIDs) as 'Refugees': Legal and Political Issues

One of the central debates around forced migration and displacement due to environmental changes concerns the use the term 'refugee' to refer to displacees. As has already been discussed in Section 2.2.3, since the introduction of the term 'environmental refugee' by El-Hinnawi, several analysts have embraced it uncritically and advocated for defining environmentally induced displacees as "refugees." Of course, such usage generates many political and legal reverberations.

Jacobson (1988) advocates that EIDs should be treated as refugees and defined in terms of habitat. She makes a distinction between 'permanent' changes in people's habitat and the processes that undermine their livelihood. This notion, in general, calls for the protection and preservation of livelihood by meeting basic human needs and conserving ecological resources necessary for human habitat. In the same line of thought, some legal theorists argue in favor of a definition of a 'refugee' based on human rights. For instance, persecution can be defined in terms of human rights violations (Coles, 1988; Hathaway, 1991) which include the right to an adequate standard of living and the right to food (Lawyers Committee, 1991). Shacknove (1985) asserts that if the state is negligent or indifferent to its obligation to protect its citizens' basic needs, this breach of contract could be the basis of international assistance to such 'refugees.'

In legal terms, the definition of 'refugees' depends on the degree of support and protection the individual or group receives from a host country or the international community (such as in the current case of the Kurds in Iraq; see Newland, 1981). Since the early 1980s, UNHCR has received authorization from the United Nations to assist people, along with recognized groups of refugees, who are displaced within their own country's borders (Kibreab, 1991). Intervention is now made to assist those displaced by a combination of war, drought and shifting entitlements to food and basic resources (McGregor, 1993; 1995; Goodwin-Gill, 1988). By 1992, UNHCR began to recognize broad groups eligible to obtain assistance which included 'internally displaced,' 'war displaced' and even the vague designation, 'other need groups' (UNHCR, 1992). Nonetheless, it is important to note that the class of beneficiaries has expanded without any corresponding broadening of state legal obligations (see Goodwin-Gill, 1988; McGregor, 1995).

In sum, the view described above concedes that people who flee life-threatening circumstances, irrespective of whether they are the product of government action (or inaction) or some other natural cause, should be regarded legally as *refugees* rather than as *migrants* (Islam, 1991; 1992; Lazarus, 1990; Hathaway, 1991; Jacobson, 1988). That is, whether a displaced person is a migrant or refugee should be determined not by the need of protection required outside the country, but by the intensity of the causes of displacement (Zolberg *et al.*, 1989). Moreover, if the circumstances at the place of origin have had life-or-death proportions, the victims are also to be regarded as refugees. The concepts of coercion and violence are here defined broadly to encompass the maximum involuntary migrants.

Still, many critics put forward a strong opposition to defining EIDs as refugees (Kibreab, 1994; Lassailly-Jacob and Zmolek, 1992; McGregor, 1993, 1995; Stoett, 1995). This critical stream of thought argues from many perspectives—conceptual, logistical and tactical—that it is wrong to form EIDs into a legal category. Lassailly-Jacob and Zmolek (1992) find counterproductive the attempt to place particular emphasis on the most desperate of the population by promoting them to the status of refugees. That process simply runs up against the blinkered UN definition of refugee, which requires the crossing of international borders. They warn further that formulating definitions that can be effective as conceptual tools is clearly necessary, but there is a risk of developing hierarchies that may be employed "for political manipulation and bring about nightmarish difficulties for people trying to find refuge" (Lassailly-Jacob and Zmolek, 1992: 3).

Those who work with refugees from a legal-human rights perspective express their concerns about the implications for the UN refugee network of shifting emphasis away from 'political persecution' to environmental exodus. Roger Winter (1993), the Director of the US Committee for Refugees (in 1993), admits that the concept of "economic/environmental migrants" is "too simple a term but is better than using the word 'refugee' [which] would be harmful to the protection and assistance efforts of those working for people fleeing *persecution*" (cited in Stoett, 1995: 36). Stoett (1995) predicts that if refugee status automatically implies the right to asylum, it will then be extremely difficult to find governments to accept the proposition.

McGregor attempts to further unravel the limitations and dangers of promoting the concept of 'environmental refugees':

> Whilst it is important to highlight environmental problems and their association with migration pressures, in so far as the term "environmental refugee" conflates the idea of disaster victim and refugee, its use brings with it the danger that key features of refugee protection could be undermined and the lowest common denominator adopted (McGregor, 1995: 8).

The strengths and merits of this argument cannot be denied. Avoidance of the term refugee in cases of environmentally and/or internally displaced people is essential both for minimizing conceptual ambiguity and for helping the cause of people who need protection.

2.2.3. NATURAL DISASTERS-INDUCED DISPLACEMENT (NDID)

Population movements can be found both as an *effect* and a *cause* of environmental change and disruption. Although the ecological impacts of mass forced migration have received serious attention only in recent years (i.e., in the 1990s; e.g., Black, 1994; Bloesch, 1995; Jacobson, 1994; Lonergan, 1995; McGregor, 1993; UNHCR, 1995), the volume of literature on the topic is considerable. Overall, the findings reveal a converging trend, namely that large influxes of people can cause severe damage to ecosystems. These ecological disruptions may cause further uprootment of people and damage to their livelihood. The present study primarily deals with the latter issue: the impact of environmental disasters upon people's livelihood and displacements.

The studies on environmentally induced displacement unanimously agree that *natural disasters* are one of the primary sets of causes of forced population migration (El-Hinnawi, 1985; IFRCRCS, 1995; Lonergan, 1995; Myers and Kent, 1995; Jacobson, 1988). In this book the term "natural disaster" connotes an integration of natural and human factors. This implies that although environmental events are sudden and triggered by natural (i.e., geophysical, climatological, and biological) forces, these forces are not necessarily the ultimate determinants or root causes. As Cannon (1994) explains, the risks inherent in nature consist of a wide range of hazards which pose threats to livelihood, safety, material production and other aspects of social, cultural and economic resources. Hence, within the nature-society interface "hazards are natural [in terms of trigger mechanism], but that in general disasters are not, and should not be seen as the inevitable outcome of a hazard's impact" (Cannon, 1994: 13). Human contexts then make it possible for a hazard to be 'realized' as a disaster. For Cannon, natural disasters in no way imply a 'natural' causality since:

> explanation of disaster causality is only possible by understanding the ways in which social systems themselves generate unequal exposure to risk by making some groups of people, some individuals, and some societies more prone to hazards than others. In other words, disasters are not 'natural' (not even sudden ones) because hazards affect people differently *within* societies, and may have very different impacts on different societies (Cannon, 1994: 14).

For this reason the last few decades have shown a generalized pattern of very high human mortality due to natural disasters in the developing world and an immense loss of capital resources in the developed world (Degg, 1992; 1993; Burton *et al.*, 1993).

The Centre for Research on Epidemiology of Disasters (CRED), in its attempt to create a global database for disasters and their management, applies specific "parameter definitions" to disasters. According to CRED's classification scheme, disaster types include those with natural and non-natural triggers. Natural-triggered disasters include avalanches, earthquakes, cold waves, high winds such as cyclones, storms, floods, insect infestation, landslides, tsunami, epidemics, volcanic eruptions, drought and famine.

Using a different classification system, Woehlcke (1992) explains environmentally induced displacement with three triggering events. These include: (1) natural catastrophes without an anthropogenic component or cause (e.g., volcanic eruptions, earthquakes, hurricanes, landslides, floods as a result of storms, forest fires as a result of lightning or self-ignition); (2) natural catastrophes with an anthropogenic component or cause (e.g., the permanent destruction of arable land and grazing land, increased heavy flooding, landslides, avalanches and forest fires as the direct or indirect result of human activities); and (3) other kinds of damage to the common property resources (environmental pollution and contamination; destruction through military activities; reduction of resources with respect to agricultural products, fuels, fresh water, and the like). Woehlcke's taxonomical groups are useful as they relate to the evolving and changing relationships in the domain of the nature-society interface. He adds that a natural catastrophe implies an extensive and far-reaching deterioration of living conditions for existence in the affected biosphere. Until a few centuries ago, human activities only rarely had an influence on the development of natural catastrophes; "as a rule, the latter were proverbially an act of God" (Woehlcke, 1992: 290). In quantitative terms, although the "classic" natural catastrophes have not been the pivotal causes of environmentally displaced persons on a global scale, an increase in the number deaths, homeless people and environmentally displaced persons can be expected in the future.

According to CRED (IFRCRCS, 1995), disasters with a non-natural trigger include conflicts such as civil strife, riots, displaced people, technological accidents and fire.

Ambiguities are clearly visible in the above classification scheme. For example, famines are as equally a result of political and social interventions as they are of droughts or floods causing food shortages. A more serious inconsistency exists in the statistics cited from the CRED databases in the World Disasters Report (IFRCRCS, 1995) on 'affected people' and 'homelessness.' While the displaced people are considered only as a component of disasters with a non-natural trigger, how then were the data on 'homelessness' generated within the category of disasters with a natural trigger? The effects upon populations from disasters, as cited in the World Disasters Report hence should be read critically.

The categorization of disasters based on whether these events are triggered by natural or human activity is a step forward. This approach addresses the practical needs of determining where and what type of interventions are required. The term natural disasters-induced displacement (NDID) is applied here in order to underscore the fact that the nature and characteristics of natural-triggered disasters and their consequent population displacements are different than other types of disasters (e.g., anthropogenic) and displacements (e.g., development-induced). As such, natural disasters-induced displacees are persons forced to leave their traditional habitat and usual home, temporarily or permanently, because of an extreme natural-triggered environmental event that undermines their existence and seriously affects their socioeconomic and cultural welfare.

2.3. Displacement, Relocation and Resettlement

2.3.1. THE MAGNITUDE AND CHARACTERISTICS OF NDIDs

Generally, natural hazards with either a slow or rapid onset are triggered by climatic, geologic and biologic forces. The former types of disruptions include deforestation, desertification, and waterlogging, whereas the latter types, which are called "elemental disruptions" (IOM and RPG, 1992), include cyclones, earthquakes, volcanic eruptions, floods, pests, and pathogens. Along with the physical parameters and severity of a particular natural hazard, its destructiveness depends on the state of vulnerable people and resources in the potentially affected area. As reviewed in Section 2.2, the categorization of 'natural-triggered disasters' is problematic, and therefore the specific numbers of affected and homeless people should not be readily accepted. However, the relative and absolute magnitudes of natural disasters-induced population displacements are immense and their associated impacts need in-depth analysis. The process of impoverishment, an effect of disasters-induced displacement, needs particular attention.

The dimensions of the impact of natural disasters are important to note. Globally, the average economic losses per year in the 1990s were US $39 billion,[16] which rose from US $3.7 billion in the 1960s (Bras and Prinn, 1993). Today, most of the economic losses (i.e., three-fifths of the total) due to natural disasters are being experienced by the developed world. In contrast, the human loss and suffering related to natural disasters are concentrated in the poor developing countries (IOM and RPG, 1992: 11; Degg, 1992). Due to diminishing biophysical and social resources in many of these developing countries people are compelled to live in areas prone to natural hazards. Then the impacts of disasters are further aggravated by poverty that leaves the victims without the means to avoid or cope with the disaster. For example, consider the disparity between the 1993 earthquake in south-central India that caused the deaths of more than 10,000 people and the recent earthquake in Los Angeles, USA, with a similar Richter-scale measurement that caused the loss of only 50 human lives (Myers and Kent, 1995: 25). Although the absolute value of economic loss due to 'elemental disruption' in developing countries is limited, their relative loss in terms of GNP is remarkably high (Wilson, 1993). The recovery of such loss is often not possible by many developing countries. Hurricane Hugo in 1989 caused losses of less than half of one percent of the USA's GNP, whereas the 1982-83 losses from El Niño in Ecuador, Peru and Bolivia cost them close to 10 percent of the GNP.

Evidence of the forced migration of people from their normal place of residence for fear of, or at the onset of extreme natural events is widely available for most parts of the world. One major reason for such wide documentation and citation is the drama of sudden catastrophic events which attracts immediate media attention. From such events two distinct trends are discernible. First, the decision-making process to migrate because of a natural disasters hazard depends on the characteristics of 'physical exposure.' Second, displacement may not always occur as a direct effect of the hazard or disaster; the indirect effects may also cause the uprooting of people through a *chain reaction*. That is, as natural events are physically characterized by their *intensive* or *pervasive* attributes, these attributes influence the nature and process of population dislocation. Thus, the eruption of Bali's Mount Agung in 1963, which was clearly an intensive event, caused the displacement of 85,000 people. Similarly, the 1982 Galunggung eruption in West Java forced 30,000 people to migrate to the outer islands of Indonesia (Awanohara, 1982). In such cases, the residents of the hazardous areas did not have the option of a pre-disaster move.

There are, however, many examples of displacement by pervasive types of hazards where some scope exists for people to move prior to the onset of the hazard. For instance, Bhattacharya (1978) records the dislocation of 195,000 people (52,000 homesteads) in India during 1924-1962 due to the changing course of the mighty Ganges over a 113-kilometer stretch extending from Farrakka to Jalangi. Kayastha and Yadava (1985) also estimate that about 15 million people are annually affected by floods in India, of whom 2.5 million become permanently homeless. In 1978, the Indian national government estimated that one in every 20 people in the country was susceptible to flooding (cited in El-Hinnawi, 1985: 14). Similarly devastating, in Mexico city in the mid-1980s more than 1.5 million people lived on the drained lake bed of Texcoco, land which either inundates or becomes a bog when it rains. Wijkman and Timberlake (1984) cite that between 1960 and 1980 floods affected 221 million people worldwide and caused the deaths of 80,000 people. Of these, the flash floods in Jamaica in 1979 were so extreme that the runoff swept away homes, roads, bridges, and crops (UNDRO, 1980); a total of 160,000 people were affected, with the number of homeless ranging between 35,000 and 40,000. These displacees, nonetheless, had some degree of choice whether or not to move elsewhere prior to their ultimate displacement.

The onset of a natural hazard displaces people by destroying their land, houses, and other tangible goods and assets. The degree of loss is generally highest in the case of intensive hazards, such as in the 1970s, when earthquakes killed 38,970 people each year. The average for that decade was particularly high due to the loss of 242,000 persons in the earthquake of Tangshan in China in 1976. The Guatemalan earthquake of 1976, which measured 7.5 on the Richter scale, killed 22,000 people and made more than one million people homeless. On a similar scope, Oliver-Smith (1982; 1994) records the displacement of the majority of people from Yungay, Peru following the 1970 earthquake and the subsequent avalanche from Mount Huascaran. Also, Kates *et al.*, (1973) estimate that about 294,000 people (i.e., 70 percent of the total population) were displaced from Managua as a direct effect of the 1972 earthquake.

Three explanations can be put forward with regard to rising disaster loss (Oliver-Smith, 1986). First, while the frequency of natural disasters may not be increasing, their intensity and vigor is greater than ever before, thus accounting for the greater loss of life and catastrophic economic losses. Nevertheless, according to findings of the Disaster Research Unit at the University of East Anglia, no major geological and climatological changes over the past five decades can adequately explain the rise in disaster impact. Secondly, the application of technology, while increasing our preventive and corrective capacities (such as weather prediction reducing potential mortality), also encourage and induce society to take more risks than ever before, especially in developing countries. Thus, it can be concluded "that disaster mortality has always been high and that our technology is not as effective in protecting us as we thought" (Oliver-Smith, 1986: 6). The third explanation, as already discussed above, is that the impact of human systems on the environment and the increase of population in disaster-prone areas increase the risk and scale of impact of nature-triggered phenomena. In other words, through social processes the vulnerability of poor masses in developing regions has increased.

Scientists and social analysts are moving towards a consensus on the issue of the alarmingly increasing number of NDIDs. As explained above, irrespective of the effects predicted by climate change models (see Myers, 1995; Myers and Kent, 1995), the loss of human lives and resources due to natural disasters, along with other events of environmental degradation, is likely to rise. The future could see large-scale displacements in both rich and poor nations; the difference in severity being that the rich will be 'susceptible' but likely able to recover in the long-term if not sooner, while the poor—that is, most of the world's population—will be extremely 'vulnerable,' unable to afford protection or the resources to cope with and recover from disaster impact (Leiderman and Hall, 1995).

Generally speaking, every year about five million people are being made homeless by natural disasters, the vast majority of whom remain permanently displaced. Their cumulative numbers are unknown. The magnitude of NDIDs has probably crossed an important threshold that renders conventional relief assistance (tents, blankets, biscuits, water bottles, medicine, etc.) completely and "embarrassingly inadequate for both the magnitude and the complexity of this crisis as it is now developing" (Leiderman and Hall 1995: 6). While the very existence of environmentally displaced persons has always been a major feature of the nature-society relationship, the increasing number of NDIDs points to significant flaws in human links with natural life-support systems. Contemporary displacees are products not of events or occurrences; they are the by-products of systemic social processes which increasingly make humans into reactive agents. That said, the most effective attempt to understand and analyze the phenomena will turn from individual events to put more focus on processes and structures.

2.3.2. UPROOTMENT AND RELOCATION OF DISPLACEES

Despite the availability of literature on population displacement due to ecological factors in different parts of the world, attempts to develop theories and conceptual frameworks for understanding involuntary migration have been limited. Nonetheless, with the advancements of general theories and models of migration since World War II (e.g., Stoufer, 1960; Todaro, 1969; Lee, 1970; Kritz *et al.*, 1981; Zolberg, 1989), there has also been a trend towards exploratory studies of involuntary migration, particularly with regard to international refugee migrations (see Adelman, 1980; Rogge, 1985; 1987; Orovwuje, 1989). In the limited volume of literature on ecological displacees, the issues of transitional movements and resettlement processes have received considerable attention.

The choice of a destination is not often available for ecological displacees. Their decision instead involves either an acceptance of institutional supports and options offered to them or a turn to socioeconomic linkages, such as friends and relatives. Brody (1970: 16) suggests in this respect that the multiplicity of factors, and their continuation over time, imply that such migratory behavior "may be most usefully considered not as an *act*, but as a *process*" [emphasis added]. A displacee takes part in this process by *reacting* to pre-existing attributes, as well as by *acting* in making choices from a given number of alternatives.

The critical resources of displacees include their past experiences, their best guess about the future, and their sense of responsibility for each other's welfare. Their responses stem from a complex web of facts and possibilities: what is known for certain about the consequences of not leaving, what is not known for certain but is suspected to be true, what is not known for certain but is overridden by the fear of staying or being left behind, and finally the knowledge of what can moderate, reverse, eliminate or help to cope with disaster impact (Leiderman and Hall, 1995). These issues form the fundamental predicament of displacees.

Governmental or institutional interventions also create pressures to evacuate a potentially hazardous place. Such relocation of evacuees may be temporary or permanent. Two examples of a temporary abandonment include the evacuation of more than 35,000 people from the city of Darwin, Australia, to avoid Cyclone Tracy (Parker, 1975), and the evacuation of 1.5 million people in Florida, USA, to avoid Hurricane Andrew (Leiderman and Hall, 1995). For evidence of permanent relocation, we can look to the resettlement schemes of the Government of Peru following upon the 1970 earthquake (Oliver-Smith, 1982), or the relocation of inhabitants of Belize City to mitigate the impact of future hurricanes (Palacio, 1982). Palacio's study of three different relocation schemes formulated by the Government of Belize following the disastrous Hurricane Hattie in 1961 demonstrates that a successful relocation and resettlement program depends upon, first, the relative power of the government in enforcing displacees to move to a certain predetermined location. Secondly, the existence and continuation of a supportive structure encourages the resettled people to remain in their new settlement.

Post-hazard forced migrations and resettlement processes in most instances entail a *suddenness* of movement and an *uncertainty* about immediate strategies and responses. The role of family and other social institutions in such post-hazard situations has been examined by several investigators, mainly sociologists and anthropologists. Quarantelli (1960: 362), for one, asserts that "the extended family is the main source to which disaster victims turn for help." By examining flood disasters in the Netherlands, the Committee on Disaster Studies shows that people tend to evacuate to friends and relatives; and the more relatives and friends one has in society, the easier it is to adapt oneself to the host community (Institute Voor Sociaal Onderzoek Van Hel Nederlandie Volk Amsterdam 1955; cited in Mileti *et al.*, 1975: 128). Kates *et al.*, (1973) similarly find that about 75 percent of the displacees settled in and around the homes of relatives following the 1972 earthquake in Managua. They observe that in the absence of an adequate and efficient civil service to counterbalance the highly centralized government, the role of family in coping with the impact of disasters is more prominent than any formal institution. Thus, individuals in Nicaragua recovered in large measure because of their strong tendency to seek out, help, and protect members of their own family.

Furthermore, Dacy and Kunreuther (1969) observe in the Waco and San Angelo communities of Texas, USA, that if families are forced to leave an area, they insist on remaining together as a unit while they are in temporary quarters. Also, whenever possible, families not only reside with close friends or relatives, but attempt to restore or repair their own dwellings rather than to look for new ones. Wallace (1957) explains this attitude among ecological displacees by arguing that in choosing to remain close to their place of origin, they attempt to reduce the amount of cognitive restructuring necessary to adapt to a new environment. Oliver-Smith (1982: 102) demonstrates empirically that the survivors' attitudes were 'conservative' towards spatial mobility—a predominantly defensive stance against further stress. It is evident that the continuity of socioeconomic support and the availability of a resource base (Scudder, 1973a, 1973b;

Scudder and Colson, 1982; Merryman, 1982; Palacio, 1982), as well as the retention of their cultural identity (Smock, 1969; Oliver-Smith, 1982), are of pivotal importance to ecological displacees in their relocation and resettlement processes.

In the specific context of resettlement of natural disasters-induced displacees, Oliver-Smith (1991), reviews three case studies from Turkey, Iran and Peru. Experience from resettlements after the 1970, 1971 and 1976 earthquakes in the Bingol province of Turkey suggests that three factors are critical in determining their success or failure: (i) the ecology of the new area; (ii) the relationship to the old village; and iii) the capacity of reconstruction of the community (see Coburn *et al.*, 1984). A certain level of economic resources, along with proximity to employment and social services, are also vital for the longer-term sustainability of resettlement projects, (Aysan, 1987; Aysan and Oliver, 1987). In addition, the success of resettlement schemes depends on the extent of a "transfer of responsibility [power] from settlement agencies to the settlers themselves" (Cernea, 1988: 28). Glittenberg (1982), based on the investigation of four reconstructed urban settlements in Guatemala, identifies that the high level of organization, the dynamisms of leaders, effective outside assistance, and the participation of all community members in decision-making and activities were the key factors in the success of the projects.

Oliver-Smith (1982; 1991) evaluates the issues that arise when "people living in contexts made unacceptably dangerous by natural or technological hazard refuse to relocate" (Oliver-Smith, 1991: 20). In these situations policy options available to the concerned authorities become very limited. Forcible relocation motivated by threats of violence, explicit or implicit, puts people under even greater stress and trauma than that posed by the risk associated with potential disaster (see Colson, 1971). Oliver-Smith draws on the case of the people of Yungay, Peru, to suggest that people faced with a choice between ecological threat and social hazards tend to accept the former types of risks. The odds and perhaps even the impact of the alternative social risks may be much worse (Oliver-Smith, 1982). Overall, an integrative approach to post-disaster reconstruction and resettlement, encompassing sociocultural and material needs, enhances the probability of successful relocation.

2.3.3. ADAPTATION PROBLEMS FOR NDIDs

The adaptation of migrants is only a recent problem in those disciplines within the social sciences that share an interest in the consequent applied and planning questions arising from migrations. Still, a growing volume of adaptation and adjustment literature has come out of the fields of psychology, anthropology and sociology, with a more modest volume from geography, social work, and economics. Generally, *adaptation* is a generic term in this literature which refers to both the "process of dealing with acculturation and the outcome of acculturation" (Berry, 1989-90: 215); *acculturation,* in turn, is defined as cultural change that results from continuous, firsthand contact between two distinct cultural groups (Redfield *et al.*, 1936). In recent decades, this group-level focus has shifted towards the individual level and more in the direction of psychological acculturation (Graves, 1967; Berry, 1989-90).

Much credit must be given to psychiatrist Eugene Brody (1970) who has extensively addressed the problems of migrants' adaptation. Brody asserts that movement over any geographical distance implies the crossing of social boundaries, whether such systems are defined in terms of national entities, regional subculture, or immediate friendship and kinship networks. Environmentally induced displacees usually lose the support provided through the social and geographical familiarity of their long-term relationships and value-systems. At the same time, they must deal with resistance to their presence in a host system into which they move. Thus displacees are under stress due to the disaster impact and are simultaneously fearful of the new threat of the unknown at their prospective destination. Between departure and destination they cope with a series of transitional factors (i.e., intervening obstacles and opportunities) which also influence their perceptions, attitudes, and capacity to deal with the host environment (Brody, 1970: 15).

In the context of the physiological adaptation and acculturation of natural disaster-induced displacees, Gerrity and Steinglass (1994) assess the nature and extent of relocation stress following floods. The Tucker County of West Virginia, USA, was selected for an investigation of the impact of home loss and relocation of more than 400 families to flooding in 1985. The findings of this longitudinal study confirm that because homes have enormous psychological and real-life importance in families, the loss of a home during a catastrophic event can have a lasting impact on family members. Fried's (1963; 1982) work underscores the fact that the depth and quality of loss experience of a home closely resemble profound grief, and for many, may generate long-term negative effects on coping ability. For such individuals, 'home' refers to a much larger entity than the physical or social environment, and for many of them represents an extension of the self, of family identity and a symbol of the future. These attributes contributing to the meaning of home create the basis for "place attachment" for individuals and families, an attachment experienced as the "positive affective bond that develops between individuals (or groups) and their residential community (Shumaker and Taylor, 1983: 233). The properties of home which distinguish it from a mere 'house' include *order* (spatial, temporal, and sociocultural), *identity* (spatial and temporal), and *connectedness* (with people, place, past, and future). A sudden disruption can destroy these qualities. Theorists working with the ideas of home and attachment commonly agree that individuals can become attached to their sociophysical environment with differing degrees of connection, and this attachment then has significant bearings upon both adjustment and loss (Heller, 1982; Shumaker and Conti, 1985; Stokols *et al.*, 1983).

In recent decades, research on disaster impact has dealt extensively with the hypothesis that home destruction leading to involuntary relocation is associated with considerable short- and long-term psychosocial maladjustment, or even major psychopathology (Erikson, 1976; Logue *et al.*, 1981; Rossi *et al.*, 1983). One well-cited case is the study of the 1972 flood in Buffalo Creek, West Virginia, USA (Gleser *et al.*, 1981). Here, the lack of consideration of prior communal ties or extended family relationships in relocating survivors led to almost uniform short- and long-term psychopathology in both adults and children. The short-term reactions in adults to the flood caused by the sudden bursting of a dam included psychic numbing, sluggishness in thinking and decision-making, anxiety, grief, despair, and severe sleep disturbances. Long-term responses (two years post-disaster) included physical complaints, survivor guilt, listlessness, apathy, decreased social interaction, and chronic depression.

Contributions from social and personal attributes research provide some interesting observations on mental health after relocation following disasters. Milne's (1977) findings on the long-term effects of Cyclone Tracy, which struck Darwin, Australia on December 25, 1974, indicate that the respondents who stayed in the community, rather than being evacuated, fared best in their post-disaster recovery; evacuees who never returned to Darwin were worst off. Milne concludes that disaster victims who stay within a familiar community with an intact social support network cope much better in the long-run with the psychosocial challenges than do those who leave their communities.

As noted earlier, in his study of earthquake displacees in Yungay, Peru, in 1970, Oliver-Smith (1982) focuses on the post-hazard resettlement process and on problems of social and cultural adaptation. He finds that the Yungainos survivors faced severe stress and physical adjustment problems including physical ailments, headaches, stomach disorder, and upper respiratory difficulties, some even after one year. As a consequence, strong opposition developed against the government's decision to relocate survivors to a distant town being developed as the new provincial capital (i.e., Tingua). Oliver-Smith views this resistance to further resettlement as a sign of sociocultural adaptation as it indicates the ability of survivors to regain a sense of participation and control over their own lives and society. Thus, the hazard victims indicated by their actions a preference to take risks within a known hazardous environment rather than risking "everything at the hands of a society in essentially strange hands" (Oliver-Smith, 1982: 103).

Du Bois (1974), Merryman (1982), and Palacio (1982) also show that problems of adaptation are faced by displacees uprooted by natural disasters. Du Bois, in investigating the flight of Malien Tuareg to Niger following the great drought in the early 1970s, observes that the refugees had serious problems in adapting to a new diet during the transitional period of migration. "The Tuareg," Du Bois writes, "have traditionally been accustomed to a diet primarily of meat and diary products. But with the death of their herds and the desiccation of the oases, the Tuareg had to shift to coarser foods, such as sorghum, to which they did not easily adapt" (Du Bois, 1974: 8). He further notes that the Tuareg were historically reluctant to being integrated into the Malien nation because of their independent spirit and their refusal to give up their traditional nomadic way of life. In a similar vein, Palacio (1982) attempts a diachronic study of the resettlement scheme in Georgetown, which was undertaken by the Belize government to rehabilitate the hurricane victims of a coastal village. The main objective of the scheme was to develop a new area as a permanent agricultural community. But because of the lack of continual technical and financial support during the transitional period, as well as because of sociocultural problems among relocatees in adapting from fishing to agriculture, the resettlement scheme did not succeed. Within ten years the resettlement area was abandoned—"in the absence of occupants the bush had reclaimed the village site" (Palacio, 1982: 129).

With respect to problems of stress and adaptation, Smock (1969) suggests that certain methods of relocation can lessen its painful character. In dealing with a drought situation in Nigeria, he finds that the resettlement process was less stressful when relocatees were substantially involved in the decision-making process. By being sensitive to the needs of relocatees in terms of their proximity to kin, land-use patterns, and ancestral rituals, the administrators, equipped with their anthropological tools, were relatively successful in implementing the relocation project. Nevertheless, an economic base was also necessary in the new setting to fulfil the expectations of the relocatees. Merryman (1982), in examining a Kenyan sedentarization project for Somali nomads who had lost their livestock and resource base due to drought, suggests that the social trauma associated with a sudden shift in subsistence can be minimized through the recognition of traditional values, the introduction of local technical and social innovations, and by maintaining a tie to the greater Somali-Islamic community. In contrast to the stereotypical Somali as a conservative, culture-bound nomad, these people were found to be receptive to change when they were able to perceive its benefits. Thus, evidence suggests that success in relocating and resettling displacees corresponds with the extent to which they are able to participate in a decision-making process which allows them to express their expectations, solutions, and sociocultural identity.

CHAPTER 3

HUMAN COPING RESPONSES TO NATURAL HAZARDS

A Survey and Critique of Approaches

Natural hazards depend on the interface between the physical environment and human society. Thus, human aspects are essential and integral components of natural hazards analysis. The ways in which human agents, as individuals and members of society, attempt to cope with risk in the environment and loss due to extreme events change with both time and space. Such variation in human behavior is largely influenced by the worldview of a particular individual or group, which then determines their approach to nature-society relationships. But how should we view the environmental events that exceed the coping ability of a given society? What are the causes of natural disasters? To what extent is the 'human-use system' of resources directly related to hazards? Furthermore, how should an individual or group respond to extreme natural events? Different answers to these questions mirror the diversity of beliefs and knowledge about the human relationship with nature.

In this chapter we will assess two problem areas in natural hazards research: (1) conceptual approaches to natural hazards, and (2) explanatory factors of human coping responses to extreme natural events. A critique of the related literature is presented in three separate sections. In the first section, approaches and definitional problems are considered. In the second section, methodological issues in hazards research receive attention. In the third section, factors affecting human perception and behavior are examined in terms of environmental risk and hazards. Thus, the goal of this chapter is to outline the strengths and shortcomings of prevailing approaches and indicate the scope and contribution of the present study to the current body of knowledge, particularly in the area of natural hazards research.

3.1. Perspectives on Natural Hazards

3.1.1. APPROACHES TO NATURAL HAZARDS

The study of natural hazards in the contemporary social sciences, particularly in geography and anthropology, has been honed from a wider interest in the nature-society relationship. One of the principal themes in the current literature on natural hazards is the causes of disaster and possible means of its prevention or mitigation (Burton *et al.*, 1993; Alexander, 1993). Three major viewpoints, differing largely in their content and emphasis, are discernible. The first approach (i.e., the nature-focused approach) views environmental hazards as external nature-generated phenomena, and hence, as essentially natural events (Foster, 1980; Frazier, 1979). This view not only puts stress on "scientific research on natural phenomena *qua* 'cause' of disaster" (Burton *et al.*, 1978: 213), but also advocates technologically biased measures in choosing means and ways to mitigate the impacts (Hewitt, 1983). The second viewpoint (i.e., the social-scientific approach) is a social engineering version of the nature-focused approach; it emphasizes the use of social science-based research and measures relating to human behavior along with technological means (Whittow, 1980; Maybury, 1986). Burton, Kates, and White (1968; cited in Burton *et al.*, 1978: 213) observe that the social-scientific approach emerges from the crisis in "man's apparently growing susceptibility to injury from natural hazards during a period of enlarged capacity to manipulate nature via technological adjustments." Based on neoclassical economic growth theory and Weberian functionalist theory of social progression, the approach further asserts that the pattern of hazard loss and cost is directly linked to social type within a unilinear sequence of stages—folk or preindustrial, mixed, industrial, and post-industrial. The third viewpoint (i.e., the social-focused approach) derives from a critical appraisal of the earlier approaches; it argues that hazard causes are functionally related to the type of social formation and historical processes of a particular society (Hewitt, 1983; Cannon, 1994; O'Keefe *et al.*, 1994). Thus, it claims that the nature of disasters and their impact in the Western developed nations can be traced to human adjustment methods within the capitalist relations of the socioeconomic structure. However, hazards in the least developed or developing societies are related to the unique social formation of these underdeveloped societies and the 'historical-structural' process under which these societies are exploited and influenced by the exogenous force of the world capitalist system (Copans, 1983; Wisner, 1980; Wisner *et al.*, 1976). Nonetheless, a caution regarding the latter explanation is necessary since an overemphasis on exogenetic factors may steer us away from the need to understand the internal dynamics of a given society in terms of its own class structure, institutions, and sociocultural traits (Brenner, 1977; Dupuy and Fitzgerald, 1977; Street and James, 1982); additionally, an external view often fails to explain ecological perspectives in these developing societies (Haque and Parvin, 1986; Daly and Cobb, 1989).

By making a distinction between micro- and macro-level phenomena, Palm (1990) critically examines various approaches to natural hazards. She considers the subjects that are indivisible as functional units (i.e., the human individual) and small aggregates that function as units (e.g., the household) as the micro level. The macro scale includes the sociopolitical-cultural environment under which individuals and households function (i.e., the nonphysical environment or milieu). Thus, Palm's definitions follow the conventions of several social science disciplines (excepting economics) in which these boundaries are common.

Anthropologists, for one, classify the micro-macro as a continuum based on size and complexity of the study groups. Thus, a contrast is made in terms of individuals versus collectives or aggregates, "individuals versus the social and cultural milieu in which they exist" (Bennett, 1969: 7). Bennett views micro as local, contrary to macro or external, and insists that the main task of anthropology is to understand "how people solve their problems in a milieu of relationships and adaptive coping" (Bennett, 1969: 24). In sociological terms, Cicourel (1981) defines a micro process as a self-contained enterprise that is locally functional and distinct from macro social facts, those normalized and context-free descriptions of behavior that stem from routine practices of everyday life. In addition, Knorr-Cetina (1981) contrasts micro and macro in four ways: (1) the individual and the collectivity, (2) action and structure, (3) small-scale uniformity, and large-scale complexity, and (4) powerless and powerful. Palm (1990) suggests that the empirical studies of natural hazards discourse have focused either on individual decision making in the context of uncertainty or on societal structures aiding or impeding response and recovery. She insists finally that there exists a gap in the integrative analysis of these two levels of studies of natural hazards.

The vast majority of pre-1970s natural hazards research dealt mainly with micro-level dimensions, concentrating on the relationships between individuals and a hazardous environment. One stream of inquiry assumed that individuals make rational and sub-optimal decisions in a risky environment; as a proactive agent, individuals were regarded as subjective entities who choose from an array of responses while confronting environmental variability (e.g., Kates, 1962; Saarinen, 1966). Using a similar framework, other research assessed the reaction of public and other large corporate or entrepreneurial entities to extreme natural events (e.g., White, 1974a; Rossi *et al.*, 1982). However, very little effort was taken to interrelate various levels of decision-making into a coherent whole, especially recognizing the complex interactions among individuals, groups, cultural systems, and the political economy (Palm, 1990; Drabek, 1986).

The sociological literature on human responses to disasters exhibits numerous examples of micro-scale research dealing with a single cause-outcome system (Drabek, 1986; Mileti *et al.*, 1975; Quarantelli, 1977). The subject matter of these studies includes the responses of individuals or households to specific hazards or disasters, general inquiries into the responses of individuals to low-probability events, and research of risk perception and decision-making. In an extensive review (979 books), Drabek (1986) identifies various levels at which responses are made: the 'individual system level,' the 'group system level' (i.e., families, emergent groups), the 'organizational system level' (i.e., media, schools, emergency management agencies), 'community system level' (e.g., local government bodies), and 'social system level.' He notes that of the total 1,250 'specific findings' of the surveyed literature, a total of 423 (i.e., 34 percent) belong to the category of 'individual' and only 85 (i.e., 7 percent) to the category of 'societal system.'

The analysis of micro-scale individual behaviors relate to the social, cultural, and political economic context. Blaikie *et al.*, (1994) and Hewitt (1983) suggest that individuals' responses should be explained in terms of constraints set by the political economy. Hewitt (1983), in his attempt to advance an alternative paradigm of hazards research, asserts that: (1) natural hazards are characteristic of the societal context within which they take place, and thus integrate into 'ordinary life,' (2) perception of, and behavior to cope with natural hazards are less associated with the geophysical conditions that trigger environmental events than with the prevailing institutions and values of society, and (3) not all responses to natural disasters are explained by conditions unique to the disasters; rather they are correlated with the ongoing societal order and the larger historical circumstances within which nature-society interactions have evolved. The current practice of and emphasis on geophysical monitoring, prediction, and the study of relief and reconstruction are misleading as "such activity is likely to restore the very infrastructure that may have significantly contributed to converting an otherwise normal geophysical event into a human disaster" (Palm, 1990: 22).

Similarly, Blaikie *et al.*, (1994) argue that analyzing disasters from a macro perspective allows one to exhibit why they should not be separated from everyday living and to demonstrate how the risks involved in disasters should be linked to the 'vulnerability'[17] created for many people through their normal existence. They add that when solutions are segregated from the social frameworks that affect how hazards influence people, too much emphasis is placed upon the natural hazards themselves and not nearly enough on the social environment and its processes. Furthermore, Blaikie *et al.*, argue that most research on disasters stresses the significance of geotectonics, climatological, or biological 'trigger' events that arise in nature, while others place their focus on individual behavior, psychological and physical trauma, and legal, social, and economic impacts. These approaches, however, wrongly assume that disasters are departures from "normal" social functioning, and that recovery means a return to a normal state. Without denying the importance of natural hazards as trigger events, Blaikie *et al.*, (1994) advocate studying the various ways in which social systems generate disasters by making people vulnerable—they call this the 'vulnerability approach.'

Exclusive attention to either individual (micro) or societal (macro) analysis, as Palm (1990) explains, stems partly from assumptions concerning the nature of decision-making within the political economy. The micro-focused analysts assume a benign political-economic structure from which the most significant decision-making occurs at the level of the individual; other dimensions at different "scales" can be left within the *ceteris paribus* assumption. In contrast, other scholars concede that the effects of the political economy are overwhelming, and that no important choice is left to the individual. These analysts attempt to understand more fully the dynamics of the political economy, analyzing historical, social, economic, political and cultural forces. It may be argued here that Palm's classification of hazard studies into micro- and macro-scale, although useful for understanding the nature and focus of research pursuits, assumes that these scales represent distinct, easily separable entities. Exclusive attention either to individual or societal perspectives will result in an inadequate explanation of hazard response.

What I then propose is an agenda for research that requires a holistic approach— the NAPP framework that accounts for the subjective role of individuals and their relationships with societal systems and the physical environment. The objectives and hypotheses of the present study are hence derived from the underlying postulation that patterns in human coping responses to natural hazards develop based on complex social environments and relational patterns by which individuals are constrained. Changes in these relationships can only be explained through a historical perspective on the development of various social, economic, cultural, religious, and other local institutions. The replication of models and assumptions on natural hazards issues taken from the experience of the Western capitalist world may have serious limitations when applied uncritically to non-capitalist social contexts.

The present study intends primarily to demonstrate that the impacts of natural hazards are not only dependent upon generic causes of natural events systems, but also upon internal social structures that transform these events into differential hazard effects among social classes. A diagnosis based on the internal complexities of nature-society relationships within a broader social perspective is necessary to determine the viable options to mitigate the hazard impacts, particularly in developing nations.

3.1.2. DEFINITIONAL PROBLEMS FOR NATURAL HAZARDS

An overview of the definitions of 'natural hazard' reveals a lack of consensus. As a result of disparate concepts and goals in different disciplines or fields, there are varied meanings and criteria posed for defining natural hazard. Yet, on closer examination the elements and orientation inherent in the given definitions are usually related to three causal agents: biological, geophysical, and human activity. It is, however, often difficult to isolate the causal agents from each other when dealing with hazards within the framework of nature-society relationships. For instance, a number of hazards are caused by people's activity in transforming nature for their utility, but the harmful effects of these actions are transmitted through natural processes, e.g., air-pollution, acid rain, and radioactive fallout. Geographers, indeed, usually restrict themselves to hazards generated from geophysical sources (Burton and Kates, 1964).

As mentioned in Section 3.1.1, among a variety of conceptual orientations in natural hazard definitions, one major approach regards these phenomena to be essentially natural geophysical events. For example, the official British definition of 'absolute drought' considers it to be a period of at least 15 consecutive days with 0.01 inches or less of rain per day (Meteorological Office, 1963; cited in Burton and Kates, 1964: 413). Likewise, UNESCO (1970) publishes data on various physical agents to indicate hazard frequencies and magnitudes. Kates' observation in this respect is particularly worth noting here. He finds that:

> varying definitions of disaster arise from the sociological structure of science and technology with its curious historical anomalies in the professional and disciplinary organization of scientists and engineers. These disciplinary territories combined with an ideological bias towards neutral nonhuman scientific study and abetted by the relative ease of some geophysical measurements leads to . . . absurdities of observation and definition (Kates, 1975: 1).

There is a general acceptance among technocratic-systematic advocates that natural disasters are consequences of 'extremes' in geophysical processes. The events and essential features of disasters are seen to depend chiefly upon the nature of natural parameters, e.g., cyclones, floods, and drought. Hewitt (1983: 5) observes that although it is accepted that 'hazard' refers to the *potential* for damage that requires the presence of a human component, actual usage "almost invariably refers to an objective geophysical process, such as hurricane or frost, as 'the hazard'." Thus, the direction of explanation still flows from the physical environment to its social impacts; this is reflected well in the fact that the scientific literature primarily deals with and recommends geophysical monitoring, forecasting, warning system development, and direct engineering in relation to natural agents (Ang, 1978; World Bank, 1990). Evidently, according to this—one of the dominant views—disaster is attributed to nature. Here, the structure of the problem is regarded as dependent upon the ratios between given forces of nature and the advanced institutional and technical counterforce (Hewitt, 1983: 6).

Consequently, the technocratic-naturalistic approach focuses on three main areas: (1) commitment to the scientific understanding of geophysical processes, (2) modification of natural processes through planning and management of human actions and activities, and (3) development of institutions for emergency measures. But these commitments only seem to create larger gaps. By subordinating other modes of understanding, Hewitt suggests:

> technocracy gives precedence in support and prestige to bureaucratically organized institutions, centrally controlled and staffed by or allocating funds to specialised professionals . . . The 'social, economic and political "people" factors involved in hazards reduction' . . . also can be and usually *are* approached technocratically . . . the 'natural science—technological fix' approach to hazards itself, essentially, a *sociocultural construct* reflecting a distinct, institution-centred and ethnocentric view of man and nature (Hewitt, 1983: 8).

Despite these issues, the dominance of the technocratic-scientific and its institutional recognition can be seen in the title of the United Nations declaration of the 1990s as the 'International Decade for Natural Disaster Reduction'. By focusing on the bahavior of nature, the declaration, its associated mandate and the follow-up processes have been solidly behind technical solutions "to the supposed excesses of the as yet untamed side of nature" (Cannon, 1994: 17).

Questions on this nonhuman view of natural hazards may be raised: is the physical event *per se* sufficient to cause hazard effects without any interaction with human elements? Obviously, a drought in an uninhabited area or a hurricane far out over the ocean cannot be called a threat or hazard. The natural event, even with high energy parameters, itself is not hazardous, but it is a necessary instigator of a particular effect on society. While neither the natural nor the social elements can be denied in natural disasters, Cannon (1994) and O'Keefe *et al.*, (1976; 1994) argue that hazardous events themselves are 'natural' while disasters are not the inevitable outcome of a hazard's impact. Cannon insists that it is the social systems which make a 'disaster' out of a situation which otherwise might not have been so serious.[18]

Another approach[19] defines natural hazard from an ecological perspective, and views it as "an interaction of man and nature, governed by coexisting state of adjustment in the human use system and state of natural events system" (Kates, 1971: 438; also see Kates, 1962; Burton and Kates, 1964; Burton *et al.*, 1978). The human use system is described as the smallest managerial unit capable of independent adjustment to the potential hazard; the natural events system denotes the measured statistical parameters of natural phenomena in terms of magnitude, frequency and duration (Smith and Tobin, 1979: 3). Following a positivist epistemology of separation between human society and nature, the ecological frameworks maintain that the natural events system and social system function independently of each other. It is the interaction between the two that creates resources on the one hand, and risks or negative resources on the other. The social and technical mechanisms enable people to find in nature what is of utility to them and to buffer what is harmful, which results in the development of complex sets of human adjustments to potential hazard impact. These adjustments are defined as the actions taken to reduce damage, to adapt to the hazard, and to remove or lessen the environmental constraints or resistances (Hunker, 1964). Failures to adjust to the natural events system cause hazardous impacts on humans and their activities.

Westgate and O'Keefe (1976) and O'Keefe *et al.*, (1976), however, express their strong disagreement with the nature-scientific and interactionist definitions of natural hazard. They argue that these approaches and definitions presuppose the interaction of a natural event and society, "but each of them is content to present the disaster occurrences—the static symbol of the disaster process—rather than the disaster process itself." These conventional approaches therefore fail to envision a continuing process of nature-society relationships and to determine the disastrous events as manifestation of that process. It is also noticeable that the interactionist definition of hazard wrongly denies the interdependence of human perception and adjustment to social processes by presupposing that adjustment decisions are related to individuals' characteristics and their free choice as independent variables (Baird *et al.*, 1975; Waddell, 1977; Hewitt, 1983; Morren, 1983; Cannon, 1994).

Based on research at the University of Bradford, with experience of field surveys in many developing countries, Baird *et al.*, (1975), Hewitt (1983), Morren (1983), and O'Keefe *et al.*, (1994) all suggest that instead of considering hazards as extraordinary events—as maintained by conventional definitions—it is more realistic to view natural hazards as extreme versions of situations in the everyday life of people in a given society. Such a concept is necessary to answer why the vulnerability of underdeveloped or developing societies to natural disasters tends to increase over time. These researchers further suggest that the analysis of marginalization and underdevelopment[20] is particularly pertinent to understand the increasing vulnerability to hazards in underdeveloped societies. 'Marginalization,' write Baird *et al.*, "is the process which leaves the underdeveloped populations more vulnerable [to natural hazards] than it was earlier to the vagaries of the environment" (Baird *et al.*, 1975: 29).

Hewitt (1983) criticizes the dominant ecological views which argue that hazards stem from intrusion into the 'human use system'—characterized by patterns of settlement and activity allowing effective and controlled use of natural resources—of unforseen, essentially independent natural processes of extreme and rare occurrence. Humanity's relation to nature, in this approach, is wrongly assigned two modes—one normal, secure, 'productive' and the other abnormal, insecure, and the occasion of losses. The dominant view also infers that persons and institutions are uniformly and unambiguously committed to eliminate known, manageable risks from everyone's life. Hewitt (1983: 23) questions this assertion, citing that there is evidence "that human groups and institutions are rather less fervent about equality and social justice than that." He insists (a) that most natural disasters are 'characteristic' rather than accidental features of the societies and places where they take place, (b) that the risks, uncertainties, pressures that bear upon awareness of and preparedness for natural fluctuations are generated within the context of 'ordinary life', rather than stemming from the rareness and scale of those variations, and (c) that the natural extremes are more expected and knowable than many of the contemporary social developments that pervade everyday life.

Over the last two decades, the search for an alternative view of natural hazards has crystallized into the *vulnerability* approach. As O'Keefe *et al.*, (1994) explain, when the probability of the extreme physical occurrence can be held as a constant, as seen in the last half century, logically then the explanation of the increasing number of disasters must be sought in an explanation of the growing vulnerability of the population to such events. In contemporary developing societies, as population continues to expand and as resources continue to be controlled by a minority, the actual standard of living declines for much of the world's population. This population is increasingly vulnerable to environmental variation as the process continues. Blaikie *et al.*, (1994) suggest that the vulnerability of people should be viewed in relation to their characteristics of predictive capacity, coping and resisting ability and recovery from the effects of a natural hazards. In this way, a temporal dimension is embodied in the concept of vulnerability as "the most vulnerable groups are those that also find it hardest to reconstruct their livelihood following disaster" (Blaikie *et al.*, 1994: 9). However, a caution is put forward here. As a rule the poor suffer more from hazards than the well-offs, although poverty and vulnerability may not be uniformly or invariably correlated in all cases. Thus, it is the social forms of disaster explanation which need significant attention.

In the context of vulnerability analysis, Cannon (1977; 1994) makes a distinction between *hazards* and disasters. As he puts it, "*hazards* are natural, but in general *disasters* are not, and should not be seen as the inevitable outcome of a hazard's impact" (Cannon, 1994: 13). The emphasis here is on the condition of the people which creates the context for a hazard to become a disaster. This includes the degree and types of their vulnerability in association with the technical issue of how society interfaces with the hazard in terms of mitigation and preparedness. Cannon insists that opportunities and risks in the environment for the society at large as well as for individual members are conditioned by the varying characteristics of social systems. This implies that there are no actual generalized opportunities and risks in nature, but instead there are varying sets of access to opportunities and exposures to risks. These features are invariably related to the socioeconomic systems. In vulnerability analysis, the key then is to interpret how social and economic systems place people at different levels of risk from extreme natural events. 'Vulnerability'—a measure of the degree and type of exposure to risk generated by different societies—explains this 'social causation'.

In view of the overall problems regarding natural hazards, I would like to assert that three major issues demand attention for an effective operational definition, particularly in the context of developing societies. First, it is remarkable that the majority (more than 90 percent) of current research efforts and resources are suggestive of technological or structural solutions to natural hazards problems (Lewis, 1975; FPCO, 1995). This emanates from an emphasis on the piecemeal and particularistic notions of the natural event occurrences. But a valid concept of natural hazards needs a comprehensive understanding of the various aspects of natural events and societal coping ability, given the level of socioeconomic development. Second, the effects of hazards depend both on the physical characteristics of the natural event and the state of social conditions. In regard to natural phenomena, it is noticeable that uncontrollable acts of nature like earthquakes, avalanches, and tsunamis, put the whole population of a certain geographical area at risk of death or loss, a situation which may be called a *general vulnerability* to a natural hazard. But the pervasive natural events like droughts, floods, riverbank erosion, coastal erosion, or snow, differentially affect various segments of a vulnerable population. In other words, two individuals with equal vulnerability within the same community may not have similar ability to adjust to hazard impacts, and thus their *specific vulnerability* or *proneness* is not equal (Westgate and O'Keefe, 1976: 62). The 'proneness' is to be measured in terms of an individual's situation and relationships within a broader social spectrum. Third, as opposed to vulnerability and proneness, a certain segment of society—the powerful interest groups and dominant social classes—could also benefit from natural disasters. This is because of their advantageous entitlements[21] to resources, and their capability to use the socioeconomic institutions and apparatuses in times of crisis or "disasters" (see Hewitt and Burton, 1971: 17). A double view of adverse effects and gains needs attention in natural hazards research in order to understand human coping response and adjustment systems, especially in the case of pervasive natural hazards.

There is little doubt that the usage of the term "natural disasters" is varied, implying different meanings and underlying postulations. Following Cannon (1994), I apply the terms 'natural hazards' and 'natural disasters' with a nature-society integrative connotation. The rationale of accepting "natural" is that the events are sudden and triggered by natural (i.e., geophysical, climatological, and biological) forces, but with the recognition that these forces are not necessarily the ultimate determinants or root causes. The NAPP framework suggests that hazards or their associated threats are natural or nonhuman, in terms of the triggering process. But the impacts of hazards after their realization in events—whether they create no adverse or disastrous outcome—require a human context. The outcome of a hazard's (natural, in a broad sense) impact necessitates a human element, and thus natural disasters imply a social construct of a naturally triggered event.

3.2. Natural Hazards Research

3.2.1. IN SEARCH OF A SOUL: THE DOMINANT VIEWS VERSUS THE ALTERNATIVES

Most research concerning natural hazards in relation to human components has been conducted during the past three decades.[22] Notably, a prominence of geographical and anthropological work has been clearly visible in the literature on natural hazards, but the issues and findings have generally been interdisciplinary in nature. Furthermore, a distinction can be made between the early studies which deal primarily with human cognition, perception and adjustment behavior in relation to hazards phenomena and more recent studies which focus on the aspects of hazard impacts in a variety of societal spheres. Crozier (1981: 106), in the context of the literature of the 1970s, characterizes these societal aspects of hazard impacts as the "marginalization" and "dependency" of underdeveloped societies. The natural hazards research discourse in the 1980s and 1990s turns more to encompass inherent socioeconomic processes rather than exogenous forces, and therefore focuses on human vulnerability and proneness to explain causes and remedies of disasters (Varley, 1994).

As an area of academic research, natural hazards analysis is still in a formative stage. The contemporary interest in the field has developed since the 1950s, when a study group at the University of Chicago under the leadership of Gilbert White promoted the 'natural hazard paradigm'—these ideas have since received global interest and have resulted in widespread research activity, especially with respect to the developed world (White, 1973; 1974a). While White's initial work on flood problems in the United States was primarily concerned with the benefits and effects of large-scale preventive and structural schemes (White, 1973), his focus has gradually shifted to the study of perception and response of individuals to hazardous environments in which risk and uncertainty exist (see Kates and Burton, 1986a; 1986b). For instance, the prime concentration of international collaborative research in the late 1960s, which involved researchers from fifteen nations and focused on nine hazard phenomena, was to determine how people perceived and appraised the occurrence of hazards, and to identify the range of possible adjustments to those hazards (Walmsley and Lewis, 1984; Burton *et al.*, 1978). Gold (1980: 213) also points out that natural hazards research in the 1970s adopted all or part of the Chicago-based "ecological approach" in which the general goal was to further strengthen behavioral studies through an accumulation of findings on a range of hazards within a well-defined paradigm. Indeed, the policy-oriented nature of the paradigm as well as its implicit support of the *status quo* politico-economic systems, facilitated a widespread acceptance of its cross-cultural settings.

In the face of a growing interest in *critical science* in human geography since the 1970s (Holt-Jensen, 1980), the neopositivist-behavioral framework of natural hazards research has come under attack from several scholars (e.g., O'Keefe, 1975; Wisner *et al.*, 1976; Cannon, 1977; 1994; Varley, 1994; Blaikie *et al.*, 1994). It has been argued that, analogous to deterministic views, the ecological framework wrongly concedes that natural hazard occurrences are the result of "natural" and inevitable operations of physical processes (Waddell, 1977; Smith and O'Keefe, 1980; Haque and Parvin, 1986). Also, the notion that disaster occurrences are interfaces between a vulnerable population and an extreme event has failed to explain the social context in which a certain segment of a population becomes particularly susceptible to hazard effects.

Despite such criticisms against the behavioral-ecological approach to natural hazards, the search for a comprehensive alternative mode of analysis of the phenomena has been limited (e.g., Blaikie *et al.*, 1994). Notably, the general emphasis of the critical or vulnerability approach turns to the relevance of social structure and related variables, including the political economy. Analysts with this critical orientation have claimed that the characteristics and intensity of hazard effects are related to the nature of social formation and historical process (e.g., Waddell, 1977; Wisner *et al.*, 1977; Cannon, 1977; 1994; Smith and O'Keefe, 1980; Varley, 1994; Blaikie *et al.*, 1994). Overall, the work from this perspective has demonstrated:

> the important extent to which natural disaster, its causes, internal features and consequences are *not* explained by conditions or behavior peculiar to calamitous events. Rather they are seen to depend upon the ongoing social order, its everyday relations to the habitat and the larger historical circumstances that shape or frustrate these matters (Hewitt, 1983: 25).

Recent reviews (e.g., Varley, 1994; Cannon, 1994) acknowledge that the 'vulnerability approach' has very limited success in offering concrete conceptual tools which could lead to the development of human actions for mitigating disasters. In addition, international agencies involved with developing disasters-related policies in general have used the concept of 'vulnerability' as a "technical" rather than a social measure (Kreimer and Munasinghe, 1992; Parker, 1992), and have in so doing 'marginalized' the vulnerability approach (Varley, 1994). Two reasons are identified for this misleading usage (Varley, 1994). First, a number of contributors to this approach write from an avowedly Marxist perspective, and call for political intervention. In Cuny's (1983: 7) words, "reducing the vulnerability of the poor is a development question . . . that . . . must be answered politically." In the perspective of governments or international agencies, the inherently political nature of reducing vulnerability is likely to render this kind of intervention problematic (Kreimer *et al.*, 1992). Secondly, vulnerability analysis has not received the full attention it deserves because of its concentration on unequal distribution of resources and poverty. Although Cannon's (1994) research shows that vulnerability and poverty are not the same, in practice it often is the poor who suffers most, in relative terms, from a disaster (Cuny, 1983; Anderson and Woodrow, 1989). But eradication of poverty, though socially desirable, cannot be achieved in the short-term. In a comparable context, Torry (1986: 13) cautions that researchers overly stressing the *ultimate* causes of disasters run the risk of "beating around the analytical bush."

By recognizing that the behavioral-ecological approach has severe shortcomings and that the alternative structural and vulnerability paradigms of natural hazards research generally have not been able to influence policy issues profoundly, the present study aims to test the hypotheses of the dominant approaches as well as their criticisms in an empirical context. My analysis then follows implicitly a conceptual bias towards the issues and postulations put forward mostly by structuralists concerning the relationship between society and nature.

An understanding of the methodological procedures and problems in hazards studies is required of us to determine the appropriate instruments to make full use of a particular research paradigm. In this way, I attempt in the following section to critically evaluate various methodologies and the extent to which they examine human responses to hazards in terms of cognition, perception and coping behavior.

3.2.2. METHODOLOGICAL PROBLEMS IN HAZARDS STUDIES

Our discussion in the preceding section indicates that there may be some serious conceptual weaknesses in the dominant approach to natural hazards studies. Several methodological problems also arise in the mainstream of research on natural hazards issues. Walmsley and Lewis (1984), in this regard, claim that problems in hazards analysis are caused by the unique nature of the field in comparison with other social sciences. The lack of predictability of natural events implies that research can only be conducted as hazards occur, or more usually, after they have occurred. Given such time frames, the major and dramatic aspects of hazards usually receive attention and often provide a distorted impression of people's actual behavior (Seehan and Hewitt, 1969; Hewitt, 1983). On the one hand, the post-event surveys of respondents' past behavior create the risk of encouraging *post hoc rationalization* since respondents are allowed time to ponder their actions. Alternatively, investigations immediately following a hazard occurrence place the researcher in a highly reactive environment "where responses are given as much to the presence of an intruder as to the questions asked" (Walmsley and Lewis, 1984: 112).

Hewitt (1983), in his seminal work, questions the benefits to a technocratic approach of classifying or dividing hazards. The problem here is handled by an extreme narrowing of the range of interpretation and acceptable empirical evidence. What is commonly cited across the various specializations of geophysical fields is the attempt to reinforce the separateness of disasters and their causes. Proponents of this approach try to determine hazardous events in step-like changes of geophysical measures that result in disastrous damages. Human susceptibility is essentially linked to these extreme 'pulses' in nature through such postulations as 'damage thresholds.' These above reductionist notions can easily convince and assist the prevailing institutions and powers, but they are both conceptually and methodologically constrained due to three features: (1) the significant extent to which a natural hazard is not explained by, nor uniquely dependent upon the geophysical processes that may trigger damage; (2) the significant extent to which human awareness of and responses to natural hazards are not associated with the geophysical conditions; and (3) the important degree to which a natural disaster, its causes, internal features and impacts are not explained by conditions or behavior peculiar to calamitous events (Hewitt, 1983: 24-25).

During the last two decades, various statistical and psychological methods and techniques have been employed to appropriately and reliably depict human cognitive processes and characteristics, consciousness, and feelings about risks in the environment, especially in association with natural hazards. The validity and effectiveness of these techniques can be questioned as to what degree they are capable of encompassing the very dynamic elements of human spirit and emotion, subjectivity, values and culture. For example, questions arise with regard to the differential meaning of hazards to the people they affect (Saarinen, 1974; Golant and Burton, 1976). Golant and Burton (1976) employ the psychological technique of 'semantic differential analysis'[23] to explore preconceived ideas and feelings held by individuals concerning hazard phenomena.

Dissatisfaction with the application of this sort of technique, however, has been expressed by a number of analysts (Hudson, 1980; Gill, 1982; Walmsley and Lewis, 1984); among them, Hudson is most critical of the behavioral methods. In his extensive review of behavioral techniques used on environmental images, Hudson (1980) argues that the semantic differential technique fails to relate individual images to the structure of a group or to the generalized image in a valid or meaningful way. As a result, he says, the technique ignores the social constraints which influence learning. He cautions that this does not imply that such constraints wholly determine behavior and completely eliminate an element of individual action, but that they set limits which constrain the degree of choice open to individuals (Hudson, 1980: 355). Gill (1982: 25) adds that the major drawback of the behavioral technique is that a preferred judgement of factors is inherent in the procedure because the rating scales are preselected by *a priori* notions of the researcher.

With a strong emphasis on the individual's cognitive-learning process as central to understand his or her adjustment strategies, Saarinen (1966) and Brooks (1973) adopted a modified 'apperception test' to isolate *personality* traits.[24] Saarinen, however, confirmed only a limited degree of relationship between personality and perception of hazard compared with geophysical characteristics of an area and past experience. Mitchell (1979: 212) draws attention to a number of weaknesses of these psychological methods applied by Saarinen and his adherents. Particularly, Mitchell notes that the standardized photographs of a presupposed ambiguous scenario often are not found to be ambiguous in the vision of the respondents. This occurs mainly because of the familiarity of respondents with similar landscapes and environments shown in standardized photographs. Thus, the effectiveness of the identification of personality characteristics by this method is doubtful since the photographs are not amenable to a wide variety of interpretations.[25]

Following the ecological concepts of nature-society relationships, several investigators attempt to adopt a general systems framework into natural hazards research. In this way they attempt to explain the various elements and relationships involved in human adjustments to natural hazards (e.g., Kates, 1971; Kirkby, 1973; Saarinen, 1976). Although the general limitations of the ecological approach have been succinctly noted in the preceding chapter, some methodological problems are worth reiterating here. Johnston (1983b: 44-5), for instance, demonstrates that systems raise numerous problems in calibration or fitting the linked equations to data; specifically, problems arise if modelling of the system is incomplete or data are not available for all the links. The latter is more common in most underdeveloped countries. Also in this regard, Johnston argues that:

> The systems approach makes more explicit the naturalistic assumption underlying the positivist conception of social science. Systems of interest to human geographers are treated *as if they were machines*, comprising linked working parts which have little or no independent existence and whose position in the system are predetermined (Johnston, 1983b: 45).

Although the underpinnings of behavioral techniques bear the conceptual burdens of positivist conceptions of humanity and nature, "it would require substantial testing before vindicating Kates's belief that this *could* be an operational model" (Gold (1980: 210).

Some serious reservations are thus given to the application of universal concepts and methods to natural hazards studies. The issues in this regard center around the effectiveness of adopting the findings of research in developed countries to contemporary developing nations. Brookfield (1964), in attacking such a blind generalization, asserts that although the mean results of a large number of observations in a statistically controlled way may serve some purposes, they may also be very illusive. In many cases there are limits to the range of territory over which results hold their validity. Also, such ahistorical methods of functional analysis are not capable of uncovering the prevailing relationships in a given society that have evolved from historical-structural processes.[26] As such, Brookfield recommends the development of an understanding of social organization and its underlying processes, using participation-observation to overcome such shortcomings. He argues that such an alternative method is necessary because:

> when an individual human geographer is sitting down in one small corner of a foreign land, and seeks to interpret the geography of that small corner, then it is difficult for him to do so without trying to comprehend the perception of environment among the inhabitants (Brookfield, 1964: 287).

Saarinen (1974) also illustrates some practical problems in using a standardized questionnaire with an implicit universal character. Among many drawbacks, he cites the difficulty in adapting the questionnaire to other sociocultural contexts and languages. He notes that the opinion-seeking methodology of an open-ended questionnaire used in developing-world situations results in both incomprehensible dimensions and in the reflection of the investigator's ideas because of the respondent's total dependence upon the interviewer. Consequently, the findings do not necessarily touch the cognitive world of the respondent, but rather represent intentions and meanings in the interviewer's own terms. As Waddell (1977) suggests, a substantial change in the research methodology, operational procedure, and content of standardized questionnaires (based on Western experience) is necessary to analyze developing societies. The success of a research methodology to depict human responses to natural hazards in such cases depends on its ability to identify local factors, as well as to build up generalizations for a better understanding of the phenomena in cross-cultural contexts.

3.3. Human Coping Responses to Natural Hazards

3.3.1. COPING ADAPTATIONS AND RESPONSES

The everyday life of an individual is profoundly influenced by a complex set of inherent and internalized elements such as physiological, psychological, and cultural traits, and the social environment in which he or she lives. These social conditions, in association with technological level, resource and capital, and access and entitlement, chiefly characterize the nature of risk and "proneness." Human actions in dealing with both 'expected' risk and unknown risk basically stem from the conditionality of a given society—that is, the way the social system makes people aware of the hazard and the possible outcome of human action or intervention in mitigating the effects of natural processes. Human coping behavior to address perceived and often unforeseen risks—"the manner in which people act within existing resources and range of expectations of a situation to achieve various ends" (Blaikie *et al.*, 1994: 62)—is then linked to the individual's subjective intentions and social allocation of resources and accessibility. In turn, the coping responses undertaken to save lives, properties, assets and other types of resources include defence mechanisms, active measures to solve problems, and methods for handling stress (Murphy and Moriarty, 1976).

As it is often assumed, the objective of coping responses is for survival in the face of adverse events. Blaikie *et al.*, (1994) find that although the above assumption is common, it blurs other important objectives which can be assessed when applying Maslow's hierarchy of human needs (Maslow, 1970). The peak of the hierarchy is the need for self-realization; this is founded upon an acceptable standard of living—a lower-level need than self-realization. Lower needs may also include adequate shelter and food for comfortable survival, while other needs near the base of the hierarchy will include minimum security from violence and starvation. In a recent work, Doyal and Gough (1991) conclude that failure to fulfil a "core" of basic human needs means that other needs cannot be met (see Wisner, 1988). Coping responses in the face of adverse circumstances may then be found as a series of adaptive strategies to protect needs as high up the hierarchy as possible in the face of threat (see Blaikie *et al.*, 1994).

Human coping responses to natural hazards, in general terms, reflect human knowledge, perception and efforts to avoid or minimize adverse effects. The intentions and actions involved in coping and adaptive processes to any particular hazard are normally put into effect within a short span of time, and in most cases they are temporary relative to the ecological-cultural adaptive process. Coping adaptations are therefore to be separated into those that are *purposefully* undertaken, and into other *incidental* actions that have the effect of minimizing potential losses (Burton *et al.*, 1993). An example may help to explain this variation. Instead of using brick, steel, concrete, and other durable housing materials, floodplain occupants in Bangladesh use thatch, bamboo, wood, and corrugated iron sheeting. Houses made of these materials are readily moveable in emergencies, and thus are far less susceptible to loss by riverbank erosion than are permanent brick structures. I observed that none of the sample households in the Brahmaputra-Jamuna floodplains had brick or concrete structures, although the general poverty of rural communities is also a major reason for the use of local and cheap materials.

The modes of coping with hazard can be grouped into two strategies: (1) avoidance of loss, which can be achieved either through *preventive* action at the collective, state or individual level or by *change in resource-use* or location (Burton *et al.*, 1993); or (2) mitigation of loss, attained by an array of measures to minimize loss and facilitate recovery. The above strategies differ significantly between individuals or groups with different patterns of access (Blaikie *et al.*, 1994). Two broad generalizations can be made in this respect. First, the purpose of many responses is to secure requirements quite low on the hierarchy. Hence, it may be preferable to someone to improve access to a minimum level of shelter, food and physical security than to increase income and asset-inventory. Second, in a socially and ecologically risky environment, maintaining command of these requirements may need diversification of options and access to resources. Blaikie *et al.*, (1994) include six other coping strategies for adverse events that are perceived to have precedents consisting of actions prior, during, and after the event. These are: (1) creation and maintenance of labor power, (2) enlargement of food and asset inventory, (3) diversification of the production methods, (4) diversification of income sources, (5) development of social support networks, and (6) adoption of post-event coping strategies based on event-experience.

The proponents of the Chicago school of hazards research, i.e., the behavioral school, assume that individuals who have either common access to information and resources or are unaware of risks because of their ignorance or certain beliefs and faith, do not attempt to alter the causes of hazards (Burton *et al.*, 1993: 59). These individuals eventually accept and bear the impact independently and may not pass the losses on to others in overt ways. The above assumptions are questionable—individual acceptance of risk, threat, and suffering should be viewed in terms of its contextuality. In societies where state and social institutions largely determine access to resources and location of residence and work, individual ability and power are constrained. Acceptance of loss due to adverse events here is not an option but a state or condition created by external forces. However, depending on social notions and prevailing political-legal systems, a given society may redistribute the loss-bearing among a broader spectrum through insurance schemes or relief and rehabilitation measures.

Our understanding of human coping responses to natural hazards, particularly from a human geographical perspective, has developed with an evolutionary course. According to the conventional approaches, human coping responses to natural hazards are generally "related both to perception of the phenomena themselves and to awareness of opportunities to make adjustments" (Burton *et al.*, 1978: 34). Geographer's concerns about human responses to disasters usually concentrate on people's long-run persistence in occupying a hazardous environment, the process of risk perception and appraisal, and on the adjustment strategies as overt behavior.

The beginning of contemporary research into human responses to hazard was White's (1945) study of *Human Adjustment to Floods* in the United States. The study explored the possibility of alternative solutions to technological measures which created an illusion of security and consequently even greater damage (also see Kollmorgen, 1953; White, 1961; 1964). The pertinent questions posed by White and his co-researchers provided a general research problem: "How does man adjust to risk and uncertainty in natural systems, and what does understanding of that process imply for public policy?" (White, 1973: 194). The most significant contribution of White's studies was its seminal influence on subsequent research, especially in terms of attempts to develop an integrated conceptual-empirical paradigm (e.g., Burton, 1962; Kates, 1962; Saarinen, 1966; White, 1974b; Burton *et al.*, 1978).

The subsequent development of Simon's (1959) model of "bounded rationality" changed the direction of research on human response to hazards. His concept of "satisficing" replaced the conventional economic optimization mode of human behavior, which was based on the assumption of "rational economic man." Kates (1962), employing this notion of bounded rationality, investigated the perception and behavior of floodplain residents in La Follette, Tennessee, as well as five other communities in Indiana, Wisconsin, California and New York. The implication of adopting the concept of bounded rationality is that, in contrast to a more positive behavioral perspective, an individual is considered to have only incomplete knowledge, is often incapable of comparing alternatives, and does not necessarily strive for optimum solutions. In contrast, the theoretical 'bounded rational individual' systematically weighs all possible alternatives and seeks to maximize opportunities; this neopositivist thinking, however, does not explain the basis of human expectation, which is postulated as being influenced by "early socialization experiences" (Brody, 1970: 14). The limitations of the behavioral approach were well reflected in Kates's (1962) findings that expectancy of future flooding was not related to age, income, or education, but rather to past flooding experience.

The bearing of social processes upon individual behavior has been stressed since the publication of Kenneth Hewitt's collection *Interpretations of Calamity From the Viewpoint of Human Ecology*. This work challenges the dominant naturalistic-technocratic or behavioral-ecological stances, and demonstrates how macro-level social, historical, political, economic and other pertinent factors condition people and guide their options to act or react. Natural disasters and their impacts "depend upon the ongoing social order, its everyday relations to habitat and the larger historical circumstances that shape or frustrate these matters" (Hewitt, 1983: 25).

The significance of the political economy perspective in human coping strategies to natural hazards has been explained by Varley (1994). She suggests that a 'vulnerability analytical framework' is capable of answering why those who know about affordable techniques for building homes better able to withstand earthquakes fail to adopt them. People may face a limited choice of construction materials with which to build their homes, but what determines whether or not the needed materials are available? "In short", Varley writes, "the explanation can never be as simple or matter of knowledge possessed by individuals viewed in isolation from the rest of society, since society is not merely an unstructured sum of individual decision makers" (Varley, 1994: 4; also see Watts, 1983).

3.3.2. DETERMINANTS AND EXPLANATORY FACTORS OF COPING RESPONSES

The inquiry into the factors related to perception of, and alternative coping responses to hazards has received a great deal of attention during the past three decades. Earlier studies, focusing primarily on a few selected developed countries (viz., United States, United Kingdom, Canada, and Australia), have asserted that the human responses to disasters are associated with (a) the nature of resource use, (b) frequency and magnitude of past hazards; and (c) personal experience of them (White, 1964; 1973; 1974b; Kates,

1962; 1971; Burton, 1961a; 1961b; 1970; Penning-Roswell, 1972; Penning-Roswell and Underwood, 1972; Sewell, 1965; 1969; Heathcote, 1969; 1974; Brooks, 1972; 1973). Following Kates' model of human adjustment to natural hazards, much empirical research has been carried out in the West since the early 1960s (see White, 1974b), these studies have generally identified a) previous experience, (b) attachment to place, and (c) potential hazardousness of an area as the principal correlates to human coping responses to hazards.

Previous Experience

Part of the conventional wisdom of hazard studies suggests that experience of an event heightens the awareness of future events (Walmsley and Lewis, 1984: 115); such a hypothesis is supported by a wide body of evidence (Kates, 1962; Kirkby, 1974; Oliver, 1975; Irish and Falconer, 1979; Gold, 1980). Close scrutiny of the literature on this issue, however, provides a contrary perspective. For example, Saarinen (1966), Oliver (1975), Irish and Falconer (1979), and McDonald (1979), in dealing with flood and drought hazards in the United States and Australia, observe that previous experience and duration of residence have a positive contribution to the awareness of future hazards and adjustment measures. Blaikie *et al.*, (1994: 64) find that people make their decisions on the assumption that a particular risk will appear for which they have some experience; thus, all coping strategies for adverse events have a perceived precedent. However, Kates (1962: 140) argues that although people on floodplains appear to be very much 'prisoners of experience,' the effect of such experience does not consistently proceed in the direction of taking individual action to reduce damage (also see Hanson *et al.*, 1979). Also, Jackson and Mukerjee (1974) suggest that very few inhabitants of San Francisco consider the threat of earthquake seriously, despite the occurrences of six major earthquakes in California during 1872-1972. As mentioned earlier, both Saarinen (1966) and Kirkby (1974) observe that people tend to be consistently optimistic in their assessment of hazard occurrences. Kirkby, by examining the individual and community responses to rainfall variability in the Oaxaca valley of Mexico, confirms that people's memory of specific natural events tends to be limited in duration as well as confined to larger events. The above evidence, in aggregate, suggests then that the recollection of past experience of hazards by individuals may be less than perfect. We can therefore argue that awareness and action relating to hazards does not necessarily develop from the lessons of previous experience.

Attachment to Place and Potential Hazardousness of an Area

Statistics show that recent decades have witnessed a significant rise in the risk of natural disasters despite improvements in preparedness and mitigation responses. Alexander (1993) identifies four reasons for this trend. First, geographical inertia and lucrative economic returns in specific locations have led to a continued inhabitance of past and potential disaster zones. Capital may have subordinated all other concerns in our commodity-driven global order, placing economics at the top of the social need hierarchy. Second, particular risks or dangerous conditions have often been accentuated by managerial or bureaucratic malpractice. Third, the processes of physical capital accumulation and encroachment into natural regions have led to capital and population concentration in hazardous locations or regions. Finally, there has been a lack of understanding and recognition of the need for a concerted effort to address extremely adverse events.

The special interest of geographers in the process of human occupance of a varied natural environment has led them to deal with the issue of people's emotional *attachment to place*. Yet Gold (1980: 207) believes that despite its profound role in determining human adjustment strategies to hazards, this aspect has received relatively little attention. Research on this issue began with the need to understand why stressful environments, like river basins or sea shores, are relatively densely settled even though these areas are generally prone to natural hazards. In explaining the continual settlements on the Nile, Tigris, Euphrates, Hwang Ho, Indus, and Ganges, Tuan (1974: 114) points out that the flood hazard posed by these rivers has been formidable, but the economic advantages of living there, and the cultural significance with which these places are endowed—what he calls "environments of persistent appeal"—outweigh the negative features. Gold, in this respect, asserts that:

> The rivers were tyrants to be mastered; once mastered, they were symbols of that achievement as well as the 'string' that binds the society together. Attachment to place is not only a feature of such illustrious examples, but a strong force which can be witnessed whenever people seek to return to their home areas after the impact of a natural disaster and to reconstruct their lives there (Gold, 1980: 207).

Jackson and Mukerjee (1974: 163) support this view with evidence from a study on adjustments to earthquakes in San Francisco. They found that the attachment of the population to San Francisco was so deep that not even one respondent included the risk of earthquakes in their perceived disadvantages of living in the city. This situation also was reflected in the expression by more than two thirds of respondents to why they remain in the city with the typical answer that "I *love* the city." Oliver-Smith (1982) also found a similar tendency among victims of a Peruvian earthquake in 1970, in that the sociocultural identity of the Yungay city-dwellers was strongly tied to the image of the area. Above all else, the Huascaran peak was considered to be the core element in the identity of the region.

It is apparent that the aesthetic and emotional attachment to a place, along with the social and cultural identity of people, is a predominant feature not only in developing countries but also in industrialized societies. This results in individuals settling and staying even if the area is known to be generally hazardous. The interrelationships between these variables, namely, potential hazardousness of an area and people's attachment to a place, are thus complex. For example, as discussed above, Tuan (1974), Jackson and Mukerjee (1974), and Oliver-Smith (1982) all find that people are strongly attached to the image of a place even though that place may be known to be hazardous to scientists and objective observers. In contrast, Burton *et al.*, (1978), Saarinen (1966), and Brooks (1972, 1973) assert that such perception varies in accordance with the potential hazardousness of a place; for example, persons in zones of high bank erosion potential will show a remarkable sense of erosion perceptivity, while those living at more distant areas from the river will have lower levels of perception. The populations of the intermediate area will tend to fluctuate between two extremes in their perception.

Overall, it appears that both in terms of attachment to a place, as well as potential hazardousness of an area, the distance of residence from the riverbank or the zoning of the floodplain, on the basis of the potential hazard event, is of great importance in the analysis of the human response system to riverbank erosion hazard. Nonetheless, underlying the above psychosocial features of a sociocultural identity and image of a place, the economic benefits and resource base of an area are still the fundamental attractions for human occupance.

Social Stratification Versus Social Class Structure

The significance of the socioeconomic variables which correlate with human perception and adjustment behavior to a hazard has only recently been recognized in the literature on natural hazards. As mentioned above, we can identify two major trends in the earlier studies of natural hazards in the Western world. First, because of their over-emphasis on individual-personal characteristics from a behavior standpoint, these studies usually exclude an examination of socioeconomic indicators as determinants of human response to hazards (e.g., Burton, 1962; Sewell, 1965; Saarinen, 1966). Second, the few empirical studies which do incorporate socioeconomic variables commonly find no significant relationship between these factors and the human responses to natural hazards (e.g., Roder, 1961; Burton, 1961a; Kates, 1962). For example, Roder (1961), Burton (1961a), and M. Islam (1974) show no relationship between socioeconomic grouping and the expectation of further flooding in the study areas.

Several later studies, however, observe quite contrasting patterns, indicating a positive correlation between socioeconomic variables and human response to hazards (e.g., Baker and Patton, 1974; Leigh and Sim, 1983). Remarkably, with few exceptions, such evidence is concentrated in developing societies. Baker and Patton (1974), for example, notice a positive relationship between high status groups and better educated people and the attitudes towards damage prevention adjustments to hurricane hazards. The policy implications of such findings suggests that relatively intensive efforts should be directed towards the lesser-educated segments of society in order to counteract their lack of confidence in taking adjustment measures. Leigh and Sim (1983: 49-50), in an examination of attitudes and adjustments to flood hazards in a mixed community of Malaysia, confirm a statistically significant relationship between risk of future flood and educational level. People in the study area with only a few years of schooling tend to view floods as cyclical events rather than random ones, implying an attitude that "it won't happen again for a long time."

Nonetheless, opposition to the hypothesis that human responses are dependent upon some independent variables of *social stratification* (viz., income and education) has also been expressed quite forcefully in recent works on natural hazards. We can find two distinct routes in such alternative approaches to so-called Weberian social stratification and neoclassical economic growth. The arguments in the first route suggest a causal relationship between natural disasters, particularly drought in the African Sehelian region, and colonial invasion, including the encroachment of the 'world capitalist system' into pre-capitalist societies (O'Keefe, 1975; Cannon, 1977; 1994; Watts, 1983). Cannon (1977), for example, demonstrates that the transformation of indigenous agricultural and pastoral activities to commercial production displaces local communities from relatively better land. This eventually leads to over-grazing of poor land, soil erosion, and the deterioration of the physical and social environment. O'Keefe (1975: 20) further asserts that the adoption of alternative strategies to counteract drought among people in semi-arid regions of Africa was hindered by the "impress of colonialism." Watts (1983: 33) adds that while these 'peripheral capitalist' societies become vulnerable to capitalist crises by virtue of their incorporation into the capitalist world economy, their partial transformation allows them to remain vulnerable to crises usually associated with underdeveloped societies. Further, while the indigenous mechanisms have been lost, the peripheral societies are unable to acquire those characteristics of capitalism that help mitigate natural disasters.

The second analytical route that disapproves of the functional stratification method in explaining the causes and impacts of natural hazards places emphasis instead on internal *social class* structures rather than exogenous factors. Wood (1981: 4), in an appraisal of class analysis in "postcolonial semi-feudal societies" like Bangladesh, argues that the Weberian view of class in terms of social stratum and status precludes the necessity of considering a class position which primarily determines one's consciousness (also see Wiest, 1979: 182). Ahmed (1973: 211) clarifies this emphasis on class analysis, stating that "Classes exist in relation to the means of production. This relation defines the positions people occupy in the organization of production, which in turn refers to a hierarchy of a composite of social, economic, and political differences." The analysis based on social functional groupings is thus unable to reflect the dynamic aspects of "social relations of production and exchange" in which human economic activity and forms of cultural and political relations are located. Wood further points out that the concept of class can only be adopted in terms of its 'relational aspects,' and therefore cannot be defined by reference to internal attributes—a class only exists in relation to other classes either through opposition or through domination and bondage (also see Ossowski, 1969). Thus, the relational analysis "can go beyond the level of appearances and provide an account not only of *where* different categories of people are located in a system but *how* they are located in terms of these features of structural fragmentation" (Wood, 1981: 4).

By adopting the social class perspective as an analytical tool, several analysts indicate that the indirect effects of natural hazards—primarily created by social motivation and manipulation of the abnormal situations (i.e., caused by extreme natural events)—may have a more profound impact than the direct losses due to the hazard (see Sen, 1980, 1981a, 1981b, 1982; Alamgir, 1980; Muqtada, 1981). For instance, Alamgir (1980) and Sen (1981a; 1981b) demonstrate that floods, droughts, and tidal bores not only cause a loss of cattle and human lives, crops, and other assets, but they also trigger a process of starvation and famine by compounding the inequalities in resource distribution. In this connection, Sen (1980) suggests the adoption of the *entitlement approach* to investigate the human response and adjustment characteristics to natural hazards and their related features (such as starvation and famine). He recommends that we see

> what a person is entitled to in a given society is constrained by a collection of relationships covering the legal system (e.g., ownership laws), the political system (e.g., presence or absence of social security or right to work), [and] economic and cultural influences (e.g., the family system of sharing incomes and food (Sen, 1980: 615).

I give special attention in this book to the role of *social structure* in determining the human responses to extreme natural events and the magnitude of hazard effects. Following Keesing (1981: 519; cited in Haque, 1988c: 48), *social structure* is defined here as "the organization of a group or society . . . in terms of structures, of positions and roles: a formal abstraction from the ongoing social relations within communities." In contrast to the concept of social stratification, which emphasizes differential ranking or status of social groups, classes, or categories, an analysis of class structure stresses relational aspects of social groups, mainly in terms of the ownership of the means of production. In this study, particular stress is given to the relational aspects of individuals in terms of landownership patterns. As well, economic activity and the labor process are considered pertinent variables in exploring the process of human coping adjustment to hazards in a broader social context. Among demographic variables, only household size is considered especially relevant to socioeconomic structure because of its direct implications for manpower and the labor force structure of floodplain production units (i.e., households). Besides its differential economic value among social classes, household size is also of significance for sociopolitical power in the 'semi-feudal' rural society of Bangladesh.

Nonetheless, the variables which reveal social stratification, such as income, education, and social status, will be used in this study to substantiate our analysis of variables concerning social structure (Chapters 7 and 8). The potential of other variables for revealing the relationships between human coping responses and previous disaster experience, as well as human coping responses and hazardousness of an area, will also be given due attention in the following chapters.

Part Two

Riverine Hazards and Human Ecology: Bangladesh

CHAPTER 4

PHYSICAL DIMENSIONS OF RIVERINE HAZARDS IN THE BENGAL BASIN

The Case of the Brahmaputra-Jamuna Floodplain

What are the nature and magnitude of the environmental features of Bangladesh's deltaic plains and floodplains? How do the events associated with these environmental features pose threats to human lives and resources? Answers to these questions are fundamental to understanding the nature of environmental hazards, and requires us to unravel the dimensions of *physical exposure*. Specifically, since the range of environmental events and their statistical variability profoundly affect the ability of humans to cope with potential threats and actual disastrous events, the assessment of physical dynamics is an utmost necessity.

The plains—including both deltaic and floodplain formations—of Bangladesh are among the most dynamic landform systems in the world. The rivers of this country form a lifeline for her people, providing an ecology endowed with fertile land and an immense supply of water. However, the volatility in terms of the physical parameters of these riverine systems often exceeds the coping abilities of floodplain inhabitants. Extremely high floods and the associated changes in channel form and structure through either radical shift or bank erosion often pose serious threats in the active floodplains. Peak discharge (or flood) and riverbank erosion can be closely correlated. Ample literature is available on fluvial geomorphology which depicts the interrelationship between river flow or discharge, and channel dimensions (Schumm, 1972; Morisawa, 1985; Knighton, 1984). As a river channel is basically the outcome of flowing water, the magnitude and frequency of runoff events can be a major factor in determining bank stability of a river channel. Leopold and Maddock (1953) used the term *hydraulic geometry* to describe such interrelationships among hydraulic factors, especially depth, width, velocity, suspended load and discharge, which influence the shape of natural stream channels.

In identifying the role of controls for channel form adjustment, Knighton (1984) isolates discharge and sediment as the most dominant independent variables, which in turn integrate the effects of climate, vegetation, soil, geology, and basin physiography (see also Morisawa, 1985: 71). The interrelationships of these two variables with other dependent variables are not easily depicted.[27] Morisawa (1985), in dealing with the overall geometry of a river, concludes that at each reach the morphology of a river is an outcome of the interaction of the following factors: hydraulics of flow (i.e., velocity, discharge, roughness, and shear), channel configuration at that reach and immediately upstream (i.e., width, depth, shape, slope, and pattern), load entering the reach (i.e., calibre and amount), and the nature of bed and bank material. Adjustment mechanisms then include erosion and deposition in the channel (altering form, slope or pattern) and the creation and movement of bedforms. It follows that discharge, bed and suspended sediment, slope and consequent channel patterns are the most important aspects of a river system.

In this chapter, my focus turns to: (a) discharge and its variability, in terms of flood-peak, and (b) the resultant channel shifts and bank erosion processes. In order to illustrate the physical dimensions of extreme riverine events in the floodplains, the physical (i.e., geological, meteorological and hydrological) characteristics of the region are discussed. This is followed by an examination of the nature of peak discharges and consequent flooding in the Bengal basin. The final section examines the Brahmaputra-Jamuna floodplain from a geomorphological perspective as a specific case of river channel shifting and bank erosion problems.

4.1. Physical Characteristics of the Bengal Basin

4.1.1. GEOLOGICAL FORMATION AND TECTONICS

The Bengal basin—a geosyncline in geological terms—occupies the entire political territory of Bangladesh with the exception of the Tertiary hills of the eastern and southeastern regions. The remaining portion of the basin lies within the Indian territory of West Bengal. The Bengal basin is bordered on the north by the Shillong Hills or the Assam Plateau, which consist mostly of Eocene sandstone and limestone. The elevations of the plateau approach a range of 1,372 to 1,829 meters above sea level. On the northwest and west, the basin is bounded by the peninsular shield of India or the Rajmahal Hills, consisting mostly of Lower Jurassic traprock, and averaging 152 to 244 meters above sea level. The eastern boundary is formed by the Naga-Lusai folded belt or the Chittagong-Tripura Hills, ranging in elevations from a few meters to about 610 meters. The southern fringe of the basin is open towards the Bay of Bengal (Figure 4.1; Morgan and McIntire, 1959; Coleman, 1969).

In light of existing knowledge, the geological formation of the Bengal basin can best be described in the broader context of the Plate Tectonic theory. The Plate Tectonic theory suggests that the formation of the Himalayan mountain ranges was the product of a northerly movement and collision of the Indian Tectonic plate with the East Asian and Eurasian plates (Gordon, 1972; Molnar, 1986). Since this collision, which took place about 40 million years ago, the Indian plate continued subducting underneath what is now Tibet, the southern edge of the main Eurasian plate. This process moved the Indian plate under the Tibetan at a rate of five centimeters a year, and elevated the central Great Himalayas about one centimeter a year (Molnar, 1986: 154), creating the

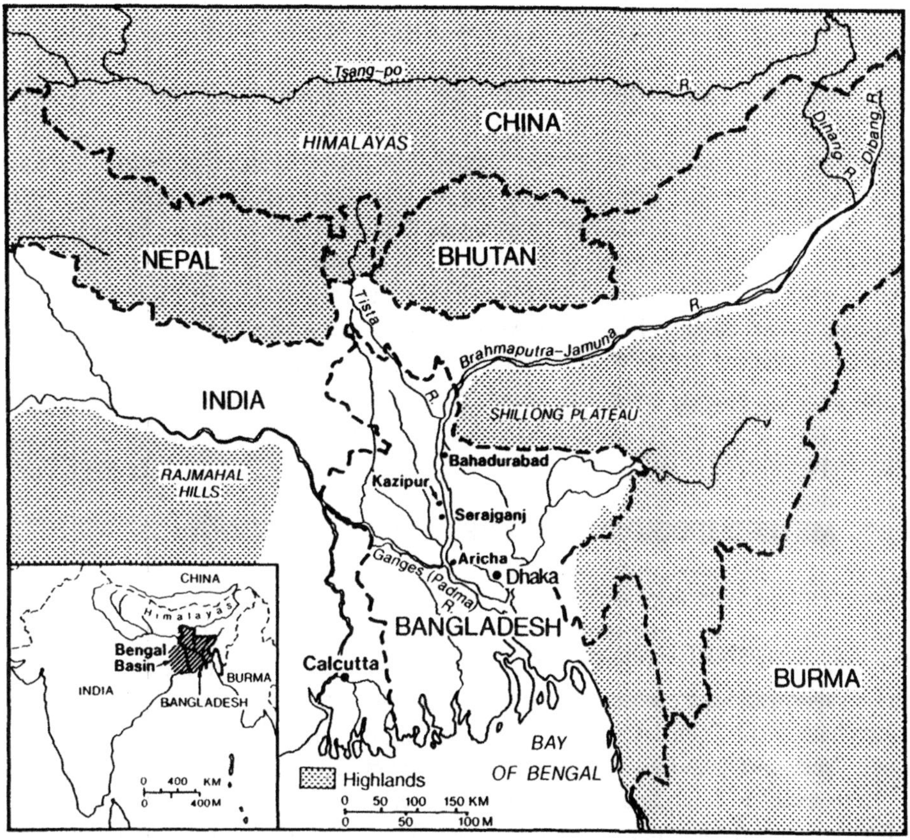

Figure 4.1. Location map of the Bengal basin and the Brahmaputra-Jamuna River.

highest landform on earth. The northeastern part of the Indian plate fractured and sank below sea level during the Oligocene period, shaping a great geological trench. During the last 37 million years, massive alluvial sediments from the very young Himalayas were carried down to the valleys and the sea by a dynamic river system. The complex geological and hydrological processes subsequently filled the subsided basin with sediment and resulted in an ever-changing floodplain and deltaic system. Based on a recent marine seismic expedition in the Bay of Bengal, Curray and Moore (1971) registered that the undersea fan of sediment from the Himalayas deposited by the Ganges-Brahmaputra rivers was 1,000 kilometers wide, over 12 kilometers in depth, and 3,000 kilometers long. Today, the surface of the Bengal basin is composed of the three great rivers in the world: the Ganges, the Brahmaputra, and the Meghna.

The active tectonic nature of the region is associated with the subduction faults in the north and transform faults in the east. Molnar (1986: 154) asserts that slippage in the tectonic processes in the Himalayan region generates "a roughly annual moderate earthquake and the rare, but all too frequent great earthquakes." The severe earthquakes of 1762 and 1897, and more than 20 other earthquakes since 1900 support this assertion of frequent tectonic activity. Several studies confirm that the tectonics of northern and eastern India controlled the courses of the main rivers of the region (Murthy and Sastri, 1981). Some historical studies of the 1897 and 1950 earthquakes in Assam, India, argue that these events caused a chain of effects involving landslips, rockfall on hillslopes, subsidence of ground and eventually changes in the course and morphology of some tributary channels (La Touche, 1910; Hayden and Pascoe, 1919; Chatterjee, 1947). The details of the relationship between earthquakes and channel changes of the Brahmaputra will be discussed in Section 4.3.4. Poddar (1952) and Goswami (1985) note that the 1950 earthquake, which was of Richter magnitude 8.7, caused temporary blockage of the Subansiri River in Assam, India, and of the Dihang and Dibang in Arunachal Pradesh, India. Subsequent release of the ponded water by sudden bursting of the dams resulted in devastating floods downstream. The process also is believed to have triggered an accelerated pace of sediment transfer through the river drainage systems to the sea.

4.1.2. METEOROLOGICAL FEATURES AND THE MONSOON

A well-defined hydrological cycle contributes to the physical processes of the region. From June to October, the summer monsoon months, the South Westerlies bring with them evaporated water from the Indian Ocean, the Arabian Sea and the Bay of Bengal. This evaporated water feeds the permanent ice and glaciers that cover 30,000 square kilometers in the Himalayas. The glaciated heights, including those of Tibet to the north, feed the Ganges and the Brahmaputra rivers. The snowpacks in the Himalayas begin to melt in March and continue up to June, often causing pre-monsoon floods.

Rainfall during the summer monsoon months plays a major role in supplying the characteristic abundant water of the region. Overall, the Ganges-Brahmaputra watershed receives an average of 1,500 millimeters of rainfall a year. The gradient of this annual rainfall decreases from west to east: from 3,000 millimeters at the Bengal coast, to 600 millimeters in Rajasthan, India, in the extreme west, and down to 450 millimeters in Lasha, Tibet, in the rain shadow of the Himalayas (Rogers *et al.*, 1989a: 9). More than two-thirds of the total precipitation fall in the watershed during the summer monsoon, which generates highly variable water discharges in the surface and subsurface drainage systems. A considerable portion of the water precipitated by the monsoon—almost close to 50 percent of the total volume—returns to the sea as surface runoff. The remaining contributes to the groundwater runoff in the unconsolidated alluvial soil to a depth of 6,000 meters or more.

Marked spatial variation over the catchment area of the major rivers of the Bengal basin is visible (see Johnson, 1982; Haque, 1988c). The extreme northern and the western areas of the region receive precipitation in the range of 450 - 600 millimeters annually, while the eastern regions have recorded maximum precipitation that ranks among the highest in the world. The mean of the 50-year (1910 - 1960) record of annual rainfall in Darjeeling (West Bengal, India) is 3,099 millimeters (Schwartzberg, 1978: 5). Holeman (1968: 742) also recorded the extreme case of Cherapunji in the Khasia Hills of Assam, India, just north of the Sylhet district of northeastern Bangladesh. In the month of August of 1841, 1,036 millimeters of rainfall were recorded in one day in this area. The average annual rainfall at Chrerapunji is 10,820 millimeters.

Rogers *et al.*, (1989b) analyze rainfall data recorded in 11 sites for 150 years (1830 - 1980) in the lower reaches of the Bengal basin. They find "no long-term trend up or down in the regional monsoon rainfall nor any statistically significant cycles or periodicities" (Rogers *et al.*, 1989a: 12). The implication of this finding is that the patterns of monsoons have been consistent historically in South Asia. However, Shukla (1987; 1989) reports that there exists an east-west alternation within the Ganges-Brahmaputra drainage basin. He suggests that droughts over central and northwest India occur in years of excess rainfall over Bangladesh and northeast India and *vice versa*.

Within Bangladesh, the range of total annual rainfall is from under 1,500 millimeters in the west to over 3,000 millimeters in the east (i.e. Sylhet and southern Chittagong). The rainfall regimes are, however, remarkably similar throughout the country: the dry season (November - March), the 'little rain' season (April - May), and the rainy season (June - October; see Johnson, 1982: 28-32). Although only less than eight percent of the Ganges-Brahmaputra drainage basin is occupied by the country, fluctuation in local or regional rainfall can affect the discharge and flooding of an extensive area. The marginal additions of precipitation to existing bankfull runoff thus can create a great havoc for the weak economy of Bangladesh. Similarly, excessive rainfall over central, western and northern India can cause flooding in Bangladesh's river systems, even though the rainfall may be below average in Bangladesh. This susceptibility to both greater-than-average local and/or extra-territorial rainfall makes the country one of the most flood-prone regions in the world.

4.1.3. SURFACE WATER SYSTEMS AND GEOMORPHIC SETTING OF MAJOR RIVERS

As stated above, the Bengal basin consists primarily of Quaternary alluvial sediments deposited by two of the largest rivers in the world—the Brahmaputra and the Ganges. The region is also drained by another major river of the subcontinent—the Meghna—and numerous streams and distributaries. The major rivers of Bangladesh altogether drain a 1,559,400 square kilometer catchment area, with only 7.5 percent of this area in Bangladesh itself. Because of its alluvial character, the relief of this plain is almost flat. With the exception of Tertiary hills in the southeast and Pleistocene terraces in four enclaves,[28] the elevation of the Bengal plain ranges between five and six meters above sea level. Such topography suggests why more than one quarter of

Bangladesh experiences a regular annual flood. The country receives an average of 870 Million Acre Feet (MAF) of river flows from India, China and Nepal. The net flow to the Bay of Bengal from Bangladesh's rivers (adding an average rainfall of 203 MAF and subtracting 120 MAF that accounts for the estimated evaporation, evapotranspiration and deep percolation) accounts for 953 MAF. This immense volume of water passes over an almost flat topography, characterized by a very gentle slope towards the Bay of Bengal. The cumulative results of these features have created one of the most complex surface water networks in the world, with a total of more than 250 perennial rivers (Zaman, 1988: 33). The dominant physiographic features of Bangladesh thus consist of the Ganges delta, and the floodplains of the Brahmaputra-Jamuna and the Meghna river systems, along with a network of hundreds of tributaries and distributaries (Ahmad, 1968; Spate, 1954; Rashid, 1977).

The Ganges and its tributaries drain an area of about 960,000 square kilometers (Table 4.1). The magnitude of water flows at a magnitude of roughly 446 MAF (Zaman, 1989: 99). The three trans-Himalayan distributaries—the Karnali, Sapt Gandaki and Sapt Kosi—originate in Nepal and contribute about 71 percent of the natural dry season flow and 41 percent of the total annual flow of the Ganges system. The main stream of the Ganges originates at Manosh Sarovar in the lower Himalayas and rises west of the Nanda Devi range; it has a length of 2,500 kilometers from head to mouth (Table 4.1). On rounding the Rajmahal Hills, the Ganges enters Bangladesh at the western extremity of the Rajshahi district and flows east-southeast towards its confluence with Brahmaputra-Jamuna near Aricha. From the point of confluence with the Brahmaputra-Jamuna River to the confluence with the Meghna River near Chandpur, the Ganges is known as the *Padma*.

The delta of the Ganges is bounded by its western-most distributary of the Bhagirathi-Hooghly and to the east by the Meghna River. This triangular shaped landform is drained by several large tributaries: Ichamati, Kabadak, Bhadra, Bhairab, Nabaganga, Gorai-Madhumati, Arial Khan, and Tetulia. Spate (1954) divides the Ganges delta into three zones: the *moribund*, *mature* and *active* delta. The Moribund, the oldest part of the delta, occupying the northwestern and western-most areas, was formed by silt from the shifting main streams and major distributaries. The slope of the moribund zone is very gentle, causing quick silt deposition in riverbeds. This process induces occasional floods prompted by the decreased carrying capacity of the river channels, particularly during monsoon periods. It also causes a gradual or sudden shifting of channels to a more low-lying area, often destroying hundreds of village settlements and cultivable land (Nicholas, 1962). The mature delta zone includes the districts of 24-Paraganas (West Bengal, India), the northern and southern parts of Khulna and Jessore respectively, and the northwestern portion of the Faridpur district in Bangladesh. At the present time this zone does not receive extensive silt or diluvian deposits (Rashid, 1977: 43; Zaman, 1988: 34).

TABLE 4.1. Characteristics of selected rivers of the world

River and Location	Length, Head to Mouth	Area of Drainage Basin	Average Discharge at Mouth	Extreme Discharge During 1969-1975	Average Annual Suspended Load
	Km	Sq Km ('000)	Cumec	Cumec	(Million Metric Tons)
Amazon (Brazil)	6,300	5,800	180,000	249,000	360
Brahmaputra-Jamuna (Bangladesh, India)	2,900	670	19,000	81,000	478
Congo (Zaire)	4,700	3,700	39,000	-	-
Ganges (Bangladesh, India)	2,500	960	12,000	74,060	1,450
Irrawaddy (Myanmar)	2,300	430	14,000	-	300
Mekong (Vietnam)	4,200	800	11,000	-	170
Mississippi (USA)	6,000	3,300	18,000	55,500	296

Sources: Oberlander and Muller (1987: 405); and UNESCO (1974; 1979).

The active delta covers most of Bangladesh's coastal plains. The tidal plains of Barisal, Patuakhali, Khulna and the *Sundarbans* (the mangrove forest) in the southern part of the country characterize this active zone. Often included in the larger deltaic region, the coastal plains extending from Noakhali to the south of Cox's Bazar are also subject to the action of tides. The dynamism of the deltaic plain has made its landform ever-changing. Most of the land is subject to annual inundation, and new land formation from sediment deposits is an important feature of deltaic processes, particularly in the Meghna-Tetulia-Sahbazpur distributary systems. There are numerous *chars* (mid-channel sandbars) of varying ages and sizes in the active delta region. As new silts fill up the channels, water tends to make its own space, causing erosion of riverbanks, encroachment into *chars*, and the eventual disappearance of *chars* and islands. The inhabitants of the active delta thus face the challenge of continuous adjustments to changing environmental conditions.

The Brahmaputra-Jamuna River presently flows through the Garo-Rajmahal gap, particularly through a zone of subsidence between the two Pleistocene tracts: the Barind and the Madhupur (Rashid, 1977: 26; Morgan and McIntire, 1959). The Pleistocene sediments of the Bengal basin are presumed to be floodplain deposits of the former Brahmaputra-Ganges system (Geddes, 1960: 580; Coleman, 1969: 133). The present river flow traverses a distance of about 2,897 kilometers through three countries—China, India, and Bangladesh. In the upper valley of the river, it is known as *Tsangpo*, meaning 'the purifier.' As the river enters India near Dadiya, the main channel, which is known here as *Khiang*, receives two major tributaries—Dibang and Lauhitya—and takes the name of Brahmaputra, which means the son of the Creator.

The Brahmaputra, which is by then several miles wide, flows through the Assam valley of eastern India. By providing the Brahmaputra with the only masculine name among Indian rivers, the language of the region emphasizes its unique character (Desai (1968).

Traversing round the spurs of the Garo Hills near Goalpara of Assam, the river enters Bangladesh and flows southwardly for a distance of about 270 kilometers across the valley between the Garo-Rajmahal gap before joining the Ganges[29] near Aricha. About 16 kilometers north of Bahadurabad, the Old Brahmaputra takes off from the main channel and begins to flow southwardly as one of the principal distributaries. This distributary once formed the main channel of the Brahmaputra river system. The river channel changes its name to the *Jamuna* after its bifurcation with the Old Brahmaputra. A dual name—*the Brahmaputra-Jamuna*—is used in the present study to refer to the comparatively new channel that occupies the reach from the Bahadurabad Ghat to Aricha, and to distinguish it from that of the older Brahmaputra (Rashid, 1977: 24).[30] Between Nunkhawa and Aricha, the Brahmaputra-Jamuna River receives several large tributaries. Of these, the Raidak, Dharla, Tista, Ghaghat, and Hurasagar are the major western tributaries. The Bangali and Ichamati are major distributaries which flow off from the Brahmaputra-Jamuna to the west, while the Jenai, Lohajang and Dhaleshwari are major distributaries on the eastern bank. The total drainage area of the Brahmaputra-Jamuna down to Aricha is 528,176 square kilometers, of which 293,687 square kilometers are in Tibet, 241,707 square kilometers in India and only 46,782 square kilometers in Bangladesh (Rashid, 1977: 64).

Morgan and McIntire (1959) identified faults on both sides of this mighty river in Bangladesh. Six *en echelon* faults on the western margin of the Madhupur tract, which cause contrasting relief between Pleistocene and Recent sediments, are noticeable. Although the history of the structural activity associated with the Madhupur region is not precisely known, geomorphic features such as a high degree of dissection, higher local relief, and lack of deep valley filling, altogether indicate a rather recent origin (Coleman, 1969: 137). Fergusson (1863) asserted that the Madhupur region had been uplifted and tilted in recent times. Referring to the earthquake of 1762, he insisted that it was accompanied by uplift and subsidence of large tracts of land. This structural activity had supposedly contributed to the abandonment of the old channel of the Brahmaputra to its present course west of the Madhupur tract.[31] A large fault, 64 kilometers long and upthrown to the southwest, is also identified along the northeastern edge of the Barind region. Coleman (1969: 137) claimed that this fault had a significant influence on the course of the Karatoya-Bangali River. He postulated that although structural movements in the Barind area have been active, they possibly consisted of broad block uplift or of slight doming. This resulted in a flat terrace surface and the absence of any remarkable dissection in the local relief. The fluvial origin of these sediments is supported by the presence of several large abandoned stream scars on the surface of the terrace.

Besides the Pleistocene deposits, the remaining landforms of the basin consist predominantly of Recent alluvial and deltaic sediments. In areas above tidal inundation, which includes the Brahmaputra-Jamuna floodplain, the deposits are composed primarily of silt and sand (Huq, 1974: 17-18). Morgan and McIntire (1959: 17-18) examined the grain size and water content and found that the Pleistocene sediments of the area are almost identical with those of Recent floodplain sediments. The Recent sediments, however, are more loosely compacted and have a high, although variable water content. The Pleistocene sediments contain calcareous nodules and a lower proportion of water, which results in them being more compact. The combination of high water content and loose cohesion of Recent sediments along with the great amount of material carried within the channels of the Brahmaputra and Ganges rivers results in their high susceptibility to erosion.

The Meghna is the third largest river in Bangladesh, and originates in one of the rainiest hillslopes of the world, the Shillong Plateau in Assam, India. The headwater of the Meghna consists of several streams that meander through Assam for about 400 kilometers and enter Bangladesh from the northeast in the form of two major tributaries: the Surma and the Kushyara. These two tributaries rejoin to form the main channel of the Meghna in Bangladesh. The principal floodplains of the Meghna are located along its lower reach.

The general topography of the Ganges, the Brahmaputra-Jamuna, and the Meghna is flat plain. However, the common variation in local relief is largely influenced by three dominant morphological features. First, the *natural levees*, alluvial ridges with one to two meters height above the floodplain surface, are formed by sediment deposits on the riverbank by the flood waters. Second, *backswamps* or *levee-flank depressions*, are areas of minor relief which commonly drop one to two meters lower that the floodplain surface. In the lower Ganges and upper Meghna floodplains, deeper and more extensive swamps are found, their origins relating to structural depressions in the floodplains (Johnson, 1982). Third, a common feature in all major floodplains of the country are the extensive abandoned channel systems, developed over centuries by the lateral shifting of river channels; related geomorphological features include *cutoffs*, *oxbow lakes*, and channel scars (Rizvi, 1955; Rasid, 1966; Rasid and Paul, 1987; Rahman, 1969).

Overall, the physiographic landforms and features in areas southwest of the Ganges-Meghna confluence represent more pronounced deltaic characteristics. This active zone relies on a complex network of distributary channels and associated island systems. The rest of the plains may be generally considered formations associated with riverine floods and the gradual deposition of sediments. The effects of the tide further play a significant role in the fluvial processes of the distributary channels up to a distance of about 150 kilometers upstream from the estuaries. A considerable proportion of land in Bangladesh is subject to either coastal inundation related to high tides or to riverine floods caused by seasonal runoff, which raises the water level in the channels and on the floodplains.

4.2. Floods in Bangladesh

4.2.1. THE GENESIS OF FLOODS

A distinct seasonal flow is common to all the mighty rivers of Bangladesh. For comparison, the combined average peak flow of the Ganges, Brahmaputra-Jamuna and Meghna is about 2.5 times that in the Mississippi River (Brammer, 1990a: 14). As soon as snowmelt commences in the Himalayas and pre-monsoon rainfall occurs in Assam, India and the northeastern part of Bangladesh, the water levels in the Brahmaputra-Jamuna and the Meghna correspondingly respond and begin to rise during March - April period. The water level of the Ganges rises much later in May due to a relatively drier catchment area, delayed precipitation and a longer distance from the catchment area to the delta region (Table 4.2). With the onset of monsoon rains, water levels rise rapidly during the June - July period. However, the peak levels of the Brahmaputra-Jamuna and the Ganges usually do not coincide: the former reaches its peak water level in July - August and the latter about a month later, in August - September. The pattern of water level in the Meghna system resembles the one for the Brahmaputra-Jamuna, although the peak level of the Meghna often extends into September due to "backing up of water above the confluence into the Ganges" (Brammer, 1990a: 14). Throughout the September - November period, all the rivers fall rapidly as dry winter approaches the region.

TABLE 4.2. Discharge data for Bangladesh's major rivers (in cumec) (based on 1934 - 1988 period)

Discharge parameter	Ganges (Hardinge Bridge)	Brahmaputra-Jamuna (Bahadurabad)	Meghna (Bhairab Bazar)
Mean low flow	790[a]	37,550	1,332
Mean peak flow	51,625	65,491	14,047
Mean annual flow	8,544	19,557	6,748
Highest peak flow	72,300	99,500	19,800

[a] Mean for 1975-1988 (since commencement of the Farakka barrage). Mean for 1934-1974 was 1,806 cumec.

Sources: Brammer (1990a); and Matin and Husain (1989).

Based on cross-sections surveyed by the International Engineering Company (IECO) during the 1960s, Coleman (1969) estimated that the bankfull discharges are close to 56,000 cubic meters per second (cumec) in the Ganges and the Brahmaputra-Jamuna, and 8,500 cumec in the Meghna. Rasid and Paul (1987), however, find these figures overestimated due to the fact that the floodplains of these rivers experience overbank flooding almost every year for a duration of 60-120 days in the summer months. They estimate, using the mean monthly discharges, that the average combined bankfull discharge for the Ganges and the Brahmaputra-Jamuna is approximately 35,000 cumec and for the Meghna only 5,000 cumec. A major limitation in collecting and generating flow-duration data from a cross-sectioned survey is created by the uncertainty in determining bankfull capacity—specifically, unstable alluvial channels may vary considerably from one flood season to another.

The empirical data compiled by the Master Plan Organization (MPO, 1990) and Matin and Husain (1989) provide a relatively reliable and precise hydrological perspective on Bangladesh's major rivers. These studies reveal that the average annual flow of the Ganges and Brahmaputra-Jamuna, based on the mean for 1934-1974, is around 28,100 cumec; for the Meghna River it is 6,748 cumec (Table 4.2). The mean peak flow of the Brahmaputra-Jamuna alone, however, is as high as 65,490 cumec at Bahadurabad reach, and its recorded highest peak flow reached 99,500 cumec. Similarly, although mean peak flow of the Ganges at Hardinge Bridge is around 72,300 cumec, its highest peak flow was recorded at 76,000 cumec. Overall, these parameters confirm a very high degree of variability in river discharges and suggest the difficulty in predicting the flow regimes of these fickle rivers.

4.2.2. TYPES OF FLOODS

The types of floods in the region are examined by several investigators (UNDP, 1988; Brammer, 1990a). A normal annual flooding, which is termed *barsha* by the local people, benefits paddy rice and other major crops. This flood usually arrives with limited severity that the local conditions—such as, settlement and housing, agriculture, and physical infrastructure—can normally withstand. However, when the timing, duration, and magnitude of floods depart from the characteristics of *barsha*, they cause damage and loss. If water rises earlier than normally expected, or excessive water accumulates and rises more rapidly than crops can withstand, agricultural losses become inevitable. Extremely high and sudden floods also breach transport infrastructure, industrial and residential sites, and flood embankments. The local people call these unusually high floods *bonna*.

Riverine floods can be classified into three main types: (i) *flash floods*; (ii) *rainwater floods*; and (iii) *monsoon-seasonal floods*. Since flooding associated with tropical cyclones in the Bay of Bengal is not directly related to floodplain fluvial processes, this type is omitted from further discussion. *Flash floods* take place chiefly in the eastern and northern floodplains, and originate from exceptionally heavy rainfall in the neighbouring elevated areas. Brammer (1990a) cites that the lower Atrai basin and lower Purnabhaba basin could be affected by a steady rise and a relatively rapid recession of water from the Barind tract. The regions characterized by a steady slope in their valleys are subject to flash floods. The northern and eastern Piedmont plains and the Chittagong coastal plains thus experience flash floods. The actions of flash floods are more liable to breaching embankments and rapidly silting streambeds and natural levees than to damaging crops and properties.

Rainwater floods occur as a result of copious rainfall within the territory of Bangladesh. The high rainfall intensity and duration of the monsoon may cause water volume in excess of the local drainage capacity. Local runoff is accumulated in low-lying areas and depressions due to heavy pre-monsoon rainfall during April and May. In the June - August period, local rainwater is increasingly concentrated on the arable land by the rising water level in neighboring rivers. The groundwater table even rises above the surface (Brammer, 1990a: 17). Thus, a series of local floods may be generated, with little propensity to siltation on the ponded land as the rainwater is generally devoid of sediments. The magnitude of rainwater flooding varies within the rainy season and from year to year, according to the conditions of local rainfall and present water levels in the major rivers. Areas susceptible to precipitation floods include the major floodplain basins and the meander floodplains.

Monsoon-seasonal floods stem from snowmelt in the Himalayan heights and plentiful seasonal rainfall during the monsoon throughout the upper catchment area. The resultant runoff causes the major rivers to rise over a period of one to three weeks. Monsoon-seasonal floods are prevalent in active river floodplains. In unusual flooding years, river water not only spills over natural levees, but also encroaches into distant land areas and eventually may merge with neighboring meander floodplains. Monsoon-seasonal floods are also known as 'river floods' (Brammer, 1990a) as they occur primarily from river runoff and carry huge volumes of alluvial sediments with them that spread out over the inundated floodplain zones.

4.2.3. FREQUENCY AND MAGNITUDE OF FLOODS

Since time immemorial the floodplains and the deltaic plains of the region have experienced normal and unusual, damaging floods, yet systematic records of such flooding only began after the great floods 1954. Evidence of earlier devastating floods can be found in government gazetteers, literary works, and folk tales; for instance, a detailed description of the severe floods of 1787, which reportedly changed the course of the Tista, is given in the Imperial Gazetteer of India (1908; cited in Miah, 1988: 1). Ahmad also cites from this source that:

> The great Tista floods occurred in 1787 and diverted that river into a new channel to join the Brahmaputra. As a result, the Brahmaputra itself shifted into a new main channel westward through the Jenai, to develop into the Jamuna which in turn found confluence with the Ganges near Goalundo (Ahmad, 1968: 33).

Reference can be found to recurrent floods during 1887-1930 over which the shift of the old course of the Brahmaputra to the Jenai, now known as Jamuna or Brahmaputra-Jamuna, was progressing (Morgan and McIntire, 1959).

Attention to the devastating floods in the northwestern region of Bangladesh in 1922 by the Government of Bengal Presidency turned into a commission, headed by P.C. Mahalanabis. The commission examined flood occurrences between 1870 and 1922 and submitted its report in 1927. Based upon statistical inferences from the available data, the commission calculated the recurrence interval of four categories of floods in order of intensity: 2.25 years for normal flooding, 4 years for moderate flooding, 7 years for severe, and 33 to 50 years for catastrophic flooding (Miah, 1988: 2). Chowdhury and Bhuiya (1990) reviewed the historical records of the major floods in the western Bogra and Mymensingh districts, concluding that the timing of floods is a more significant determinant of flood damage than magnitude since the damage and loss mainly take place due to a lack of anticipation and unpreparedness of floodplain inhabitants. Rasid and Paul (1987) compiled historical records, finding at least six major floods in the 19th century: in 1842, 1858, 1871, 1875, 1885, and 1892. The number of comparably severe floods in the 20th century was 14: in 1900, 1902, 1907, 1918, 1922, 1954, 1955, 1956, 1962, 1968, 1970, 1974, 1987, and 1988 (Coleman, 1969; Tarafdar, 1978; Brammer, 1990a).

The normal floods *(barsha)* inundate about 26,000 square kilometers of the land surface of Bangladesh (i.e., 18 percent of total area). During a "catastrophic flood", the areal extent may reach over 82,000 square kilometers (57 percent of total area; MPO, 1985; Miah, 1988). In the 1950s, severe floods occurred successively, in 1954, 1955 and 1956, when more than one quarter of Bangladesh was inundated. Devastating floods were frequent throughout the 1960s and early 1970s (Table 4.3), engulfing an area ranging between 35,000 and 52,000 square kilometers. Two very exceptional years were 1987 and 1988 when more than 40 percent of the country was devastated by consecutive floods. Such a succession of two or more catastrophic floods is a rare phenomenon, although in recent experience it has happened twice—in 1954, 1955 and 1956 and in the recent years of 1987 and 1988. But does this evidence show any significant trend towards increasing flooding? The World Bank concludes that despite such consecutive floods in recent decades, there is no scientific evidence of any significant trend (Rogers *et al.*, 1989). However, the extent of flood-induced damage and loss has increased significantly as population and new development activities have become more established on the floodplains than ever before. These aspects of human occupance of floodplains will be discussed in Chapter 5.

Out of a total of 32 annual available records, on 21 occasions one-fifth or more of Bangladesh was flooded (see Miah, 1988). The long-term time series data indicate that on average 22 percent of the area of the country is flooded each year, with up to 60 percent flooded by a 100-year return period flood (Table 4.3). Severe damage to crops, infrastructure and urban settlements generally takes place when floods cover more than one quarter of the area of Bangladesh (Figure 4.2). The French Engineering Consortium suggests that when more than 35 percent of the territory is inundated, the floods cause catastrophic damage and loss of life (Rogers *et al.*, 1989).

TABLE 4.3. Return period according to areas affected by floods

Return Period (Year)	Flooded Area (Percentage of the Country)
2	20
5	30
10	37
20	43
50	52
100	around 60
500	around 70
mean	22

Source: Rogers *et al.* (1989: 37).

Quantitative information on depth of floodwater is meagre. Rasid and Paul (1987), and Chowdhury and Bhuiya (1990) have attempted to overcome this limitation by relating flood height to some of the morphological features of the floodplains. *Shallow floods* imply a depth of floodwater on the floodplains of less than one meter, thus the natural levees stand above the flood level. *Moderate floods* refer to depths ranging from one to two meters. The depth exceeds two meters and levees are submerged during *deep floods* (Abbas, 1963). Since backswamps and abandoned channels form the most low-lying areas, the standing water in these areas is often deeper than three meters.

4.2.4. THE FLOODS OF 1987 AND 1988

The two most catastrophic floods of the 20th century, in both physical and socioeconomic terms, occurred consecutively in 1987 and 1988. The severity of the 1988 flood disaster alarmed the global community through widespread coverage from the international media. Meanwhile, the response from the international community to provide immediate relief was prompt and substantial (Boyce, 1990; Haque and Zaman, 1993). Technical assistance to help formulate longer term solutions to the country's chronic flood problem was also offered by many developed countries and international development agencies. In the following pages the physical characteristics of the 1987 and 1988 floods will be assessed to determine any possible differences between them in terms of their causes and effects.

The 1987 Floods

The 1987 floods inundated 54,491 square kilometers, that is 39.92 percent of Bangladesh, the highest areal extent recorded up to that time (Figure 4.3). Overall, the 1987 floods were largely characterized by *rainwater floods*, although *flash floods* were also registered in eastern foothill areas in late June and August - September and along the Tista, Atrai, and other northwestern rivers in July and August (Brammer, 1990a:

Figure 4.2. The eroded Asian Highway near Aricha, Manikganj district, during the 198l flood. Damage to valuable infrastructure and disruption of vital transportation by unusually high floods create severe setbacks for national economic progress.

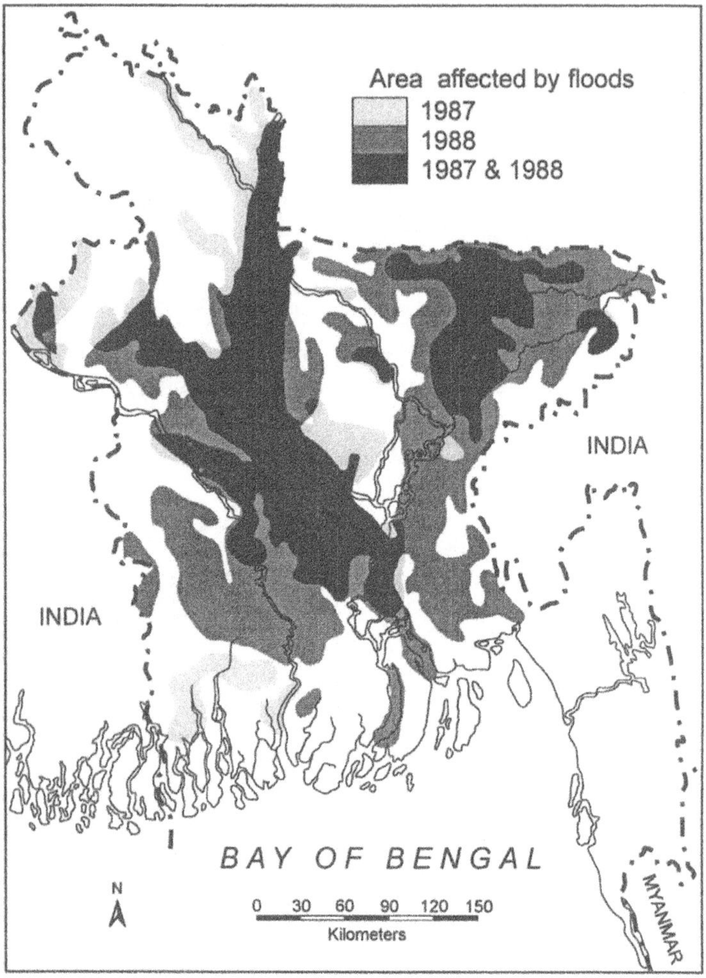

Figure 4.3. Areas flooded in 1987 and 1988 (based on Landsat images).

118). During June - September, several periods of very high rainfall were recorded in northern Bangladesh and adjacent parts of India, namely, the states of Meghalaya and Assam. Consequently, most stations in northwest Bangladesh showed an excess of 300-700 millimeters of rainfall above long-term averages for the four monsoon months. In July-September, northeastern border areas of Bangladesh and the adjoining Meghalaya state of India had excesses of 500-1000 millimeters of rainfall. Overall, rainfall totals were highest in the humid northeastern districts but the largest percentage excesses were in the northwestern districts.

An examination of the chronological events associated with the 1987 floods reveals that in the last week of June the northeastern part of the country was struck by the onrush of water from the Meghalaya Hills. The Surma River at Sunamganj crossed the danger level on June 25 and stayed above this level (local bankfull stage) till July 8. This flash flood was chiefly the result of an increase in the discharges of the tributaries of the Surma River, such as Gowain and Piyain, submerging an area of about 1,600 square kilometers. During the first two weeks of July (i.e., July 3-17), the southeastern foothill areas became affected by another flash flood, inundating Chakoria, Cox's Bazar and Ramu *thanas* (sub-districts).

An early flood fed by the river runoff commenced on July 20 in the northwestern districts of Dinajpur, Rangpur, Kurigram, Gaibandha, Nilphamari and Thakurgaon. Local rainfall during late July and early August was extremely high and incessant. At Dinajpur station a total of 674 millimeters of rainfall was recorded between July 23 and August 3; for the same period at Kaunia, the recorded amount of rainfall was more than 1,200 millimeters. Miah explains why the floods of 1987 in the northwest can be termed as rainwater floods:

> Although Thakurgaon town itself and the areas adjoining it were submerged, the Karatoya at Panchagarh didn't overtop its banks at all. And the Purnabhaba at Dinajpur flowed only slightly (the maximum being 77 cm on 14 August) above the danger level and that too for a very brief period of 4 days, from 12 to 16 August. This simply means that the flood waters didn't come via the river-routes but rushed down overland from the northern slopes (Miah, 1988: 37).

Overland flow increased because minor rivers could not accommodate the rainwater due to silted and raised beds and the raised water table. In early August, a large flash flood in the northwestern region of Bangladesh added to the overland flow. Eventually, the high water levels in the major rivers aggravated rainwater flooding. The Brahmaputra-Jamuna crossed the danger level at Serajganj on July 28 and constantly remained above this level till August 28. The water levels of the Ganges at the Hardinge Bridge rose to the danger mark in the same week (i.e., July 30) as did the Meghna at Bhairab Bazar on August 2. The respective peaks of these rivers were reached on August 15, 19 and 7 (Brammer, 1990a: 18; Miah, 1988: 39-41). The water levels of the Ganges and the Meghna remained above the danger levels till the end of September; the Ganges reached its highest recorded level (i.e., 14.7 meters) at Hardinge Bridge.

Some characteristics of the 1987 floods deserve mention. The floods occurred in several spells, up to a maximum of four in a particular area. Also, since the floods exceeded any previous recorded geographical extent, infrastructures built above the normal floods were damaged by the incursion of floodwater.

The 1988 Floods

Contrary to the 1987 floods, the floods of 1988 were predominantly monsoon-seasonal floods. Hydrological and meteorological records indicate that the water levels of the

Meghna and the Brahmaputra-Jamuna rose above the danger mark during the second week of July with runoff aggravated by heavy, early monsoon rainfall in the northeastern region of Bangladesh and the state of Meghalaya of India.

The situation deteriorated further as the entire northern and northeastern parts of the subcontinent experienced a period of continuous rainfall for ten days in late August. The resultant river runoffs from the major systems in the region created the most catastrophic floods of recorded history as the parameters superseded the widest areal extent of the 1987 floods (Figure 4.3) and other previously recorded peak flood levels (Table 4.4). In the Brahmaputra-Jamuna basin for 8 out of 10 stations water levels exceeded the 1987 records; in the Ganges basin for 5 out of 9 stations the 1988 flood level exceeded the previous year's level. The total flood discharge through Bangladesh was in the order of 192,000 cumec (Matin and Husain, 1989: 6).

Analysis of rainfall patterns over Bangladesh in August 1988 reveals that precipitation was not concentrated only within the country, rather it was extensive, covering Assam, India, Nepal, Bhutan and northern West Bengal, India. Also, the intensity of rainfall in Bangladesh was less vigorous compared to the same period of 1987. Several places in the northwest and northeast experienced daily rainfall amounts exceeding 100 millimeters and ten-day totals exceeding 500 millimeters. However, the border region and adjoining areas in India received rainfall exceeding previous records, with return period frequencies of 50-100 years (Brammer, 1990a: 21; Islam, 1989).

TABLE 4.4. River reaches crossing the previously Recorded Peak Flood Level (RPFL) during the 1988 floods

River Reach	Station	Recorded Peak Flood Level (RPFL)		1988 Peak Level	
		Date	Water Level (meter)	Date	Water Level (meter)
Brahmaputra-Jamuna	Bahadurabad	08/06/74	20.28	08/30/88	20.62
Brahmaputra-Jamuna	Serajganj	08/20/84	14.62	08/31/88	15.12
Buriganga	Dhaka	08/08/55	7.09	09/04/88	7.58
Lakhya	Narayanganj	08/12/74	6.23	09/05/88	6.71
Ganges	Goalando	08/08/74	9.60	09/02/88	9.83
Gorai	Gorai Railway Bridge	09/19/48	13.58	09/02/88	13.65
Kumar	Faridpur	09/09/54	8.64	09/02/88	8.76
Surma	Sylhet	06/02/59	1.92	07/07/88	11.95
Khowai	Habiganj	10/03/59	10.90	07/08/88	11.00
Meghna	Bhairab Bazar	09/08/74	7.65	09/10/88	7.66

Source: Matin and Husain (1989: 74).

TABLE 4.5. Channel-bed gradient of the major rivers in Bangladesh

River/Channel Reach	Gradient (cm/km)
Ganges (250 km reach within Bangladesh)	
Subreach 1 (upper 90 km)	5.0
Subreach 2 (middle 80 km)	5.0
Subreach 3 (lower 80 km)	4.0
Brahmaputra-Jamuna (275 km reach within Bangladesh)	
Subreach 1	10.0
Subreach 2	7.0
Subreach 3	6.5
Lower Meghna (the main channel within Bangladesh)	3.0

Sources: Latif (1963: 103); Rasid and Paul (1987); and Haque (1988c).

The flood peak of the Brahmaputra-Jamuna was attained on August 30, and was the highest on record with an estimated return period in the order of 100 years. The Ganges at the confluence with the Brahmaputra-Jamuna (Aricha) and the Meghna at Bhairab Bazar reached record high water levels on September 2 and 7 respectively, with a return period in the order of 50 years (Rogers *et al.*, 1989a). A total of 82,000 square kilometers were inundated (57 percent of the country), covering 53 of the country's 64 districts, as well as two-thirds of the capital city of Dhaka (Figure 4.3).

4.2.5. GENERAL CAUSES OF FLOODS

In hydrological terms, *flood* is defined as the instantaneous peak discharge of a river channel during a given study period (Linsley *et al.*, 1982). For the operational and common-use purposes, the overbank flow is emphasized. United Nations (1972) and Rasid and Paul (1987) offer a definition of flood as the occurrence of water levels whose magnitude exceeds the natural banks along the channel, thus allowing water to flow over adjacent low-lying areas of the floodplains.

Coleman (1969), Rasid and Paul (1987) and Miah (1988) examine the general causes of floods in Bangladesh, concluding that three major factors cause both normal and damaging floods in the country: (1) accelerated seasonal runoff from the drainage basin, (2) the hydraulic and physiographic characteristics of the river channels and floodplains respectively, and (3) local precipitation. The three great river systems drain an extensive area of 1.56 million square kilometers, most of which receives an exceedingly heavy monsoon rainfall. As Bangladesh covers only 7.5 percent of the total basin area, most of the catchment areas for the major river systems of the country lie beyond her borders. By and large, when intensive rainfall occurs simultaneously over the entire catchment, the combined runoff from the tributary systems generates a colossal volume of discharge draining out to the Bay of Bengal through Bangladesh's river network.

Notably, the upstream areas of this network lack large reservoirs, either natural lakes or human-made, that would assist in the rapid transmission of floodwater. Consequently, Bangladesh's river systems strain to efficiently discharge the flow that originates in an area 12 times greater than the area of Bangladesh.

Several hydraulic factors exacerbate the problem of discharging this enormous volume of water. A very low channel gradient (Table 4.5) and loss of channel capacity due to siltation results in rising water overtopping the riverbanks and submerging the vast floodplains. Extremely low gradients of all three major river systems create inefficient channels for discharging the excess water during high floods. Although the Brahmaputra-Jamuna River has a slightly steeper gradient than the Ganges and the Meghna, the larger volume of sediment load in its water outweighs the potential efficiency. A low hydraulic efficiency of the Brahmaputra-Jamuna is apparent in the plan views, particularly for low flow season. The plan views show numerous channels, sandbars (*chars*), and shoals, indicating the overall low hydraulic efficiency of the river. A similar braided pattern is also apparent in the low flow season plan views of the Ganges.

A critical factor for a flood occurrence is the coincidence of monsoon peak discharges in the main streams with prolonged localized precipitation within Bangladesh. Johnson (1982), in this context, explains that rainfall alone can account for the flooding of extensive areas of near-level land as the peak monsoon daily precipitation in July averages about 13 millimeters.

4.2.6. CAUSES OF THE 1987 AND 1988 FLOODS: HUMAN ACTIVITIES VERSUS NATURE

Generally, research supports the stated contention that floods in Bangladesh result primarily from snowmelt in the high Himalayas and rainfall in the adjoining drainage basins; localized precipitation contributes mainly to the inundation of the many basins contained within the deltaic plain. Three major thrusts are adopted, however, to explain the consecutive catastrophic floods of the late 1980s in Bangladesh. The first of these, an argument which has become widely cited in the popular media, attributes damaging high floods to the impact of increasing human interference with the natural ecosystem. In particular, deforestation in the upper catchment areas of mountainous regions is blamed for the floods. International development agencies such as the World Bank, the United Nations Development Program (UNDP), as well as much of the Western media, argued in 1988 that the rapid elimination of all accessible forest cover in Nepal is accelerating direct runoff while also reducing stream flow during the subsequent dry season. These processes, in turn, result in rapid siltation of river beds and concomitantly increase the frequency of high flooding in the lower reaches of the rivers.

Several recent studies, based upon longitudinal observation, question these assertions. Ives and Messerli (1989) and Hamilton (1987), for instance, suggest that deforestation in the Himalayas could not have anywhere near as significant an effect on the floods in the farthest downstream zone of the Bengal basin as is generally purported. They also insist that the terraced cultivation that replaces the forests prevents excessive soil erosion in the mountainous regions (Ives, 1989). In addition, based upon an analysis of rainfall data, Brammer (1990a) and Miah (1988) establish that the 1987 floods were mainly exacerbated by localized rainfall, rather than by runoff generated in the upper catchment. Similarly, Miah (1988), Islam (1989), and Haque and Zaman (1993) confirm that some 90 percent of the 1988 floodwater originated in areas adjacent to the Bangladesh border instead of in the watershed areas in Nepal and China. Such data lend support to the proposition that deforestation in the upper Himalayan region was not a dominant cause of the 1987 and 1988 floods.

The second argument also blames human activity for the increased frequency of high floods. It is argued that embankments, dykes, and other flood-related engineering schemes along the upstream reaches of the rivers reduce storage capacity in the basin and thus induce higher flood peaks and a steeper rise in downstream water levels (Shaw, 1989). In specific terms, major political parties and the national newspapers charged that the opening of the Farakka barrage on the Ganges at the height of the monsoon flow was the prime cause of the 1988 floods. While some studies produced by the International Engineering Company (IECO, 1977) note that there was a substantial reduction in the winter flow in the lower distributaries of the Ganges, no scientific evidence has been produced which links the effects of the barrage with high floods during the monsoon (Crow, 1981; Crow *et al.*, 1995). As such, it is difficult to determine which of these factors may have caused, and to what extent repeated the catastrophic floods in Bangladesh.

A more plausible explanation of the 1987 and 1988 floods can be found in the physical factors governed by natural forces. As discussed earlier, the 1987 floods were predominantly rainwater floods aggravated by heavy localized precipitation, and to some extent by flash floods in northwestern districts and eastern foothill areas. Northeastern border areas had excesses of 500-1000 millimeters in the July - September period, but in absolute terms, the largest percentage excesses were in the northwestern districts. Brammer finds that:

> [T]hese periods of heavy rainfall caused widespread flooding, especially in the north-west. The many minor rivers in that region were unable to accommodate the huge run-off and the out-flow from the raised water-table because of silted channels and constrictions caused by fish traps and by road embankments lacking adequate bridges and culverts. A large flash flood early in August along rivers entering Bangladesh from the north-west added to the overland flow. High levels in the major rivers aggravated rainwater flooding where they blocked drainage outfalls (Brammer, 1990a: 18).

The geographical spread of the upper catchment areas of all the major rivers flowing through Bangladesh is among the largests in the world. When copious rainfall occurs simultaneously and for a protracted period over several tributary systems, the combined runoff translates into extremely high flood levels in the main channels downstream. An analysis of rainfall during July - September 1988 reveals that the floods that year were caused by a combination of exceptionally high monsoon rainfall in the northern and eastern zones of the Indian subcontinent (i.e., most of these areas are located outside Bangladesh) and a rare coincidence of the peak flows of the Ganges and the Brahmaputra-Jamuna. Normally, the interval between the flood peak of the Brahmaputra-Jamuna at Serajganj and the Ganges at Hardinge Bridge is about four weeks. Even in some of the damaging flood years of 1954, 1974 and 1987, flood-peaks were experienced in intervals of 30, 27, and 34 days, respectively. In 1988, however, the interval was less than three days. Even when comparing water levels between normal flood years and those of unusual floods, an interval of three to four weeks appears to be the normal condition (Table 4.6). Moreover, between June 1 and August 23, 1988, fifty percent more water flowed into Bangladesh through the channels of the Brahmaputra-Jamuna compared with the same period in 1987 (Haque and Zaman, 1993: 97). As a result, out of 34 hydrological stations along the river, ten recorded flood peaks higher than at any previously recorded time and 22 recorded peaks higher than during the 1987 floods.

Besides the coincidence of the peaks of the Brahmaputra-Jamuna and the Ganges, the factor of spring tide aggravated the 1988 flood levels (Islam, 1989). The backwater effect of spring tides in the large estuaries of southern Bangladesh added to the severity of the 1988 floods. Islam (1989: 60) observes that, "due to the new moon coming in conjunction with the solar eclipse on September 11, 1988, the coastal water inflated and blocked the passage of floodwater downstream." This caused an extensive spread of floodwaters across the floodplain as well as a prolongation of the flood-stay period.

4.3. River Channel Shifting and Bank Erosion Problems in the Brahmaputra-Jamuna Floodplain

4.3.1. CHANEL CHARACTERISTICS OF THE BRAHMAPUTRA-JAMUNA RIVER

To illustrate the bank erosion and channel migration problems within the Brahmaputra-Jamuna floodplain, our focus will now shift to an examination of channel changes. The first part of this section describes the significant variables related to channel *hydraulic geometry*, with a presentation of fundamental data on the Brahmaputra-Jamuna river system. The second part focuses on the effects of morphological factors on channel forms and characteristics.

Hydrology and Channel Characteristics of the Brahmaputra-Jamuna River

To date, a great deal of research has been carried out to investigate the principal characteristics and channel morphology of the Brahmaputra-Jamuna River.[32] The investigations of Morgan and McIntire (1959), Coleman (1969) and Khan (1995) have

TABLE 4.6. Timing and the length of the gap between flood peaks of the Brahmaputra-Jamuna and the Ganges

Year	Water Level (in meters)		Length of the gap between flood peaks (in days)
	Brahmaputra-Jamuna	Ganges	
A. Typical (normal) flooding years[a]			
	(at Bahadurabad)	(at Hardinge Bridge)	
1977	18.4	14.2	25
1978	18.3	13.8	20
1979	18.6	14.3	22
1980	19.0	14.0	29
1981	18.9	12.8	32
1982	18.7	13.1	10
B. Unusually high flooding years[b]			
	(at Serajganj)		
1954	14.2	14.9	29
1974	14.2	14.4	26
1987	15.1	14.8	33
	(at Bahadurabad)		
1988[c]	20.8	14.9	3
	(at Serajganj)		
	15.1		

[a]Obtained from the plotted data prepared by Rahman (1984).
[b]Source: Miah (1988).
[c]Source: Islam (1989).

already provided the most comprehensive and analytical viewpoints about the lower reaches of the river. The following discussion of the Brahmaputra-Jamuna is based primarily on the findings of these three major studies and on recent data from the Water Development Board of the Government of Bangladesh.

Discharge Regime. As described earlier, the Brahmaputra basin receives abundant rainfall in the upper valley area of the Indian states of Arunachal, Assam and Meghalaya, mostly during the summer monsoon period (i.e., June to October). The extent of spatial variation over the catchment area of the Brahmaputra River is noticeable. For instance, although the central basin of Assam receives 2,125 millimeters of rainfall annually (Viz and Shenoy, 1968: 267), the long precipitation records exhibit maximum precipitation that ranks among the highest in the world. Most catchment areas of the river receive more than 3,175 millimeters of annual rainfall.

Although the average annual rainfall in northern Bangladesh is as high as 2,271 millimeters (see BBS, 1984: 24-26), its contribution to the total discharge of the Brahmaputra-Jamuna River is nominal (only 5 percent). The runoff of the Brahmaputra is thus the outcome of the combined effects of spring snowmelt and summer rainfall in the Himalayan mountain system of Tibet and northeastern India.

Historical records of water and sediment discharge of the major rivers of Bangladesh are limited, with relatively reliable data available only from the early 1950s onwards (see Morgan and McIntire, 1959: 323). It was only after the heavy floods of 1954 that discharge observations were recorded at some sites on the Brahmaputra-Jamuna (Tarafdar, 1974: 26). Presently, water levels are measured at 16 stations and water discharge and suspended sediment loads are measured at four stations by the Hydrology Division of the Bangladesh Water Development Board.

Notably, while the length of the Brahmaputra-Jamuna River and the area of its drainage basin are smaller than many of the selected rivers, only the maximum discharge of the Amazon exceeds that of the Brahmaputra-Jamuna for the specified period (Table 4.7). In terms of the average discharge at its mouth, the Brahmaputra-Jamuna River ranks fourth in the world. On the basis of the average discharge per unit of drainage area, which amounts to 0.0345 cumec per square kilometer, this river supersedes all other rivers in the world (Khan, 1995: 83). Based on water level records since 1949 and discharge measurements taken since 1956 at Bahadurabad station, the maximum discharge of the Brahmaputra-Jamuna was recorded at 99,500 cumec on August 30, 1988. The highest flood level of 20.62 meters was also recorded on that date. The minimum water level was measured to be 11.40 meters in February 1952, while the minimum discharge was recorded at 2,860 cumec in March of 1971. The river's average discharge of around 20,000 cumecs and its bankfull discharge of about 44,000 cumecs (Consulting Consortium 21/22, 1993) make it the fourth largest freshwater flow in the world. Based on records for the 1972-1985 period, the average annual runoff of the Brahmaputra-Jamuna at Bahadurabad is estimated to be around 613.3 million cubic meters (Table 4.7). Khan (1995) cites that the total annual runoff can vary between 25 billion and 763 billion cubic meters.

The extent of flooding of the great rivers of Bangladesh is primarily influenced by the amount of peak discharge and the duration of high flood stages. This in mind, it is interesting to compare the annual discharge cycle of the Brahmaputra-Jamuna with that of the Ganges. The Brahmaputra-Jamuna carries down its floodwater from the last week of May to the first week of October, with July and August as the months of maximum discharge (see Table 4.7); the flood peak of the Brahmaputra-Jamuna usually occurs prior to peak flooding of the Ganges (see Coleman, 1969; Rashid, 1977). As Coleman (1969) notes that, 58 percent of the Ganges River experiences flood peaks in the later part of August or the first part of September. As a result of variations in the flood peaks, the Brahmaputra-Jamuna floodplain is prone to two or three flood peaks in a year. The first flood occurs during the May - July period, due to spring snowmelt and monsoon rainfall in the upper valley of the river. Discharge often increases from about 8,000 cumec in May to 70,000 cumec in July. The second flood most often strikes during the August - October period, aided by the strong backwater effect of the flooding Ganges. The analysis of the discharge regime of the Brahmaputra-Jamuna River demon-

TABLE 4.7. Average water discharge of the Brahmaputra-Jamuna (in 1,000 cumecs) at Bahadurabad station by month and annual runoff (in billion cubic meters), 1972-1985

Water-Year	Apr	May	June	July	Aug	Sept	Oct	Nov	Dec	Jan	Feb	Mar	Annual Runoff
1972-73	7.2	17.8	33.0	43.4	37.0	29.9	16.3	8.7	5.0	3.9	3.4	4.0	554.5
1973-74	7.1	14.3	37.1	31.9	45.3	30.5	22.7	10.4	5.5	3.9	3.6	4.1	572.8
1974-75	6.8	17.2	30.9	56.1	49.9	48.4	25.9	9.3	5.5	3.9	3.4	3.8	690.3
1975-76	7.1	10.9	24.5	43.0	39.4	37.0	26.0	11.5	6.9	5.0	4.2	5.2	584.2
1976-77	7.9	12.4	30.0	46.0	38.0	28.7	16.7	9.5	7.0	4.9	4.2	5.8	557.0
1977-78	11.0	20.6	40.0	45.2	55.5	42.0	28.5	13.7	8.2	5.8	4.8	4.8	739.4
1978-79	6.2	13.6	33.5	42.8	39.6	32.9	19.1	9.8	7.3	5.3	4.5	4.6	579.0
1979-80	7.5	13.0	17.5	45.0	40.0	47.2	36.6	13.6	9.9	5.4	4.6	5.3	645.0
1980-81	8.6	17.5	29.7	38.2	48.8	33.3	23.0	10.7	6.7	5.0	4.5	5.3	611.2
1981-82	8.0	11.3	20.2	41.1	38.5	36.7	16.7	8.4	5.3	4.3	3.7	4.0	427.2
1982-83	7.6	13.9	29.6	58.8	35.1	31.8	17.7	8.4	5.4	4.9	4.3	5.3	562.2
1983-84	7.3	15.1	24.7	40.9	35.9	44.9	26.0	10.8	6.2	4.8	4.0	4.0	686.3
1984-85	8.3	16.5	33.2	53.3	37.8	48.6	21.4	10.5	6.5	4.9	4.0	5.0	763.7
Mean	7.7	14.9	29.5	45.1	41.6	37.8	22.8	10.4	6.6	4.8	4.1	4.7	613.3

Source: BWDB (1985; unpublished data).

strates overall the highly variable character of its yearly flood cycles. Such erratic discharge profoundly influences the channel pattern.

Bed and Suspended Sediment. The Brahmaputra-Jamuna River is one of the most heavily sediment-laden rivers in the world. Holeman (1968), and Milliman and Meade (1983) consider this river as second only to the Yellow River in China in terms of sediment transported per unit drainage area; Goswami (1985), on the basis of total load measurements at Pandu, Assam, India, ranked the Brahmaputra in fourth place. Goswami observed that more than 95 percent of the total sediment is transported by the river during the rainy season (May through October) with the major portion of the sediment load (i.e., 95 percent) being carried by the relatively frequent moderate floods. These findings conform with assertions made by Wolman and Miller (1960).

The sediment transport regime in the Brahmaputra-Jamuna is extremely complex. Data on sediment discharge, particularly bed-load sediment, are meagre. Because of high costs, the measurement of suspended sediment load in the Brahmaputra-Jamuna is not obtained on a daily basis; instead, between one and five measurements are taken in a month by the Hydrology Division of the Water Development Board. During low flow months, sediment discharge measurements are often not attempted. Data on suspended sediment discharge of the Brahmaputra-Jamuna therefore are limited in scope and should be considered as only close approximations.

The sediment transported by the Brahmaputra-Jamuna consists of chiefly medium to fine sand and silt, largely the same materials found in the banks. Suspended sediment and bed material samples are presently collected at Serajganj, Bahadurabad and Nagarbari stations. Data from Serjganj station showed that 70 to 90 percent of suspended load samples were composed of silt and clay (0.001 to 0.02 millimeters). Among the remaining fraction, only traces of fine and very find sand (0.355 millimeters) could be found in the composition of suspended sediment samples taken in Serajganj. The bed material at Serajganj station was found to be characterized by 95 percent fine sand, with negligible proportions of coarse sand, silt, and clay (Latif, 1969: 1697; Rahman, 1984: 10). The maximum size of bed materials was recorded at 0.83 millimeters.[33] From grain size analysis of bedload material, Coleman (1969: 178) also reported that the range in median size was observed to be from 0.34 millimeters (fine sand) to 0.03 millimeters (silt), with a mean of 0.17 millimeters. Overall, since the bank materials rarely contain any clay and are characteristically non-cohesive, the riverbanks are very susceptible to erosion.

Besides the effects of sediment size and cohesiveness on bank erosion, the seasonal variation and the overall magnitude of the sediment discharge regime have significant bearings upon bank stability. Nearly all sediment is carried by the Brahmaputra-Jamuna River as bedload and suspended load, with only very nominal wash load. The depth-integrated samples that are obtained for measuring sediment loads contain the suspended load and some of the bedload. The recorded daily sediment transport rate of the reaches of the Brahmaputra-Jamuna River in Bangladesh ranges between a high of 2.26 million tons and a low of 0.02 million tons (Khan, 1995: 72). Coleman (1969), however, reported a much higher sediment transport rate based on measurements made in the mid-1960s: 5.0 million tons per day and a total of 600 million tons annually. Upstream, at Pandu, Assam, India, Goswami (1985) observed an average transport rate of 2.12 million tons per day with a maximum of 26.5 million tons per day during the peak flow season. Notably, the measurements of sediment concentration in the upstream reaches, namely in Assam, are taken more frequently and in more locations than those in Bangladesh.

The daily sediment discharge data by month for selected water-years (i.e. 1976-77, 1977-78 and 1983-84) at the Bahadurabad station indicate a high degree of variation between low water and high water seasons (see Haque, 1988). The low water period records have a maximum value of 20,000 tons of sediment discharged per day in the selected years. A far greater range of variation occurs during the high flood water season. For example, the highest level of daily sediment discharge was recorded in the summer months of 1977 (i.e., at more than 2 million tons per day). Sediment discharge begins to rise in the early summer monsoon period (in April) and maintains high magnitudes until the end of September, which is indicative of its high correlation with the water discharge regime. Khan (1988) describes the relationship between water and sediment discharge as:

$$Q_s = 0.38 Q_w^{1.36}$$

where,

Q_s is the sediment discharge in tons per day, and

Q_w is the water discharge in cumec.

He adds that the regression line underestimates sediment discharge at higher flows, as revealed by a logarithmic plot. Overall, the high degree of correlation between sediment discharge and water discharge is expressed in the r^2 value of 0.95.

Slope. Khan (1988) plotted the distance-elevation data for the reaches of the Brahmaputra-Jamuna River in Bangladesh—a distance of about 200 kilometers (Figure 4.4). This bank elevation plot with distance for the entire river is modified from Goswami (1985). In general, the river can be grouped into two reaches of different slopes. The upper reach is much steeper, with a slope of 0.000105, compared to the lower reach which has a slope of 0.000053. The average slope of the river in Bangladesh is 0.0000819. During the flood stages the water surface slope is 0.000066 (Coleman, 1969). The disaggregated data show that the water surface slopes of the reaches of the Brahmaputra-Jamuna River in the Bangladesh territory range between 0.000055 and 0.000093 (0.29 and 0.49 feet per mile) during high flow season and between 0.000053 and 0.000089 (0.28 and 0.47 feet per mile) during low flow season (Table 4.8).

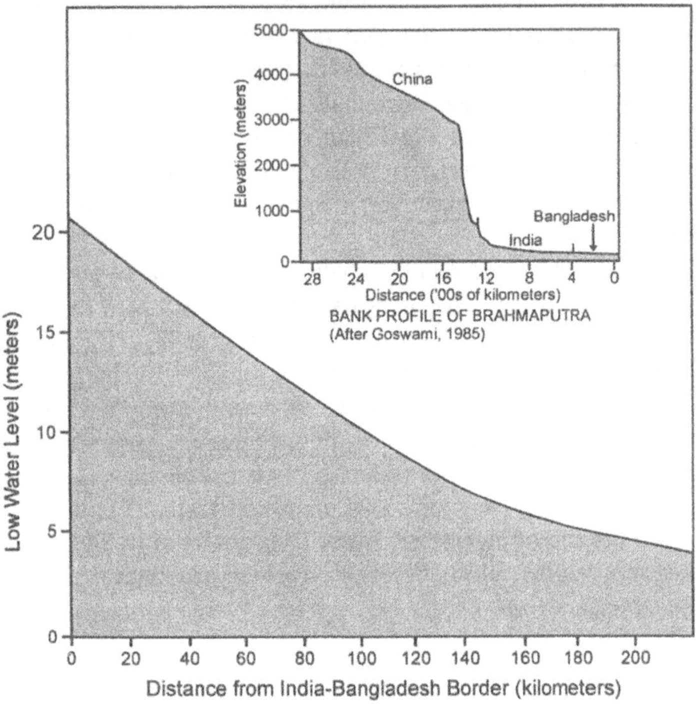

Figure 4.4. Distance-elevation plot of the Brahmaputra River (after Khan, 1995).

TABLE 4.8. Water surface slopes at different reaches

River Reach	High Flow Season Slope	Low Flow Season Slope
Nagarbari to Serajganj	0.000055 (0.29 ft/mile)	0.000053 (0.28 ft/mile)
Serajganj to Milanpur	0.000070 (0.37 ft/mile)	0.000064 (0.34 ft/mile)
Milanpur to Golna	0.000072 (0.38 ft/mile)	0.000066 (0.35 ft/mile)
Golna to Porarchar	0.000093 (0.49 ft/mile)	0.000089 (0.47 ft/mile)

Source: BWDB (1978: 4-5).

A focus on the slopes of the Brahmaputra in the upper valley reaches may be necessary here to provide an overall profile of the gradient of the river. From its origin in the south of Lake Gunkyud in southwestern Tibet (elevation 5,300 meters), the river—here called the *Tsangpo*—flows along the bottom of a longitudinal graben parallel to the Himalayan ranges for 1,100 kilometers (Goswami, 1985: 962). The river then enters a deep narrow gorge at Pe (elevation 3,500 meters); the gradient in the gorge section ranges from about 4.3 to 16.8 meters per kilometer. In its first 450 kilometers through Assam, India, the river, now called the *Dihang*, follows a slope of one meter per kilometer (Desai, 1968: 432). Between Kobo and Dibrugarh, where the Dibang and Lauhitya join the Dihang, it maintains a slope of 28 centimeters per kilometer. The slope between Dibrugarh and Dubri ranges from 10 to 18 centimeters per kilometer. In Assam, as well as in Bangladesh, the Brahmaputra flows in a highly braided channel characterized by the presence of a multitude of bars and *chars*, and with a relatively low gradient relative to its upper reaches.

Channel Pattern of the Brahmaputra-Jamuna River

In simple terms, *channel pattern* means, "the configuration of a river as it would appear from an airplane" (Leopold *et al.*, 1964: 281). Knighton (1984: 124) elaborates upon this definition by mentioning that it represents a mode of channel form adjustment in the horizontal plane, which is additional to but linked with transverse and lengthwise modes. Channel pattern affects the resistance to flow and can be considered an alternative to slope adjustment when valley slope is regarded as constant in the short time scale.[34] The importance of analyzing the channel pattern of a river lies in the fact that the planform of a reach of an alluvial river reflects the hydraulic processes of flow within the channel, as well as the related processes of sediment transfer and energy dissipation (Richards, 1982: 180).

Despite some problems with the tripartite classification of straight, meandering and braided patterns,[35] this convention continues to dominate as the basic means of channel classification. Leopold and Wolman (1957) assert that river pattern is a continuum and that each pattern is associated with a particular combination of variables.[36] As Leopold *et al.* succinctly explain:

> River patterns represent an additional mechanism of channel gradient and cross-section. The pattern itself affects the resistance of flow, and the existence of one or another pattern is closely related to the amount and character of the available sediment and to the quantity and variability of the discharge (Leopold *et al.*, 1964: 281).

This concept presumes that streams are able to form their planimetric dimensions freely in response to flows and sedimentations, yet the possibility of a transition between channel patterns is not directly addressed. Transitional patterns, however, have been observed in nature, albeit over relatively short river reaches. Some scholars assert that anastomosis in certain cases represents a transitional channel pattern (Knighton and Nanson, 1993), which seems to conform with Leopold and Wolman's view.

In the continuum of channel forms, each channel is associated with a particular combination of controlling variables. The numerical values of such variables are used by several studies to explain the three basic patterns (Table 4.9); note that the slope-discharge relationships received the most attention. A considerable difference in the form of the relationships and the values of the coefficients are apparent in Table 4.9. These variations are likely due primarily to the different environmental conditions under which the relations were developed. For instance, some of the relations were developed

TABLE 4.9. Slope, discharge and bed material size and channel pattern

Source	Equation S = Slope Q = Discharge D = Bed Material Size (SI unit)	Channel Pattern Characteristics
Leopold and Wolman (1957)	$S = 0.012Q^{-0.44}$	Braided channels plot above the line, meanders plot below. Natural streams.
Lane (1957)	$S = 0.0007Q^{-0.25}$ $S = 0.004Q^{-0.25}$	Meanders. Braided with sandy bed.
Henderson (1963)	$S = 0.00012D^{1.15}Q^{-0.46}$	Slopes greater than this braid, and less than this meander.
Ackers and Charlton (1970)	$S < 0.001Q^{-0.12}$ $S > 0.0014A^{-0.12}$ $S = 0.00085Q^{-0.21}$ $0.001Q^{-0.12} < S < 0.0014Q^{0.12}$	Straight channels. Meanders. Meanders plot above the line, straight plots below. Straight channels, alternating bars.
Osterkamp (1978)	$S = 0.001Q^{-0.24}$ $S = 0.0019Q^{-0.31}$	Includes all patterns. Coefficient value increases as sinuosity decreases or material size increases. Braided channels.

from natural streams while the others result from laboratory experiments. Overall, all relations indicate that the braided pattern occurs at a steeper slope than either the meandering or the straight pattern. This implies that a threshold value may exist for the slope beyond which the braided pattern prevails.

A *braided channel* is divided into several channels by bars or islands, and the divided channels subsequently join and redivide. A distinction between braided and anastomosing patterns can be made on the basis of their respective stability (Schumm, 1963). A braided pattern contains many unstable divided channels which change their location frequently, while an *anastomosing pattern* possesses many stable channels that retain their identities despite changing discharge and time. Based upon these criteria, several studies have confirmed the channel pattern of the Brahmaputra-Jamuna reach as a braided one (Coleman, 1969; Chowdhury, 1973; Tarafdar, 1974; Rahman, 1984).

The braided channels in small mountain streams cannot be expected to follow precisely the same patterns as those of large rivers such as the Brahmaputra-Jamuna. Still, the suggestions put forward by studies of small streams are worth reviewing as a basis for a general understanding of the braiding process. We will now turn to the findings of such studies. Basically, four specific aspects are important in summarizing the diversified conditions and characteristics of braided planimetric form (see Knighton, 1984: 144-7): (1) abundance of bedload; (2) easily erodible banks; (3) highly variable discharge; and (4) steep slopes.

An Abundant Bed Load. Braided channels occur in river systems having an overabundance of bedload, although it is generally assumed that braiding is not symptomatic of overloading. In addition to large amounts of sediment, the load contains size fractions that the stream is locally incompetent to transport—as such, they form initial deposits. The formation of bars due to concentrated deposition diverts the flow against the channel banks, which contributes to the bank erosion required for the development of a wide, shallow channel. Hoey (1992) suggests that in complex braided systems, fluctuations in bedload transport and changes in channel configuration may be autogenic, that is, inherent, and thus may not need external forcing.

The process of braiding is aggravated by the occurrence of a coarse fraction in the sediment load. This material forms the initial deposits due to local incompetency. The deposits, in the form of bars, cause the flow to strike the channel banks. The resulting bank slumping and erosion then promote a wide and shallow channel pattern (Khan, 1995: 22). Church (1972) argues that braiding occurs when the resistance to flow of the single channel becomes too high to transport the water-sediment load. The bars are formed due to deposition of excess load, and in turn divide the flow into channels around them. These channels are steeper, relatively deeper and more efficient, even though their combined resistance may be higher than that of the single channel upstream. Consequently, both the velocity and energy of the river are enhanced, allowing it to transport the incoming sediment more easily.

Schumm (1980) indicates the importance of sediment availability with evidence from the changes in the channel pattern of the Rangitata River of New Zealand. In contrast to the braided rivers of the Canterbury Plain, the Rangitata adopts a meandering planform below a gorge as a result of the reduction in sediment supply. Further downstream where the river has cut into less consolidated materials, it rapidly converts from a meandering to a braided form. Similarly, as opposed to the Ganges, the plentiful supply of sediment from unconsolidated Holocene formations and the abundance of water from summer monsoonal precipitation and spring ice-thaw contribute significantly to create the braided forms in Assam (see Goswami, 1985) and in Bangladesh (see Coleman, 1969).

Presence of Easily Erodible Banks. The availability of erodible banks is necessary for maintaining the braiding pattern. Banks composed of readily erodible material provide an abundant source of sediment and permit the channel widening so characteristic of braided reaches. The scope for localized deposition is greater in a wider channel, where a heterogeneous flow tends to affect part, rather than all of the cross-section (Knighton,

1984: 145). Also, any incipient bar deposits tend to be eliminated, rather than added to without the presence of erodible banks. Miall (1977) demonstrates that rivers with resistant banks meander rather than braid, even though bedload transport may be dominant, which too is a characteristic of a braided river. Mackin (1956) analyzes how the channel patterns of the Wood River in Idaho, USA, responded to changes in bank resistance. He attributed a sequence of meandering-braided-meandering to a corresponding sequence of flow through forest-prairie-forest environments. Nevins (1969), examining the Turandui River in New Zealand, indicates that it has converted over a period of years from a braided to a meandering pattern as a response to the planting of willow shrubs at appropriate places. Coleman (1969), and Khan and Rashid (1985) also argue that the existence of non-cohesive bank material along the Brahmaputra-Jamuna River is one of the prime factors which influences the braiding of the river.

A Highly Variable Discharge. A high rate of sediment supply is often caused by rapid fluctuations in discharge. These, in turn, contribute to bank erosion and irregular bedload movement. Morisawa (1985: 107-8) argues that, hydrologically, as compared to non-braided streams, the regime of a braided river is one of an overall highly variable discharge with higher flood peaks, higher total discharge range, and higher monthly variability. As shown above, both rising and falling river stages have important roles upon bar and braid formation. These highly variable hydrologic characteristics function as controls of bar formation, flow diversion, and new channel creation (c.f. Leopold and Wolman, 1957; Hong and Davies, 1979).

Steep Slopes. Several theoretical and empirical studies indicate that braiding develops in a situation where the slope is above a threshold value (Leopold and Wolman, 1957; Schumm and Khan, 1972). Parker (1976) and Chang (1979) further indicate that, theoretically, the degree of braiding is likely to increase with increases in the slope gradient.

Knighton (1984: 146), however, points out that *stream power*—rather than simply steep slope—is possibly more important for the persistence of braiding. Such a proposition may be supported with a number of empirical examples. The Ganges at Patna, India, has a braided pattern while its slope is only 0.000066 (i.e., 0.35 feet per mile; see Leopold and Wolman, 1957). Similarly, the reaches of the Brahmaputra in Bangladesh have an average slope (during high water period) between 0.000055 and 0.000093 (i.e., 0.29 and 0.49 feet per mile), but the channel characteristically represents a braided pattern. Coleman notes that:

> From the studies completed to date, it appears that braided channels occur in river systems having a relatively steep slope, an overabundance of bedload, or a combination of the two. In mountain streams, slope is the dominant factor causing braiding, but in large alluvial rivers such as the

Brahmaputra [Brahmaputra-Jamuna] extreme bedload is the more important factor (Coleman, 1969: 146).

The evidence of braiding in low gradient streams suggests that slope itself may not be the dominant factor in braiding. A river must be sufficiently potent to erode its banks and achieve the high bed mobility which is a fundamental requirement for braiding. None of the above factors alone is sufficient to cause braiding, but a combination of abundant bedload, easily erodible banks, and relatively high stream power, are probably necessary. The Brahmaputra-Jamuna typically represents such a combination.

4.3.2. RIVER CHANNEL MIGRATION: A HISTORICAL ACCOUNT

A fluvial system can be regarded as a physical system with a history (Schumm, 1977)—its present form is the product of past and present processes and conditions.[37] The fluvial system thus possesses a memory of temporal progress. Notably, different components of channel morphology adjust at different rates and reflect the influence of past conditions to varying extents. Richards (1982: 252), in this respect, insists that channel morphology adjusts in the long, medium, and short term to changing control variables such as water and sediment discharges. It is suggested that long-term effects, such as climatic, hydrologic and gradual tectonic effects, cause gradual and progressive adjustment of alluvial channels. Medium-term adjustments are generally related to human activities which create temporary disequilibrium within the channel, which then approaches a new equilibrium. The third category, that of short-term changes, relates to the influence of individual random extreme events, such as high-magnitude, low-frequency floods (see Richards, 1982: 252).

The primary clues for detecting changes in a fluvial system are morphologic remnants provided by, for example, abandoned channels and sediment sequences. However, these features usually provide a complex and incomplete record of changing conditions. Two methods are then generally adopted by geomorphologists to reconstruct river histories. The first method provides direct evidence of change through instrumentation, photographic records, and ground survey. The second method, involving a more indirect approach, uses historical documentation, morphologic and sedimentologic data, and "dating techniques" to provide information on spatial and temporal changes in river courses. Notably, direct observations are rarely maintained for a long period of time (i.e., usually confined to less than two decades), and often fail to provide continuous information. Photographic and ground survey records are usually limited in number and, therefore, are inadequate for long-term change detection of river channels.

In the Holocene alluvial plains of the Brahmaputra-Jamuna River, where the society is still predominantly agrarian, the use of the 'historical record' for detecting long-term channel changes appears feasible (see Grove, 1972; Hooke and Kain, 1982; Miller and Wendorf, 1958). There may, however, arise a number of difficulties in using historical documentation. For instance, historical records are often available for only specific dates, necessitating that a uniform change be presumed between each pair of surveys.

These gaps can lead to serious limitations in the historical study of channel changes as significant alterations may take place during the intervening period (see Knighton, 1984: 165).

Hooke and Kain (1982: 68-72) provide a comprehensive analysis of the causes of error in using historical records. They note that errors may arise during collection, transfer, storage, and subsequent use of documents. Common causes of error are identified as: (i) lack of awareness of *date conventions*; (ii) *frequency* and *timing* of observations; (iii) *perception* and *attitudes* of observers; and (iv) *styles* and *conventions* of writing and drawing. They add that the use of historical records can be very time-consuming and requires meticulous examination.

The changes in river courses of the Bengal basin are recorded in early scriptures, travel accounts, ancient maps, and folk literature. Each of these sources can provide only a general estimate of the direction and rate of channel changes, thus providing only two-dimensional perspectives. Another major limitation of using such material in the subcontinent is that although a reasonable amount of information can be explored to recollect the past, maps and charts based on a scientific scale did not develop until after India's contact with the outside world during the late Medieval period (Chowdhury, 1977: 145).

Despite the limitations of using historical and cartographic documentation, the accuracy of the information can be assessed in a number of ways. The principal means consists of checking the characteristics (such as the date) of the source, and comparing its information with other evidence (see Hooke and Kain, 1982: 73). To assess the validity of historical sources attention must be paid to the following aspects: (i) originality; (ii) closeness in time of documents to reality portrayed (i.e., contemporaneity); (iii) closeness in space of document to the situation described (i.e., propinquity); (iv) generalization and level of detail; and (v) purpose of the record or the reason it was compiled. The reliability of historical documents can also be verified by using external corroboration, which means finding an agreement on a specific point between two or more independent sources. Such corroborations involve comparisons of similar historical sources such as two maps or two descriptions, comparisons of evidence from different types of historical sources (for example, between map evidence and evidence in written sources), and comparisons of historical maps with a modern equivalent.

As Knighton (1984: 165) indicates, the potential use of historical sources in geomorphological study has not been fully exploited, although many researchers have successfully applied historical evidence to studies on coastal change, river channel movement, reconstruction of catastrophic events such as floods and landslides, discharge records of streams, and soil erosion (see Hooke and Kain, 1982: xix). Some specific examples are worth citing here. Grove (1972) uses land rent records for the Jostedasbre region in Norway to demonstrate how the incidence of mass movement and floods has varied since the 16th century. Hooke (1979), and Hooke and Kain (1982) use five historical surveys to determine the distribution of course changes along Devon streams and estimate rates of bank retreat.[38] Both Burkham (1972) and Schumm and Lichty (1963) effectively study channel widening and other changes over historical time in rivers of the United States. Notably, among a variety of historical materials, maps have proved to be the most important source. However, one needs to keep in mind that:

> These [maps] vary in date, scale and coverage and thus in amount of information they can provide but generally they can show by comparison whether or not there has been a change in river planform. With the aid of a series of detailed maps the sequence of changes in particular parts of a river can be established, measurements of direction, distances and rates of change can be made and processes of movement and change analysed (Hooke and Kain, 1982: 119).

In the following pages, an attempt is made to reconstruct the changes in river courses of Bengal with special reference to the Brahmaputra through reviewing a variety of historical materials. Since the origin and development of the Brahmaputra channel has been intrinsically related to other great river systems of the Bengal delta, the discussion focuses on the broader regional perspective. For our purpose, the documents, travel accounts and notes, scriptures, inscriptions, literature, books, and maps relating to rivers of Bengal and their migratory nature can be treated in terms of three historical phases: (i) ancient and early records of the Hindu and Buddhist Dynasties (from 300 BC to 1200 AD); (ii) the written and cartographic documents of the Muslim period and those from European exploration (1200 to 1757); and (iii) the British and post-Independence surveys, records, and maps from 1757 onwards.

The Ancient and Early Documentation

The earliest reference to the territory of Bengal are mythological rather than historical. The *Aitareya Aranyaka, Ramayana, Mahabharata, Samyatta Hikaya*, and other Hindu scriptures refer to the people of the land of the *Vangas* (i.e., Eastern Bengal), *Sumhas* (Hoogly district of West Bengal), *Pundras* (Northern Bengal), and other inhabitants of the delta of the Ganges or the territory east of it (Ray Choudhuri, 1943). The description of the ancient Brahmaputra River is found in the *Puranas*.[39] The *Lauhitya* or *Lohini* (the ancient name of the Brahmaputra-Jamuna River) is described, "as giving a sanctity to Kamarupa which the river crosses after traversing the Himalayas" (Mukerjee, 1938: 200), and then flowing into the southern sea on the coast of *Udayachala* (southern Bengal). These old Hindu scriptures indicate that the Atrai and Karatoya were two great rivers flowing southward parallel to each other, and that the lower part of the Karatoya was named the Brahmaputra (Mukerjee, 1938: 208; see Figure 4.5a). The greatness of the Karatoya is specifically described in the *Puranas*; it is mentioned in these scriptures that the river once formed the western boundary of the Kingdom of Kamrupa and followed an independent course to the sea.

Alexander's historians were familiar with the accounts of the Gangaridai-the people of the Ganges delta—as such, a number of Roman sources document the region. Ptolemy drew the first map of the area in 150 BC and referred to the five mouths of the

Ganges: Kambyson, Mega, Kamberikhon, Psedostomon and Anibole. In Ptolemy's description, Sounagoura (Sonargaon), the ancient capital of central Bengal is shown near the right bank of the Brahmaputra (McCrindle, 1885; cited in Kamaluddin, 1973: 2).[40] The Chinese pilgrim, Huen Tsang visited India during the seventh century AD and prepared an intriguing account of the delta (see Walters, 1905; Cunningham, 1924). From his description, it appears that *Pundra Vardhana* (northern Bengal) was one of the independent divisions of Bengal, with its capital at Mahasthana on the banks of the great river Karatoya. Interestingly, this locational delineation of *Pundra Vardhana* as lying between the Ganges and Karatoya coincides with the description of the *Varendra* in the *Rama (pala) Charitra*—the biography of the Buddhist King Ramapal.[41]

The ancient and early scriptures, travel accounts, maps and other documents of the Bengal territory thus provide ample evidence that the present day Brahmaputra-Jamuna River flowed southward through the then mighty river Karatoya in the early Christian era. The lower reach, with its name as the Brahmaputra or Lauhita, flowed to the sea through central Bengal (Figure 4.5a).

The Records of Muslim Period and European Maps

During the Muslim regime in Bengal, European contact resulted in the production of Portuguese, Italian, Dutch and French maps of the Bengal region. The maps of de Barros (1550) and Cantelli da Vignolla (1683) indicate that the Caor or the Karatoya, with the name of Brahmaputra in its lower course, flowed into the Bay of Bengal near Sandwip Island (see Figures 4.5a and 4.5b). By examining the changing patterns of the rivers of the Bengal delta, Mukerjee (1938: 124-5) finds that for centuries the main channel of the Brahmaputra passed southward through the district of Dhaka or central Bengal, and "carried troops of the Mughols, Armenians and Portuguese for many sanguinary river engagements." Because of favourable waterways in central Bengal, the capital of Ballal Sen—Rampal—was established on the western bank of the Lauhitya or Brahmaputra. During Akbar's time (1556-1605), the eastern lower reaches of the Ganges or Padma flowed through the districts of Rajshahi and Pabna, continuing onwards either through the Dhaleshwari channel or farther north along the bed of the Buriganga (Chatterjee, 1947: 13).

During the early decades of the 17th century, the lower part of the Brahmaputra, particularly to the south of Agrasindhu, evidently migrated farther eastwards. The silting-up of the channel (resulting from accelerated soil erosion upstream due to human interference and deforestation), the upheaval of the Madhupur tract and the subsidence of the Sylhet basin have all been suggested as possible reasons for this remarkable channel shifting (Mukerjee, 1938; Morgan and McIntire, 1959). As a result, the Brahmaputra took a course to the Meghna first at Narsingdi through the Arial Khan channel, and then for about a century (mid-17th to mid-18th century) through the main channel at Bhairab Bazaar. A comparison of the maps of Matheus Van den Brouck (1658-1664) and Major James Rennell (1764-1776) demonstrates some remarkable changes (Figure 4.5b). Van den Brouck's map shows that the discharge of the lower

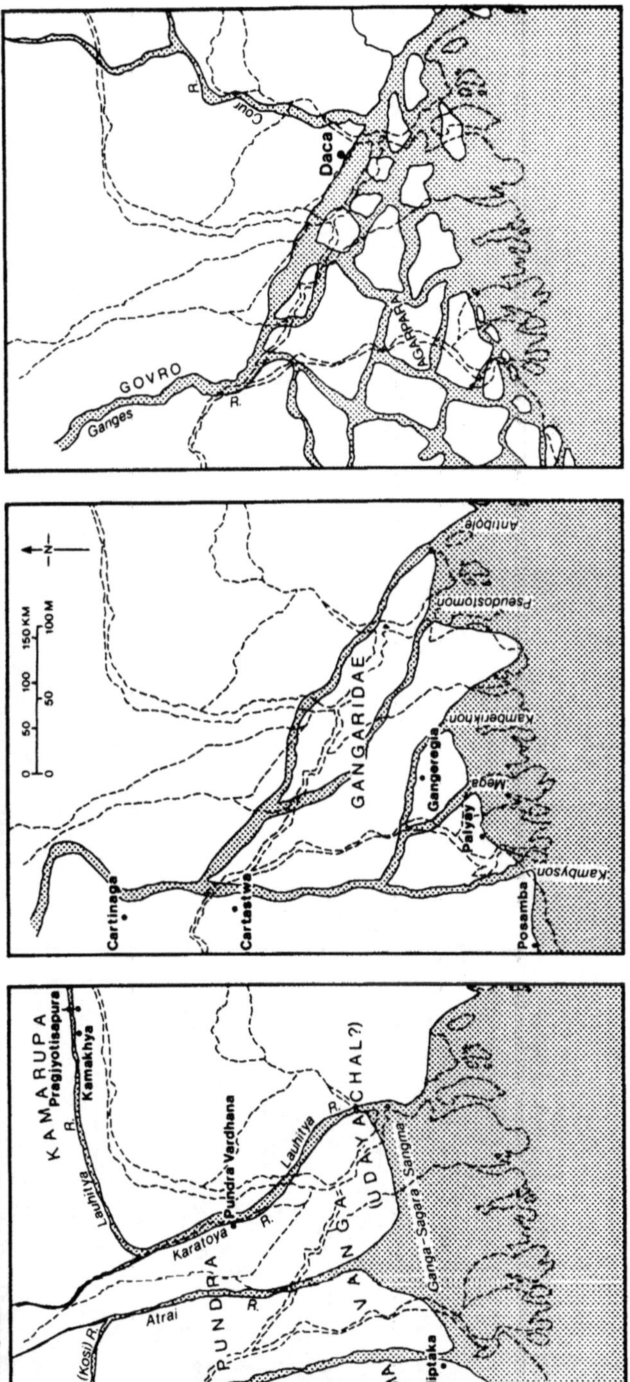

Figure 4.5a. Changes of the rivers in the Bengal delta, ancient times - 1550 AD.

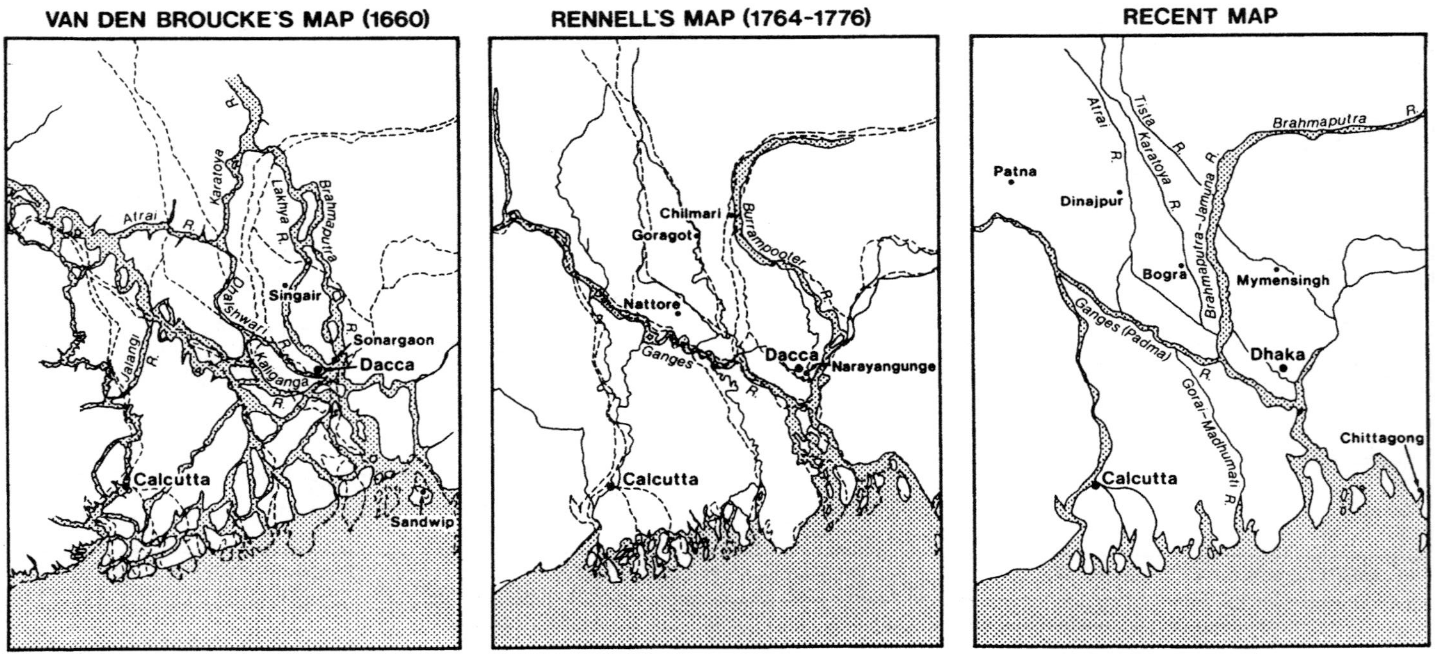

Figure 4.5b. Changes of the rivers in the Bengal delta, 1660 - present.

Brahmaputra passed through both the channel of Lakhya and the old bed of the Lauhitya across the district of Mymensingh. Major Rennell, about a century later, found that the discharge regime of the Lakhya had noticeably deteriorated. Furthermore, Van den Brouck recorded the Atrai and Karatoya channels as considerably large rivers discharging into the Ganges. The Tista was not yet a noticeable river. Rennell also noted remarkable changes in the river systems of northern Bengal during the late 18th century. The following section reviews these changes.

The British and Postcolonial Maps and Surveys

For many years after the conquest of Bengal by the English East India Company in 1757, waterways remained the predominant means of transportation and communication, especially for the purpose of internal commerce and trade. With the increasing importance of accurate geographical knowledge and navigational information of the area, the Company employed Bartholomew Plaisted as the first surveyor of Bengal, his first assignment to survey the coasts along Chittagong and the *Sundarbans*. In April, 1764, Major James Rennell received an assignment to find an alternative route from Calcutta to Dhaka to avoid the long detour through the *Sundarbans*. During 1764-1776, Rennell surveyed and compiled sets of maps of Bengal which obviously include the valleys of the Ganges and the Brahmaputra (Figure 4.5b). As Kamaluddin (1973: 5) remarks, Rennell's maps "will glow for all time as a fascinating and outstanding work of this illustrious geographer."

The most remarkable change in the river courses that occurred since the time of Rennell was the avulsion by the Brahmaputra of its eastward course through Mymensingh, and its pursuit of the narrower channel of the *Jenai* (which was locally known as the *Konai*) River following the great flood of 1787 (see Chowdhury, 1973; Ahmad, 1968). After this date the Brahmaputra flowed southward to join the Ganges (Padma) around Goalando, involving a westward retreat of the main channel of about 113 kilometers. This new channel became known as the *Jamuna*.

In studies dealing with this change there is a lack of agreement concerning the duration and causes of the westward diversion of the older course of the Brahmaputra. The arguments can be broadly grouped into two divisions. The first group postulated that the avulsion of the Brahmaputra took place at a very rapid rate, and was primarily caused by the extreme flood of 1787 or by an abnormally high volume of discharge (La Touche, 1910; Hayden and Pascoe, 1919; Chatterjee, 1947). The second group argued that the westward shift of the Brahmaputra occurred gradually in adjustment to the tectonic activities in the region (Fergusson, 1863; Hirst, 1916; Chowdhury, 1973; ISPAN, 1993).

La Touche (1910: 20-21) claims that the diversion resulted directly from a considerable increase in the discharge of the river. He concluds that the Dihang River of Assam, as a tributary of the then small channel of the Brahmaputra, beheaded the Tsangpo River of Tibet because an "enormous accession of water began to exert itself." Chatterjee (1947: 18), supporting these suppositions, insists that the main cause of the diversion of the Brahmaputra westward was the inability of the old channel to discharge the combined water volume of the Brahmaputra and the Tista. From a geomorphological standpoint, however, these views have limited scope. Evidence from recent studies of rivers subject to high flood discharge or altered hydrologic conditions indicate that the cross-sections of rivers undergo maximum adjustment particularly in terms of width (see Knighton, 1984: 114).[42] Recent studies also clearly demonstrate that both the depth and width can respond rapidly to changes in the dominant controls of discharge and/or sediment load. The development of the concept of channel adjustments and the related evidence in recent river change studies suggest quite plausibly the Brahmaputra diversion was affected by structural controls.

Hirst (1916: 14) asserts that the Brahmaputra diversion took place gradually between 1720 and 1830. He draws partly upon the observation of Buchanan Hamilton, who visited the area in 1810 and recorded that the Brahmaputra threatened to divert into its present course. Morgan and McIntire (1959: 330) note that by 1830 the diversion of low river flow down the new channel was apparently complete. Fergusson (1863) concludes that this diversion occurred chiefly due to the uplift of the Madhupur tract. Hirst elaborates on this and stated that a zone of subsidence between the two large Pleistocene blocks of the Barind and Madhupur tracts had been the prime determinant in the shift of the Brahmaputra. He claims that these two blocks had been elevated "as compensation to a line of subsidence passing approximately from Jalpaiguri to the sea, down the alignment of the present Meghna River" (cited in Rashid, 1977: 24).

A comparison of Rennell's map with successive surveys and maps demonstrates additional spectacular changes in the drainage systems of northern Bengal (Figure 4.5b). Historically, the Tista or *Trisrota* (i.e., consisting of three currents) flowed southwards through three of its distributaries—the Atrai, the Karatoya, and the Purnabhaba (see Chatterjee, 1947: 18). The maps compiled prior to Rennell's survey show the main flow of Tista through the course of the Dhepa-Purnabhaba into the Ganges in the south. Following the flood of 1787, the Tista diverted its course southeastwards through one of its old beds[43] to join the Brahmaputra (see Rennell, 1781: 95-7), which resulted in subsequent deterioration of the Atrai and Karatoya. The sudden diversion of Tista River as a tributary of the Ganges to the Brahmaputra channel in 1787 thus was the final action that triggered the diversion of the lower Brahmaputra into the Jenai channel west of the Madhupur tract (see La Touche, 1910). The Karatoya formerly flowed past Bogra, discharging itself into the Old Brahmaputra channel through Dhaleshwari. Following the 1787 flood, it took a direct route through Bangali to join the Brahmaputra-Jamuna. Similarly, the Nagar channel lost its connectivity with the Karatoya River (see Kamaluddin, 1973: 8). Overall, it is clearly evident that during the late 18th century the river systems of northern Bengal experienced remarkable changes and adjustments, of which the avulsion and consequent diversion of the Old Brahmaputra channel 113 kilometers westward was the dominant one.

Morgan and McIntire (1959) synthesize the viewpoints of the Brahmaputra diversion. They suggest that the shift took place gradually, and was influenced in part by gradual tilting of the Madhupur block. This tilting caused the Old Brahmaputra River to become antecedent in many places along its course, "necessitating [that the] river scour into slowly or periodically rising, comparatively resistant Pleistocene sediments" (Morgan and McIntire, 1959: 332). The sudden shift in the course of the Tista, with the resultant addition of its discharge to the Brahmaputra channel perhaps played a partial role in this diversion.

The process of channel widening in the Brahmaputra-Jamuna following the diversion of the old channel is explained by Chowdhury (1973). He points out that the new channel was by no means broad and deep enough to accommodate all the waters of the Tista, Jaldhaka, Torsha, Raidak, Gangadhar, Sankosh and Gadadhar, carried down by the upper or Assam Brahmaputra River. As a result of enormous loads in the discharge of the river, depending on velocity and other local conditions, a multitude of *chars* and shoals began to form. Chilmari, which presently stands on the right bank of the Brahmaputra, was situated on the right bank of a spill channel during Rennell's time. An examination of the maps of the locality indicates that between 1781 and 1865 the river had eroded the west bank approximately 13 kilometers to the west—the distance Chilmari had been from the main channel of the Brahmaputra (Chowdhury, 1973: 26). The east bank had also shifted to the east, resulting in a widening of the channel by lateral bank cutting and slumping, and by the forming of numerous *chars*. During 1866-1920, the swamp and *bils* (depressions) which were formed by the shifting of the channels westward had gradually filled up. The *chars* progressively became consolidated with new vegetation and human settlement. Chowdhury further notes that besides the process of avulsion,

> by the process of bank erosion, the old alluvium on either bank . . . has been badly affected. The eroded portion is laid down by new sediments (recent) but structural control over the course of the river exists even today. It is this structural control which remains responsible in keeping the courses of rivers of the Barind in positions or out of positions. The balance of the alignments of rivers depend more or less on the structural control here (Chowdhury, 1973: 27).

The ISPAN Charland Study (1993) supports the notion of gradual development of the Jamuna-Brahmaputra new course during the 1780-1830 period. It states that, "there is evidence that this change did not occur catastrophically, but took place gradually over a period of perhaps 30 years between 1790 and 1830" (ISPAN, 1993: 2-2).

Overall, the survey of historical maps and documents of the Bengal basin and drainage system indicate that the north-south alignment of the old alluvium of the *Varendra* or Barind largely influenced the gradient of the rivers of northern Bengal for centuries. The development of the eastern channel of the Ganges (i.e., the *Padma*) and the easterly migration of the lower Brahmaputra from Lakhya to upstream of Meghna during the 15th and 16th centuries were related to the delta building process, earthquakes, and other tectonic activities in the region. As a combined consequence of the great flood of the 1787, diversion of the Tista course and the influence of tectonic activity, the Old Brahmaputra channel took a course 113 kilometers westwards from its earlier course. This new channel, with much channel widening and *char* formation, gradually came to be the mighty Brahmaputra-Jamuna River (see Figure 4.5.b). Several studies suggest that most of these changes over the past centuries were greatly influenced by tectonic activity.

4.3.3. CHANNEL SHIFTING AND BANK EROSION

Type and Extent of Bank Erosion and Slumping

Theoretically, bank stability is primarily dependent on the behavior of the river bed during rising and falling flood stages. In the case of a *meandering river*, the changes in planform are fairly predictable. But in the case of larger *braided rivers* like the Brahmaputra-Jamuna one can find all combinations of erosion and deposition. For example, both banks may experience cut and fill simultaneously. Seven principal factors involved with short-term bank changes can be identified: (i) rate of rise and fall of river level; (ii) number and position of major channels active during flood stage; (iii) angle at which the thalweg approaches the bankline; (iv) amount of scour and deposition that occur during flood; (v) formation and movement of large bedforms; (vi) cohesion and variability in composition of bank materials; and (vii) relationship of abandoned river courses to the present-day channel (Coleman, 1969: 157). While the interrelationships among these factors are not yet precisely understood, several studies suggest a simple relationship between bank erosion and unconsolidated alluvial deposits. In other words, according to Knighton (1984: 61) and Morisawa (1985: 61), the extent and nature of bank erosion depends upon the relation between the force expended on the bank material and the resistance of the bank material. The main processes of bank cutting involve the *hydraulic action* on non-cohesive banks (see Knighton, 1984; Hooke, 1979; Simons *et al.*, 1979) and the collapse or *slumping* of large blocks of material, which is due to bank moisture and flow conditions (Knighton, 1984: 61; Figure 4.6). The bank materials, being alluvial deposits, maintain horizontal stratification, which results in greater permeability horizontally than vertically (Khan and Rashid, 1985: 4). The lateral seepage velocity henceforth becomes greater during periods of rising and falling flood waters. The seepage force is one of the main factors responsible for the rapid rate of bank erosion (see Twidale, 1964; Klimek, 1974).

Generally, the process of bankline recession in the Brahmaputra-Jamuna channel can be attributed to either the liquification and flowage, or to the shearing away of bank material (Figure 4.6). During high floodwater levels, pressure is applied to the sands below the river level, forcing water into the formation. With the fall of the water level, pressure against the channel walls decreases and water moves from the formation back into the channel runoff, resulting in lateral flowage of sand and subaqueous failure (see Thorne and Tovey, 1982; Thorne, 1982). The upper strata of silts and silty clays of the natural levee deposits often remain in place for a duration, but collapse as further undermining produces slump blocks. The flowage failures thus can occur either below

Figure 4.6. An example of the steep bank along Uttar Tekani, Kazipur during the lean (i.e., winter) season. Rapid bank slumping occurs along the Brahmaputra-Jamuna River chiefly due to the steep bankline gradient, and the non-cohesiveness and low moisture content of the soil (Photo: M.Q. Zaman).

water level (i.e., subaqueous) or in the zone between high and low water level through variations in the water pressure in the bank strata. Shear failures in the upper bank materials—the common type of bank recession—can be induced by one or a combination of factors. As noted above, most shear failures occur as a result of water pressure undermining natural levee deposits. The other major cause of shear failure is excessive steeping of the bank caused when the thalweg of the channel hugs the bank (Figure 4.6). Coleman (1969: 163) tabulates an average of "five slumps per mile over a 30 mile [nearly 50 km] stretch of bankline" of the Brahmaputra-Jamuna. He observes more than 10 bank failures per kilometer in some areas. The sizes of the slump features were also quite variable, ranging from 15 to 350 meters in length.

Three distinct types of channel shifting occur along the Brahmaputra-Jamuna River: (1) *gradual migration* of the major flow channel; (2) *rapid shifting* of the major flow channel during development of cutoffs; and (3) *sudden creation* of new flow channels on the floodplain outside the bankfull channel (Galay, 1980). Galay (1980: 546) measured rates of *gradual migration*—which refers to the shifting of the main flow channel whenever rivers flow in single defined channels—and found that they varied from 60 to 1,675 meters per flood season in the Brahmaputra-Jamuna, Ganges and Meghna floodplains. Rapid shifting of a major flow channel occurs when the main channel switches from one side of an island or bar to the other, or when a cutoff is formed. Rapid shifting generally takes place within the bankfull channel, and can take place in one flood season. The development of cutoff channels pose difficulties in predicting the rate of bank erosion; a bank that has not migrated for years can suddenly undergo rapid erosion when the main flow channel shifts rapidly. Besides the above two types of channel migration, the Brahmaputra-Jamuna experiences major changes when new channels are created outside the bankfull channel; this shifting process is similar to that of cutoff development, except for the fact that the newly created flow channels are located further away from the main flow channel. This third shifting process is more prominent in the Brahmaputra-Jamuna River in comparison to other channel shifting processes.

The International Engineering Company attempted to delineate the short-term changes of banklines of the Brahmaputra-Jamuna through compilation of historical maps, covering the period of 1830-1963 (IECO, 1964; Figure 4.7). Although the reliability of these historical maps is questionable, they do provide a valid sketch of the magnitude of bankline shifting in the area. Six maps were prepared, based on completed surveys for different years: (i) Wilcox maps (1830), (ii) Bengal Surveys (1867), (iii) Survey of the Government of India (1934), (iv) United States Army Maps (1944), (v) East Pakistan Maps (1953), and (vi) Inland Water Transport Authority Maps (1963). The maps for 1953 and 1963 were checked by aerial photography by the IECO study group and were generally found to be accurate (Coleman, 1969: 15-8).

Using the above maps, the west and east banklines, between 1830 and 1963, can be measured to determine the preferred direction of bankline migration (Figure 4.7). The center line of the 1963 channel width serves as the baseline datum. Measurements indicate that at the junction with the Old Brahmaputra, the width of the Brahmaputra-Jamuna channel increased from 7.2 to 12.1 kilometers between 1830 and 1963. The extent of bank erosion was quite extensive in several channels during the last few decades. Tarafdar (1974: 29), for instance, cites a lateral movement of bankline as rapid as 13 kilometers in a four-year period at one reach of the river. From 1830 to 1963, the junctions of the Dhaleshwari and Tista Rivers migrated upstream about 15.3 and 12.1 kilometers respectively.

Two periods (i.e., 1944-1952 and 1952-1963) were selected by Coleman (1969) for detailed study because of the greater reliability of the maps for these periods (Figure 4.7). In total, 32 stations were chosen, approximately five to eight kilometers apart; the rate of bankline migration at these locations was originally measured in feet per year, which for the purpose of this book was converted into meters per year. In consideration of the highly variable character of the river, measurements of such annual rates, however, are likely to misrepresent the actual situation. The aggregate figures of bankline movement between a base and a terminal year are hence presented in the present study.

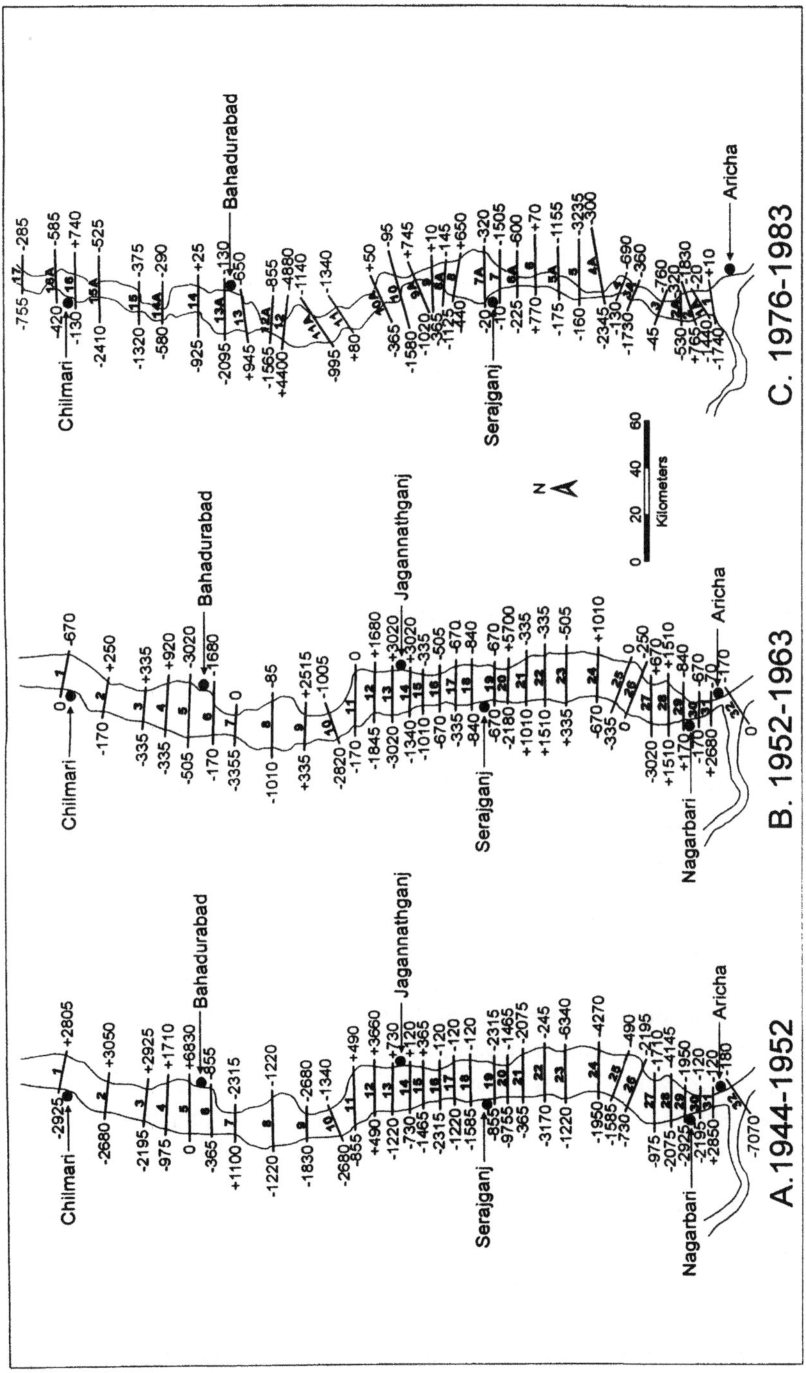

Figure 4.7. Bankline migration in the Brahmaputra-Jamuna floodplain, 1944-1952, 1952-1963 (Coleman, 1969) and 1976-1983 (Stene, 1988).

During 1944-1952, the occurrence of maximum bank erosion was recorded at 6,340 meters on the left bank, six kilometers west of Tangail Town (Figure 4.7, station 23). The maximum amount of build-out was 6,830 meters during this period at a location five kilometers upstream of Bahadurabad (station 5). At some reaches, both banks of the river were retreating rapidly (stations 10, 23, 24 and 28), whereas at some other reaches both banks were building-out rapidly (station 12). A greater number of stations along the right bank compared with those on the left bank underwent bank erosion. Similar patterns persisted during the 1952-1963 period. The maximum erosion (3,355 meters) took place in Shaghata *thana* (station 7) during 1952-1963, and maximum deposition was recorded at 5,700 meters near Porabari of Tangail (station 20) during the same period. In view of these observations, Coleman (1969: 158) made the obvious conclusion that, "thus, within the period [1944-1963] bankline migration was not predictable but, rather it was erratic." In a similar study, utilizing a Landsat data set for the period 1976-1983, Stene (1988) observed that west bankline movement averaged 120 meters per year and that nodal points exist along the river at locations where minimal bankline movement occurred (Figure 4.7). Due to the short record, however, it is not possible to establish conclusively whether these nodes are permanent or transient features of the river.

Although a few recent studies analyse continued erosion and deposition along the bankline of the Brahmaputra-Jamuna River (Stene, 1985; 1988; Bristow, 1987; Haque and Hossain, 1988; ISPAN, 1993), only the ISPAN Charland Study (1993) provides a comprehensive perspective on the contemporary extent of bankline migration. As a partner of the Flood Action Plan (FAP),[44] the Irrigation Support Project for Asia and the Near East (ISPAN) (1993) investigates the dynamics of the river in relation to *char* formation and bankline movement. This information is included in two FAP components, namely FAP 1 which dealt with the Brahmaputra Right Bank Strengthening and FAP 19 which developed a Geographic Information System to support further activities. For the latter component, a satellite image obtained on March 8, 1992 was converted to a map by geo-referencing to a standard map projection; comparisons could then be made with maps prepared in 1830, 1914, 1953 and 1973 to determine long-term trends. The study reiterates that over much of the length of the Brahmaputra-Jamuna in Bangladesh, the river is migrating westward. The overall westward shift of the river is depicted in Figure 4.8, which shows historic bankline changes between 1830 and 1989. Such an overall shifting, however, did take place through a complex process of widening and braiding, *char*land formation (medial bar building) and associated bank erosion.

Furthermore, the ISPAN *char*land study (1993) reports that the maximum westward movement of the bankline during the study period (i.e., 162 years) peaked at only 13 kilometers, just south of Chilmari and the confluence with the Tista River. Elsewhere, near Haturia in the Pabna district the bankline shifted 11 kilometers (Figure 4.11). An assessment of channel migration achieved by plotting successive center lines reveals that the overall rate of center line migration, across 20 reaches, each 10 kilometers long,

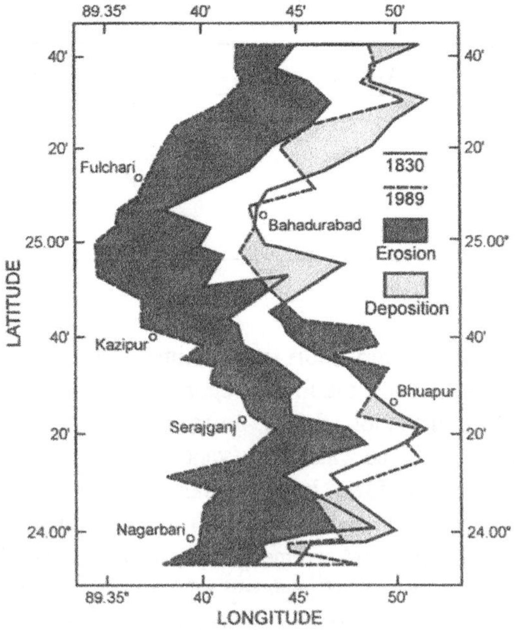

Figure 4.8. Historical bankline changes of the Brahmaputra-Jamuna River, 1830-1989 (after CBJET, 1991).

ranged between 18 and 31 meters per year since 1830. There have, however, been wide variations between reaches. The center line from the more detailed 1973-1992 bankline map series exhibits the development of channel sinuosity in the braided belt. This sinuosity was chiefly responsible for the westward migration of the channel, apparent for most of the reaches above Serajganj. The only exception to the general westward migratory trend has registered since 1973 in the southern reach, where the center line moved eastward north of Aricha (ISPAN, 1993: 2-14).

Channel widening is a significant aspect for understanding the Brahmaputra-Jamuna River morphology and its associated processes as it affects the settlement patterns and economic activities of the region. In 1830 the width of the river averaged 6.2 kilometers, and by 1992 this had increased to 10.6 kilometers (Table 4.10). With a high annual variability, the channel has widened at a long-term rate of 27 meters per year. The rate of widening, however, has accelerated since 1914, with an average annual rate of 65 meters. Even the narrowest reaches of the river have widened significantly, resulting in the 1992 minimum width between banks surpassing the mean of 1830. Overall, recent trends have shown more widening in the upper reaches than the lower reaches (FPCO, 1993: 2-14).

The FAP 1 study on the channel morphology of the present course of the Brahmaputra-Jamuna demonstrates that in geomorphological terms the two centuries

TABLE 4.10. Average width of the Brahmaputra-Jamuna, 1830-1992 (in kilometers)

Year	Average	Maximum	Minimum
1830	6.24	16.37	1.44
1914	5.55	11.96	1.06
1953	9.05	15.00	2.67
1973	8.08	15.04	2.91
1992	10.61	15.57	6.34

Source: ISPAN (1993: 2-14).

over which it has evolved is a very short span of time. Based on analysis of the long-profile and specific gauge records, the FAP 1 reports that since 1963-64 there has been no statistically significant deviation, that is, no major trends in bed levels or in water levels for specific discharges. The water conveyance capacity of the braided channel as a whole has been steady and uniform. It may be safely presumed that the courses of the Brahmaputra-Jamuna have not had sufficient time to adjust morphologically to accommodate the range of flow and sediment loads from upstream inputs. "The river channel may still be in a process of unsteady adjustment," ISPAN (1993) suggests, "with the three-dimensional geometry of the channel evolving toward a form that is in dynamic equilibrium with the flow and sediment regimes" (ISPAN, 1993: 2-12).

Formation of Chars and Impact on Channel Shifting

The Brahmaputra-Jamuna, as a large braided channel, is characterized by innumerable medial bars and islands (*chars*) that separate the flow into several channels. The initial location and formation of bars are presumably determined during flood events. As Leopold and Wolman (1957), and Krigstorm (1962) describe, the lag deposition of coarser material which the stream channel is incapable of transporting provides a locus for further deposition, causing the incipient bar to grow surfaceward and downstream. The hydraulic conditions leading to *char* development within the Brahmaputra-Jamuna are still unknown. However, it postulated that the area of deposition in the central part of the channel is basically caused by slack water conditions (Coleman, 1969: 145). This area undergoes deposition during flood periods, causing the bed to become aggraded; then because floodwaters rapidly subside, there is limited time for removal of these materials. Consequently, following the flood season, the deposited material becomes exposed as a shoal. The shoal or island is further inundated during the next flood, which commonly causes the erosion of its upstream face and deposition on its downstream side. Coleman (1969: 176) observes that in most cases the total area of an island remains the same from one flood to another, and only its location and shape change. Occasionally, whole islands are removed by erosion, or entirely new islands are formed.

With respect to the creation of mid-channel bars and *chars*, Coleman (1969) describes the development of large sand waves. These bedforms range from 7-16 meters in height and 183-194 meters in length. The average rate of movement of these sand waves is 204 meters in 24 hours, with a maximum rate of movement of 640 meters in 24 hours. The depth of water does not affect the size or height of these sand waves, although during flood stages, general lowering of the bed profile is common and applicable to all of the bedform. In such cases, the topset and bottomset bedding are removed by the migrating bedform. During the falling flood stages, the entire bed becomes elevated by a net deposition on the stoss and lee sides of bedforms. In these stages one bedform often overtakes another bedform, resulting in considerable aggradation of the bed (see Coleman, 1969: 197). These large bedforms remain as mid-channel bars and gradually build up to form *chars* (Figure 4.9). The river accordingly adjusts and widens at the point where *chars* develop.

The planform of the Brahmaputra-Jamuna River indicates that bank widths vary considerably along its entire length, with some locations showing narrow widths or nodes. The node points are characterized by narrower widths than other cross-sections and by greater stability of banklines compared to other points. Coleman (1969: 144-5) suggests that the presence of such nodes may be attributed to the existence of resistant clay as bank material. Khan (1988), however, argues that such resistant bank material is generally non-existent (or its proportion is minimal) in the Brahmaputra-Jamuna reaches. Bristow, in supporting this view, states that:

> Coleman . . . suggested that the position of these [nodes] was controlled by more cohesive banks at abandoned channel localities. Historic maps show that node points are not fixed but are transient on a scale of tens of years. One node point identified by Coleman is located just south of Sirajganj. Historic maps show that the river has previously been wider here and is not constricted by more cohesive banks but by deposition downstream from a confluence (Bristow, 1987: 65).

The presence of resistant clay, Khan (1988) maintains, may be one of the factors for node point location, although the characteristics of the braiding process may be the prime determinant in their persistence along the Brahmaputra-Jamuna River. One relevant feature is that nodes are located at certain intervals along the longitudinal profile of the river (Khan, 1988). From the examination of bankline migration for the period of 1830-1963, Coleman (1969: 160) identifies three node points in the Brahmaputra-Jamuna reach: (i) just south of Bahadurabad; (ii) immediately below Serajganj; and (iii) near the confluence with the Ganges below Aricha. The depth-width ratio at these points is found to be quite large compared to other reaches, which substantiates the proposition of river scouring at these node points. For example, the depth at Aricha is about 45.7 meters while in non-nodal locations it is within the range of 18.3 - 21.3 meters.

The ISPAN Charland Study (1993) compares historical maps to determine *char* morphology of the Brahmaputra-Jamuna River. It appears that medial bars gradually grew and evolved into island clusters, which then merged to create the large *island-chars* and *attached-chars*, features that are evident in the present braided form. The spacing of the island clusters has widened over time, and at present the width is between

Figure 4.9. An emerging *char* (medial bar) within a channel of the Brahmaputra-Jamuna River. Erosion and deposition patterns are so erratic that even the emerging *chars* may face simultaneous bank recession from the lateral movement of the channel.

20 and 30 kilometers. Although medial bar-free nodal reaches between the islands have always existed, they also have dispersed over the years. Consequently, the nodes are generally less distinct now, relative to the 1993 patterns. Overall, as the river has widened, *island-chars* have also expanded, resulting in almost constant actual width of the water surface. This process has led to a net loss of mainland floodplain area and a gain in *char* area.

Recent Satellite Mapping of the Brahmaputra-Jamuna Channel

Effective application of Landsat data has already been demonstrated in many relevant fields of land and water resources (see Lillesand and Kiefer, 1979: 543). The rationale for using remotely sensed data and mapping lies in the fact that the technique provides greater precision for mapping locations than land-based mapping techniques. The technique also allows a time-efficient assessment of land-water conditions.

Bristow (1987), Hoque (1983), and ISPAN (1993) have been using Landsat data for detection of changes along the length of the Brahmaputra-Jamuna River since the 1970s. The images used are all selected for the dry season period, because at this time of year there is nominal cloud cover and a low water level exposes the channel bars. Bristow (1987) analyses Landsat data for the period of 1972-1978 primarily to depict the temporal patterns in channel changes. He interprets the channel migration along the river in a hierarchical order: (i) the *first-order channel* migration, which comprises the entire river, has an average rate of 70 meters per annum; (ii) the *second-order channels* migrate at a rate of as much as one kilometer per year; and (iii) the *third-order channels* dissect and modify higher order bars at the falling stage. The second order channel migration in the Bahadurabad-Aricha reach is shown in Figure 4.10. From these data it appears that most channel movement occurs by lateral migration, and that the megaforms or large old mid-channel bars remain relatively stable.

The most recent attempt by ISPAN (1993), under the FAP 1 component, furthers and refines the satellite image-based study of river channel migration and its associated bankline movement. The study sequentially coveres a 19-year period, from 1973 to 1992, for a reliable detection of changes in water bodies, land, medial bars, and vegetation. Two Landsat frames, namely 148/42 and 148/43, path and row respectively, are needed to cover the study area. Considering suitability and availability, the Multi-Spectral Scanner (MSS) and the Thematic Mapper (TM) sensors are used for 1973-1989 and 1990-1992 respectively. A field check of the accuracy of the image interpretation (ground truthing) was conducted during the summer months of 1992 at 245 sites. The results reveal a 96-percent accuracy for the image classification of water and 91 percent accuracy in cases of cultivated and vegetated land.

The net shift in river banklines between 1973 and 1992 is calculated by digitally comparing bankline delineations (Figure 4.11). Over a 19-year study period, there has been an overall widening of the river in the range of 2 to 4 kilometers, resulting from consistent erosion along most of the west bank. Accretion is noted on the west bank only in two areas—one small segment in the extreme north and a 6-kilometer strip just north of the river training works established at Serajganj in 1976. In comparison, the east bank has shown a mix of accreted and eroded areas (Figure 4.11). Average rates of widening vary significantly between pairs of images, ranging from a maximum of 330 meters per year during 1987-1990 to a narrowing of 36 meters per year during 1990-1992. These patterns are indicative of accelerated bank erosion during the peaks of the 1987 and 1988 floods. They further suggest that the post-1988 flood period produced minimal erosion because discharges were below normal, probably due to bank relaxation following the exceptionally rapid retreat in the preceding period.

Interestingly, the series of peaks for channel widening during 1987-1990 exhibit evenly spaced zones of severe bank retreat interspersed with stable, less severely eroding or even accreting banks (see ISPAN, 1993: 2-20). The average spacing of erosion spikes is around eight kilometers. This corresponds to the spacing of embayments in the west bank, cut by aggressive bends in the nearest branches of the channel. The pattern of erosive spikes in general depicts some wide gaps and areas of narrowing. The 1973-1992 study period reveals a relatively higher rate of narrowing

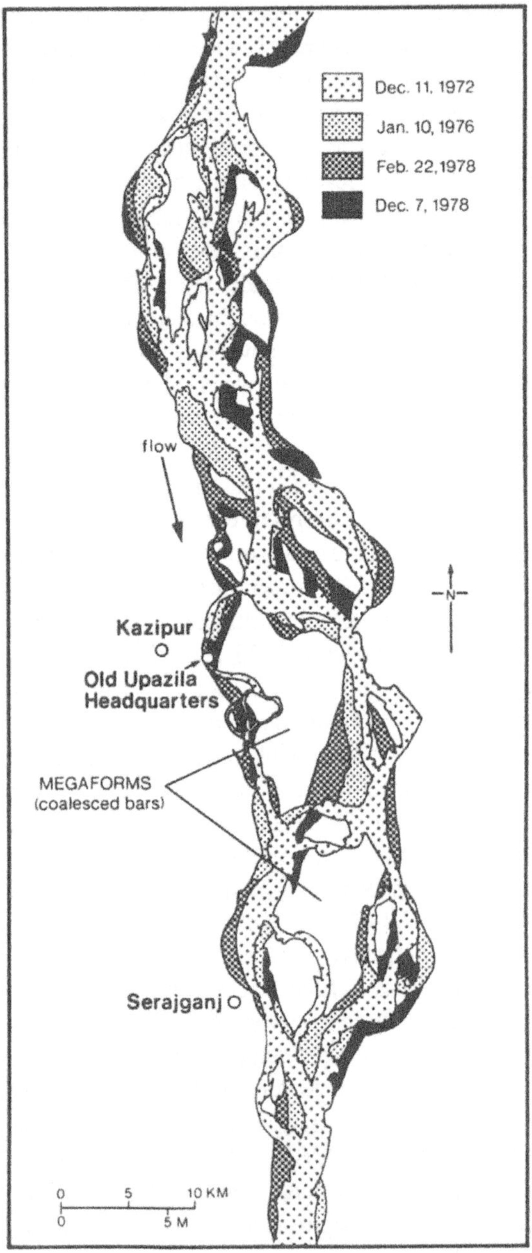

Figure 4.10. **Second-order channel locations from Landsat images, 1972-1978 (Bristow, 1978).**

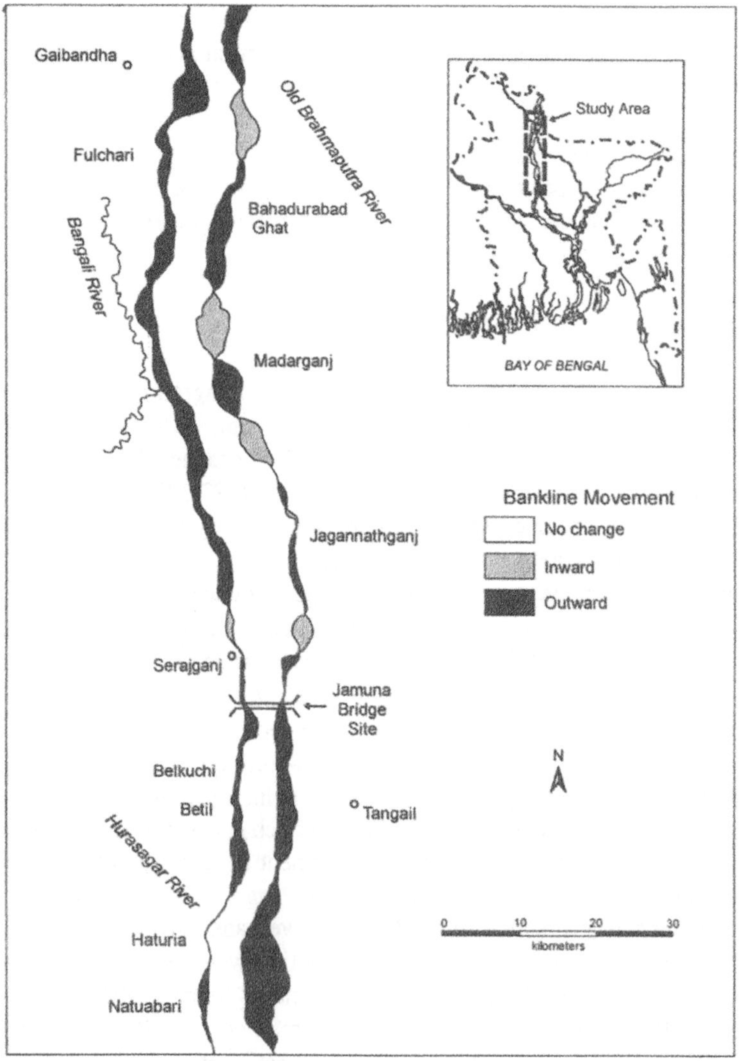

Figure 4.11. **Overall bankline movement of the Brahmaputra-Jamuna River, Gaibandha-Natuabari reach, 1973-1992 (detected from Landsat images, after ISPAN, 1993).**

and widening in the northern reach and a nominal narrowing and widening in the southern reach of the Brahmaputra-Jamuna. The FPCO reports that:

> The cause for the supposed trend in either direction and the widening is not known. The overall movement can be related to tectonic activities . . . [T]he Brahmaputra-Jamuna River has a record of avulsive behaviour, and it may well be that this overall movement has a similar background. The widening can be attributed to an advancing alluvial fan or to the fact that the river has not fully adapted itself after the shift to its new course around 1800 (FPCO, 1993: 2-16).

4.3.4. PREDICTING BANK MOVEMENT

Earlier investigations conclude that the nature of channel migration of the Brahmaputra-Jamuna is erratic (Coleman, 1969; Bristow, 1987; Khan, 1988; 1995; FPCO, 1993; ISPAN, 1993); consequently, attempts to predict river channel shifting and associated bankline movements have been scarce. Only ISPAN (1993), Khan (1988; 1995), and Khan and Booy (1991) pursue vigorous attempts in this regard to assist calculating the magnitude of risk in floodplains and *char*lands.

The Brahmaputra-Jamuna, being one of the largest braided rivers in the world, forms a complex water and sediment transport system. Thus, it can assume numerous possible states; the known laws of fluid mechanics and river behavior are quite inadequate to predict its continual morphological changes. A *single channel* alluvial river usually attains a dynamic equilibrium between erosion and deposition along its course by which it maintains a typical channel alignment and cross-sectional geometry. In comparison, a *braided river* has an extra degree of freedom, allowing the flow of water and sediment to be distributed over a variable number of different channels; the considerable changes that take place in this division of water and sediment by channel splitting and joining cannot be easily predicted. Yet, as noted above, some attempts have been made to develop short-term (e.g., annual) predictions of riverbank movement.

ISPAN (1993) uses a simple method for predicting west bankline movement of the Brahmaputra-Jamuna channel, analysing satellite images from the 1970s and 1980s. The entire reach in Bangladesh is divided into 21 sub-reaches, each 10 kilometers long and possessing 20 sampling cross-sections at 0.5-kilometer intervals along the bank. For each sub-reach, mean erosion rates for 1973-1992 are estimated. Assuming that within each 10 kilometers sub-reach over a 19-year period the distribution of bankline movement has been random, a normal distribution is applied to determine the probability of a particular erosion rate being equalled or exceeded (ISPAN, 1993: 2-20).

The above findings reveal a relatively modest average erosion rate, as opposed to many superficial and dramatic claims of rapidly retreating banklines. The maximum 19-year erosion rate for a 10-kilometer sub-reach is close to 160 meters per year, an amount which could cause serious socioeconomic loss, but geomorphologically, "would not usually be considered exceptional for this river" (ISPAN, 1993: 2-20). The median erosion rate is found to be about 80 meters per year, while bank encroachment rates greater than 400 meters per year are rare, registering at less than 5 percent of the sites. For shorter periods, erosion rates follow a common probability function. The short-term (2 to 4 year) median erosion rate range only between 50 and 110 meters per year. This range is reflective of major floods of the late 1980s and of low flow years.

For meaningful application of this information, the time and space coordinates need to be considered simultaneously. Both the historical and satellite image-based maps reveal that a very high rate of bankline retreat may occur over periods of one or a few years, but do not continue for many years at a particular location and do not take place at sub-reaches simultaneously. This observation suggests an inverse relationship between the severity of erosion and the number of years over which the erosion rate is sustained. It has been calculated that there is less than a 10 percent probability that catastrophic erosion (higher than 350 meters per year) will persist at a site for more than four years; rapid erosion (150-200 meters per year) is more likely to be sustained as the median duration is just over six years. Finally, there is a 10-percent chance that rapid erosion will continue longer than 11 years.

Khan (1988) and Khan and Booy (1991) apply the Transfer Function Noise model, suggested by Box and Jenkins (1976) in the space-time domain, to predict the location of banklines of the Brahmaputra-Jamuna River on a yearly basis. The empirical data are obtained from satellite images from 1973 and 1976-1983. From a north-south reference line, the location of banklines at 97 points are measured, and the yearly locations during 1976-1983 are linked to the 1973 bankline position. The study assumes that the position of the bankline at any location in any given year is the output of a dynamic system in which bankline positions in the previous year are the input—all other variables are lumped into the system. In Khan's (1988) attempt, the appropriate Transfer Function is determined to be a first order Moving Average model with a noise term that may be modelled by a first order autoregressive process. A typical comparison between predicted and observed bankline position is shown in Figure 4.12. The model succeeds in predicting the bankline position at the locations with a standard error of about 270 meters. The model gives poor results for the points where the changes were abrupt and massive.

Prediction of channel migration is also attempted by using a conveyance model (Goulter and Dubois, 1991; Burger *et al.*, 1988). In some studies the location of the centroid of the channel cross-section is assumed to define the position of the stream, and thus its movement with time is supposed to indicate the direction of bank migration and its future trend. Goulter and Dubois (1991) investigate the centroid of the cross-sectional area, though find it an inappropriate indicator of flow capacity, unable to account for several variables. The centroid of the channel conveyance is examined by Burger *et al.*,(1988); they proceed by dividing the cross-section into vertical strips and computing the conveyance of each strip using Manning's equation. In this study, it is assumed that the wetted perimeter of the strip is the contact line with the solid boundary. Suharyanto (1992) correctly, however, points out the error-inducing assumption that the flow in a

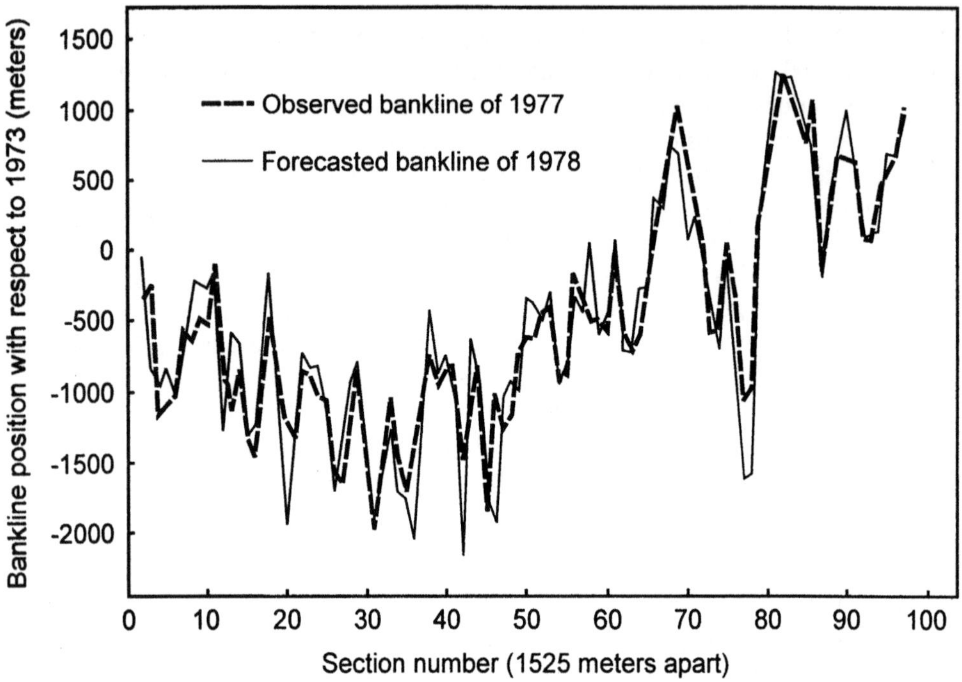

Figure 4.12. One-Time-Step-Ahead forecasted bankline as input (west bank; after Khan, 1988).

particular strip is independent of the flow in adjacent strips; monitoring the movement of the centroid of conveyance of the cross-section only gives an indication of the direction of movement of the river if it were a single-channel river. As Khan (1995: 51) elaborates, the centroid of conveyance in a bend of a meandering river indicates that the flow is concentrated close to the outer end of the bend. Monitoring the position of centroid over time, therefore, reveals the erosion of the outer bend where the erosion is most active. Earlier, Coleman (1969) chose to study the thalweg movement to identify channel migration, a study which also indicates a bias towards the outer bend (Khan, 1995: 51).

Using qualitative means, Khan's (1995) attempt to predict bankline movement one year in advance is somewhat successful, although quantitative and longer-term prediction does not appear to be feasible. This study assumes that average width-depth relations in the numerous branch channels over the recording were not far from equilibrium conditions for the given water and sediment distribution. This allows determining the degree to which the channels in each river cross-section are out of equilibrium at any given time; the expected changes towards equilibrium can then be estimated. The expected riverbank movement results from the corresponding changes in channel width. Khan (1995) uses the trends towards equilibrium conditions of individual channels in a given cross-section and superimposes a large random noise (in terms of changes in water and sediment flow over the branch channels) to predict riverbank movement quantitatively. The findings of the study overall reveal that the joining of two single channels in the following year can be predicted with a high degree of confidence using the developed qualitative tools. Similarly, the amount of outer bankline movement can be predicted quantitatively with a fair degree of accuracy one year in advance. The R^2 value based on a comparison of observed and predicted movement was on average 0.93.

4.4. Conclusions

A review of the physical dimensions of the Brahmaputra-Jamuna River reveales that the present channels are geomorphologically very young, dating back only two centuries. This braided system forms a complex water and sediment transportation process subject to rapid and erratic changes. The existing hydrological and fluvial studies of the river suggest that predicting long-term bankline movement is beyond the scope of the scientific domain, given the known laws of fluid dynamics and the considerable deviation of this young, multiple-channel river from a dynamic equilibrium state of erosion and accretion. Overall, both the conceptual and empirical studies carried out during the last four decades indicate that prediction of short-term bankline movement of the river is possible within a reasonable margin of error. This opens up a possibility of estimating the risk of a given local bank, thereby helping to minimize the potential losses to floodplain resource users. However, the inability to predict the long-term trend of bank movement due to its random or erratic distributional pattern makes the bank-erosion phenomenon a serious hazard for the Brahmaputra-Jamuna floodplain inhabitants.

The study of the interrelationships between nature and people is essential to a discussion of river-related hazards and human settlement. In the Bengal deltaic plain, the development of human habitat by a process of floodplain invasion and permanent occupance[45] is intrinsically related to the changing nature of river courses. It is undeniable that, besides their direct influence on geomorphologic features, river courses have a profound impact on human social formations and other human affairs. But an explanation of human occupance of the floodplain in terms of physical control is seriously limited and subject to extensive criticism.[46] As I have argued elsewhere (Haque, 1987c), the analysis of the human settlement in productive but hazardous floodplains should not adopt the conventional "cause and effect" model that represents a mechanistic view of the nature-society relationship. A consideration of environmental changes and social dynamics treated as interrelated variables is essential for understanding human behavior in a floodplain environment. In other words, it can be argued that neither the physical environment nor the social domain may provide independently the required explanation of the complexities of human occupance. An integrated notion of the historical development of social forces and the changes in physical systems is required to analyze the complex behavioral patterns in floodplain occupance by people who possess a variety of sociopsychological and socioeconomic characteristics. The following chapter therefore examines, from a historical-structuralist standpoint, the interrelationships between the changing river courses and the social dynamics of human habitation in northern Bengal.

CHAPTER 5

SOCIAL CLASS FORMATION AND VULNERABILITY OF THE POPULATION

A Historical Account of Human Occupance and Land Resource Management

Human existence and occupance of geographical space are obvious considerations when examining environmental disasters. Yet not all individuals will be equally prone to a particular extreme environmental event. Besides the physical parameters and characteristics of the geophysical event itself, a combination of social, cultural, economic and political variables determine the degree of an individual's *proneness* to a hazard. Thus, understanding social evolution and structure from a historical perspective in relation to resource-use and management is an utmost necessity for determining the nature of environmental threats.

In this chapter the principal dimensions of human occupance of the alluvial floodplains are analyzed from a historical-structuralist standpoint. Special emphasis is given to environmental and social changes to demonstrate how these phenomena are intrinsically interrelated. The ultimate aim is to determine the underlying factors which produce a greater *proneness* to natural hazards for the marginal and poor peasants of Bengal, relative to other social classes.

Since the *study area* for this research falls entirely within the Brahmaputra-Jamuna floodplain, the discussion, with the exception of section 5.2, focuses on the geographical area called the *Ganges-Brahmaputra Doab*[47]. However, the discussion refers to the entire territory of Bengal where information at the regional level is not available. Of the five sections presented in this chapter, the first two deal with the relationship between the physical environment and human habitat— specifically, the impact of river channel changes on human settlement and land occupance. The third section investigates the dynamics of population growth and its spatial distribution in relation to the human occupance of the floodplains. The fourth section analyses the development of land tenure and the legal system, focusing on the issue of entitlement to land. Finally, the fifth section analyses the contemporary agrarian structure and the proneness of floodplain inhabitants to natural hazards and related phenomena.

5.1. River Channel Changes and their Impact on Human Habitat

The reconstruction of the geographical, social, and economic past of Bengal and its sub-regions is a difficult task because of a dearth of reliable historical documents and statistical information. Although historians cite that *pustapala* or record keepers were attached to village and district headquarters as early as the fifth and sixth centuries of the Christian era (see Majumdar, 1971a: 641; 1971b: 332), these records, mostly on palmyra leaves, were subsequently destroyed. During successive centuries, the recording of land, settlement, and other socioeconomic statistics was generally ignored because the state had no immediate concern with land owned by individual communities (see Desai, 1959: 8; Currey, 1979: 139-152). However, numerous alternate means for historical reconstruction, such as epigraphic records, inscriptions, literature, travel documents, and other archaeological materials, allow us a reasonably clear understanding of past processes.

The literary and epigraphic evidence indicates that the first human settlements of Bengal were established on the old alluvium tracts of the Ganges-Brahmaputra Doab. The early life of the doab was characterized by a simple division of labor and redistribution, due to both the absolute dependence of the economy on fishing and to the static, caste-based social system. The standard occupations for men, *Kaivarta* and *Mahishya* (fishermen and boatmen), reflected the economic dictates of the region (see Mukerjee, 1938: 123). It is suggested that these characteristics continued to define life in Bengal throughout the Indo-Aryan civilization of the *Brahmana* (800 - 600 BC) and *Sutras* (600 - 200 BC).

Archaeological evidence suggests that from the early Christian era to the Muslim conquest of Bengal in the 12th century, northern Bengal flourished as an agricultural economy, with a few towns and cities of considerable size. During the early Hindu and Buddhist eras, a relatively complex redistribution system, founded on land and caste-based social relations, gradually evolved. New arable land or *khetra* was opened up, and pits and canals (*nala*) were excavated for irrigation purposes. By the seventh century, the majority of people in the doab had taken agriculture as the principal occupation for subsistence (see Islam, 1968: 2). The villages then grew mainly at the waterfront of riverbanks, canals, or reservoirs.

The conditions of towns in Bengal in the Early Ages were directly linked to the land revenue system in the countryside.[48] The vast scale of rent appropriation by the ruling classes and castes during the beginning of Bengal's history led the agrarian population into a societal and technological stagnancy (Pavlov, 1978: 39). The process rested on a rigid system where a relatively large proportion of the population was dependent upon appropriating surpluses from agricultural producers. With a predominantly subsistence economy based on inequitable, underdeveloped economic relationships, the Hindu and Buddhist regimes failed to mould a secure society. Therefore, when the Muslim conquerors superimposed their own system of centralized state and social bonds upon the existing systems, medieval Bengali society became stable at the village level, but turbulent in the overlying state structure (see Desai, 1959).

This transformation of Bengali society into a two-tiered sociopolitical system, combined with the eastward shifting of river courses, had a decisive bearing upon the overall human habitat of the Ganges-Brahmaputra Doab. The perennial river channels were needed for the existence and development of early settlements in the region. They were valuable on the one hand for their annual inundation and subsequent siltation of the alluvial farm lands, plus they provided easy means for the state to appropriate and transport agricultural surplus in the form of rent and tax. The later decay and silting up of rivers thus resulted in the complete ruin of many prominent cities and their surrounding rural areas. One of the most prominent examples of this was the case of the degeneration of *Pundra Vardhana Bhukti*. In the seventh century, Huen Tsang travelled from the Murshidabad area (*Kajangala*) to the capital of Pundra Vardhana, known as Mahashangarh. He recorded that Pundra Vardhana had a flourishing population "with tanks, hospices and flower groves" (cited in Mukerjee, 1938: 127). It has been suggested that the combined flow of the Lauhitya and Karatoya formed a considerably large river, and the Karatoya flowed along the eastern boundary of Pundra Vardhana (see Walters, 1905: 2). In the Tsang's account, the Karatoya was described as a "mighty river."[49]

The decay of the Karatoya and other perennial channels during the Medieval period significantly promoted the settlement of new alluvial lands in the east and south of the old alluvium tracts. The decline of human occupance of the doab is well documented in some historical materials. For instance, Barani, a 14th-century writer, records that the land between Varendra (i.e., Barind) and Madhupur was a "land of swamps" (see Mukerjee, 1938: 130). A European traveller, Ralph Fitch, also describes "few villages but almost all wilderness" during his visit through the western part of the doab (cited in Mukerjee, 1938: 131; also see Law, 1954; Ahmad, 1968: 77). On the other hand, many travellers record growing human settlement in the eastern regions since the 14th century. The famous Moorish traveller, Ibn Batuta (1328-1354) made a journey through the Meghna floodplain, which he calls "habitations along the Blue river," describing dense populations, markets, and orchards along its banks. The writings of the Chinese traveller, Ma Huan (1406), and the Italian traveller, Ceaser Fredrick (1565-1581), indicate that in Bengal only the eastern riverine tracts or seaboards were densely populated, while the interior was a region of marshes and wastes devoid of habitations.

As noted in Chapter 4, the map drawn by Van den Brouke (1660) shows that while the vigor of Karatoya had been declining, giving away its dominance initially to the Lakhya, and later to the Old Brahmaputra, the mainstream of human occupance of the alluvial land followed the changing river courses. Due to declining productivity of land in the old alluvium, along with the concentration of land in the hands of the higher castes, the capitals, towns, trade centers, as well as rural settlements, gradually moved towards the southeastern deltaic regions. Interestingly, between the 11th and 18th centuries the comparatively new alluvial plains of this central Bengal region not only attracted natives from the old alluvium tracts, but also a sizeable immigrant population from western Asia and the Arab world (Mukerjee, 1938).

Besides a steady progress in the political, economic, and demographic spheres during the Medieval period, the central and eastern Bengal plains faced several problems relating to physiographic and environmental conditions. The ruthless vagaries of nature periodically took their toll on existing settlements and the building of new habitations. Socially and technologically, village life was predominantly feudalistic in nature; as such, attempts at technological adjustments to natural hazards were largely absent.[50] Historical records on natural hazard impact and related phenomena are also not available, though it is evident from documents that village after village was washed away by the course changes and avulsions of great rivers. One such example was the eastward diversion of the main channel of the Ganges in the 15th and 16th centuries. As noted earlier, the Ganges flowed through Rajshahi and Pabna in northern Bengal, and through Dhaleshwari and Buriganga during Akbar's time (1556-1605). According to Ralph Fitch (1585), at that time the confluence of the Ganges and Meghna was near Sripur, south of Sonargaon. A comparison of Rennell's map (1764-1772) with Ralph's sketch suggests that the confluence of the Ganges shifted to around Dhakshin Shabazpur Island (i.e., present day Bhola) during the intervening two hundred years. It is recorded that the Ganges shifted its main channel further southward through the Vikrampur Parganah (i.e., southern Dhaka) during the 16th century, destroying Sripur—the capital of the Parganah—and the "far-famed temples and palaces of Chand and Kedar Roy, and later Raj Bullov" (Mukerjee, 1938: 203). The name of this portion of the Ganges, from Jaffarganj (near Goalando) to Dhakshin Shabazpur Island, became known as the *Kirtinasha* (the Great Destroyer).

As was discussed at considerable length in the preceding chapter, the drainage systems of Bengal delta underwent remarkable changes during the 18th and 19th centuries. During the period between Van den Brouke (1658-1664) and James Rennell (1764-1776), not only did the Atrai, the Karatoya, and the Old Brahmaputra degenerate, but the flood of 1787, together with a few earthquakes, caused the Tista to become a major tributary of the Brahmaputra-Jamuna. The opening up of new channels of considerable dimension through the two old alluvium tracts transformed the once prosperous lands into moribund river zones. Among many other problems, two factors affected these moribund regions: first, a decline of agriculture due to lack of a yearly flood; second, deterioration of human habitat through water logging, creation of marshes and swamps, and other related environmental problems.

Along with these changes, geographical processes also led to phenomenal alterations through the cutting and filling of a vast amount of land (Figure 5.1). Martins succinctly portrays this picture:

> On the whole it must be evident that changes in the course of rivers are attended with great loss and inconvenience. The new channel is so much land-loss, and the old one leaves behind it a marsh or a kind of lake, which for ages is rather injurious then of use . . . The towns must,

Figure 5.1. An example of ongoing river channel shifting along the Brahmaputra-Jamuna River. Riverbank erosion is a severe hazard to land, houses, crops, and trees since the disaster permanently damages most of these resources.

> therefore disappear, and the uncertainty of their place of abode . . . prevents the inhabitants of . . . Bengal from building more substantial and comfortable houses (cited in Mukerjee, 1938: 189).

As noted above, we have limited data for this historical period on land-loss and population displacement caused by channel shifting and bank erosion. Hence, the social impact of the avulsion and bank erosion of the Brahmaputra's new channel in the 18th and 19th centuries is not known precisely. Some piecemeal records, however, provide some indication of the magnitude of land-loss and other impacts of channel changes during that period. Gait, in a census report of Bengal, describes the new channel of the Brahmaputra, or the Jamuna as it was then locally called. The river, he writes:

> [is] constantly cutting away land in one direction and adding it in another. For a series of years one bank is gradually cut away and the other added to, and then for another series of years the opposite process goes on . . . the Jamuna washed away a considerable area in the Mathura thana [sub-district]. The result was that at the end of the decade the population of the

> headquarters sub-division was less by 1.1 per cent than it was at the beginning (Gait, 1902: 66).

O'Donnell (1893: 57) further records that "the destruction of many villages in the east of the district [Rangpur] by encroachments of the Brahmaputra have caused very many families to emigrate." Porter reports that during the 1920s serious erosional activities on the banks of the new channel "engulfed almost the whole of the original town [Serajganj]" (Porter, 1933: 49). The association between channel changes and the human habitat in Bengal is precisely described by Mukerjee:

> Political power, industry and trade have shifted in the history of Bengal with the constant changes that the river-courses have undergone in a delta yet in the making . . . Not less far reaching have been the effects on settlement and cultivation (Mukerjee, 1938: 149; also see Figure 5.1).

5.2. Settlement in the Delta Frontier

The earliest descriptions of the Bengal delta refer us to the classic texts of ancient India. The *Rig-Veda* (latter half of second millennium BC) and the *Mahabharata* (second century BC to second century AD), for instance, include accounts of the country of Vanga (or Bengal). Mukerjee asserts that:

> There is no doubt that the first settlement of Bengal was in the old alluvium tracts. The most ancient inhabitants were described as fishermen and uncleaned folks in the ancient literature . . . Vanga at that time comprised only the modern districts of Murshidabad, Nadia [now in West Bengal, India], Jessore, Pabna, Faridpur and parts of Rajshahi. It is probable, however, that only the riverine tract or the seaboard was populated; the interior was more or less a region of marshes and wastes devoid of habitations. Barani, writing as late as the 14th century, thus refers to Bengal as a "land of swamps" (Mukerjee, 1938: 123-130).

J.C. Jack, a British colonial administrator who spent years in the Faridpur district of eastern Bengal, observes the development of settlement in the delta in the following manner:

> It [the land] is still in the process of formation and is full of rivers which are broad and deep, heavy in the flood season with constructive silt, yet sufficiently active to work their will upon a land of plastic mud. Here, as is the fashion of the delta, the rivers are washing away mud from one bank and re-forming it on the other with a method so complete and comprehensive that in the entire tract, perhaps 900 square miles in extent, very little of the present land has been in existence a hundred years and not very much for more than fifty years (Jack, 1916: 16-17).

Several historical accounts note that the emerging delta areas and a large part of southern Bengal including the *Sundarbans*, constituted 'new frontier' in terms of extended agricultural land and new settlements (Hunter, 1875; 1877; Pargiter, 1885; Carstairs, 1895). The decay of the moribund floodplain, along with increasing population growth, caused the development of a migration stream to the delta frontier (Mukerjee, 1938; Nicholas, 1962). Migrants found motivation in the availability of new opportunities provided by rivers for transport and commerce and in the availability of fertile alluvial soil.

An interesting aspect of the early settlers of the floodplain frontier is their ethnic composition. As noted above, Mukerjee (1938: 279) traces the movement of at least three predominant Hindu castes into the delta area, the *Mahisyas*, *Namasudra* and *Pods*, whose traditional occupations included cultivation, boating and fishing. Nicholas (1962) analyses the spatial distribution of these people. He finds that the *Mahisyas* and *Namasudras* were dispersed all over Bengal, with a concentration in the eastern and southeastern portions of the delta (Bakerganj, Faridpur, Khulna, and Jessore); the *Pods* were settled in the central (24-Pargana) and south-central (Khulna and Jessore) part of the delta.

The colonization process of new land was confined to two routes: (1) land reclamation, clearing and cultivation, especially in the *Sundarbans*, and (2) claims between peasants over new accretion or depositional land in the riparian areas of the active delta zone (Zaman, 1988; Hunter, 1875). In 1783, Tilman Henckell, then judge and Magistrate of Murli (now Jessore) submitted a plan to the Governor General of British India, Warren Hastings, which would (a) grant out leases of wasteland for cultivation in order to raise revenue, and (b) settle the boundaries of estates with local landlords (Pargiter, 1885). Since the *Sundarban* forest zone was the property of the state and as such was excluded from the Permanent Settlement of 1793, the boundary areas, which were undefined, became subject to ownership conflicts. In reality, many people had cleared the jungle themselves and had established a strong sense of interest and control over the land (Nicholas, 1962; Hunter, 1875).

Similar legal problems were also registered in the active delta zone in association with settlement and ownership rights to accretion land. Here the problem was further complicated by the fact that there were always more than one claimant for the new alluvial land. It was always very difficult to ascertain ownership of newly emerged land—whether it belonged to the estate to which it may have been annexed, to some other riparian proprietor upstream who had lost his land due to erosion, or even to people on the bank of the river. Inhabitants on both banks of the river, therefore, often became very quickly involved in disputes, which at times led to battle and bloodshed for the disputed land. Colebrooke explained this situation very succinctly:

> Some of the islands, before they have acquired a degree of stability which might enable them to resist the force of the stream, are entirely swept away; but whensoever, by the repeated additions of soil, they appear to be sufficiently firm, the natives then no longer hesitate to take possession of them, and the new land becomes an immediate subject of altercation and dispute. The new settlers bring over their families, cattle, and effects . .

> these islands will continue a number of years, and may last during the lives of most of the new possessors; as they are, in general, liable to destruction only by the same gradual process of undermining, and encroachment, to which the banks of the river are subject (Colebrooke, 1803: 4-5).

5.3. Dynamics of Population Growth and Spatial Distribution

A natural hazard exists in the interface between the 'natural event system' and the 'human use system.'[51] To properly evaluate this human use system in a floodplain environment we must understand the *growth* and *spatial distribution* of population. In general, population density in combination with technological conditions largely determine the human component of an ecological threat or disaster. Moreover, demographic features are also affected by social, political, and legal systems, while the opposite is also true—these systems suggest and establish a particular demography. In the following section these aspects are analyzed in relation to floodplain occupance.

5.3.1. DYNAMICS OF POPULATION GROWTH IN BENGAL

The lack of diversity in agriculture, crafts, and industry during the Early Ages (i.e., until the 12th centuries) of Bengal's history was the consequence of a definitive caste system and a rigid division of labor (see Mujeeb, 1972: 102). This technological stagnancy led to an almost unvarying socioeconomic structure for rural life at that time.[52] While the bulk of village production was destined for direct use by the producer-community, the hands of the state snatched the surplus that then went to support the intermediaries, the sizeable armies, and the "luxurious life" of the rulers:

> The productivity of labour being low as a result of this low level of technique of production, there hardly survived for the mass people either surplus of products (after satisfying the needs of self-presentation and land revenue claims of the often rapacious government) or time for organizing a high standard of material and cultural life (Desai, 1959: 17).

The size of Bengal's population continued to follow a low-level equilibrium as a consequence of such processes. For one, high birth and death rates continued until the end of the 17th century (Obaidullah, n.d.: 33; Davis, 1951: 23). The usual demographic course was a gradual population growth for a short period of time followed by an abrupt decline due to catastrophes such as floods, cyclones, famines, epidemics, or warfare. The long-term trend was described as "one of virtual fixity of numbers" (Davis, 1951: 24).

Community ownership of land guaranteed the rural population access to food and agricultural produce in the Early and Middle Ages, yet the population of these autarchic villages was still quite susceptible to the fluctuations of environmental controls (see

Desai, 1959: 17-9). An abnormally high flood, or a crop failure as a result of drought, threatened villages with extinction because little or no external assistance could be expected. A lack of communication, transport, and inter-community social bonds insisted on self-reliance.

The Muslim conquerors from central and western Asia who brought a centralized state system to Bengal, particularly during Akbar's regime, attempted to introduce relatively intensive farming along with new agricultural practices (see Pavlov, 1978: 36; Majumdar, 1973: 176-7). As a result of Todarmal's land reform policies, the cultivation of three crops a year became common practice in the central-eastern region of Bengal (Allama, 1582; cited in Ahmad, 1968: 76). Theoretically, such a change in agricultural production would bring about a growth in population size, but this did not occur in Bengal. The increasing vigor with which surpluses were appropriated eliminated virtually any possibility of improving the living conditions of cultivators (*raiyats*).[53]

An analysis of the socioeconomic history of Bengal during the British colonial period (1757-1947) indicates that population size could not have increased substantially during the early decades after colonial conquest (c.f. Morris, 1968). War casualties probably declined considerably with the consolidation of the colonial state, but other factors causing high mortality, such as widespread famine and epidemics, were not brought under control until the early 20th century. During the rule of the East India Company (1757-1885), attempts to mitigate famine were largely left to local landlords (*zamindars*) or "native princes." Ironically, the Company was reluctant to interfere "with private trading in foodgrains by taking the responsibility for collection and distribution of food, even during the famine period" (Visaria and Visaria, 1983: 438). As a result of drastic intervention by the Company in the prevailing legal and political institutions, and a policy of non-interference in public health and food distribution, population size and growth remained stable during the 19th century.

The 1901 Famine Commission reported that the subcontinent experienced 12 famines and four "severe scarcities" during 1765-1885, noting that, in particular, great famines took place in the western and northern Bengal regions. The famine of 1770 caused the death of one-third of the total population in affected areas (Campbell, 1878). Likewise, a considerable proportion of population perished during other famines. Simultaneously the region's population was susceptible to cholera, malaria, influenza, and other epidemics. On the whole, throughout the 18th and 19th centuries death rates remained very high; the estimated crude death rate for the period 1750-1850 was 45 per 1,000 population (Visaria and Visaria, 1983: 472-6).[54] Concern by the British Raj about high mortality emerged from the problem of retaining a military and an administrative presence in areas with poor sanitary conditions, especially in the decaying moribund floodplain. A Sanitary Commissioner, for instance, in an 1873 report, mentions that although the treatment available to the military had been "fairly good and quick, nearly 20 to 60 percent of the afflicted troops used to succumb to the epidemic" (cited Visaria and Visaria, 1983: 479). Bourdillon (1883), in another case, records that in Dinajpur district mortality caused by malaria fever surpassed 30 persons per square mile. He notes:

> Out of seventeen adult Europeans fifteen had to leave the district [Dinajpur] during the year, broken by repeated attacks of fever, and official business could hardly be carried on. This terrible mortality drew renewed attention to the insanitary condition of the district, and a committee was appointed to report on the causes of this great unhealthiness and to suggest means for improving the health of the station (Bourdillon, 1883: 53).

As a result of continued efforts by the public institutions at various levels, sanitary and public health conditions in rural Bengal began to gradually improve. Plague, malaria, and small pox, diseases once reaching epidemic proportions in the region, were brought under control by the 1920s and in the succeeding decades (see Elahi and Ruzicka, 1981; Mitra, 1979). Also, due to improvements in technology to combat and avert the severe effects of flood and drought, fewer serious crop failures were experienced after the 1920s (Visaria and Visaria, 1983: 492). Despite continued environmental constraints in most regions of Bengal, the chances of survival among rural populations gradually increased.

Estimated crude death rates for the territory of present-day Bangladesh ranged from 47.3 to 11.2 per 1,000 population during 1881-1991. The rise in the death rate between 1941 and 1951 was due mainly to the Great Bengal Famine in 1943, and the effects of population displacement by the Partition in 1947 (see Robinson, 1967; Haque, 1987b; 1996a). Given the fact that international migration was nominal from the region until the mid-1970s (see Elahi, 1974; Haque, 1984), the rate of population growth, which has exceeded one percent per annum only since the 1940s (Table 5.1), has been primarily influenced by changes in mortality.

Population growth in rural Bangladesh began to accelerate in the early 1940s (Table 5.1), resulting from a number of factors, but particularly the control of endemic and communicable diseases. In contrast to the changes in mortality conditions, fertility rates stayed relatively constant—crude birth rates remained at more than 40 per 1,000 population until the 1980s. The explanation for such persistently high fertility lies not only in demographic factors (i.e., an early age for and the universality of marriage, and the traditional values towards child bearing; cf. Kabir, 1977; Khan and Ruzicka, 1981)—the socioeconomic determinants of fertility must be included.

Following Alamgir (1978: 93-100) and Cain (1977; 1981), a number of underlying social and economic determinants of persistently high fertility can be cited. First, family size is not a "decision variable in the individual welfare function" among disadvantaged rural inhabitants of the region (see Notestain, 1960: ix-x). Both Alamgir and Cain suggest that fertility-raising factors have been relatively more pronounced than have fertility depressing factors. Second, given the scale of poverty for the majority of rural people, the cost of an additional child is not generally large enough to affect a family's economic status. Furthermore, the cost of childbearing is often transferred to others or

TABLE 5.1. Population in census year and intercensal increase, 1871-1991

Year and Date of Census	Enumerated Population	Intercensal Increase	Percentage Increase	Average Annual Growth Rate (%)
1872	22,779,000	-	-	-
1881 (17 Feb)	24,926,000	2,147,000	9.43	0.90
1891 (26 Feb)	27,010,000	2,084,000	8.36	0.80
1901 (1 March)	28,927,786	1,917,786	7.10	0.69
1911 (10 March)	31,525,056	2,956,270	8.98	0.86
1921 (18 March)	33,254,096	1,699,040	5.38	0.52
1931 (26 Feb)	35,604,170	2,350,074	7.07	0.69
1941 (1 March)	41,997,297	6,393,127	17.96	1.66
1951 (1 March)	41,932,329	- 64,968	- 0.15	-0.02
1961 (1 Feb)	50,840,235	8,907,906	21,24	1.96
1974 (1 March)	71,479,071	20,638,836	40.59	2.64
1981 (5 March)	87,120,119	15,641,048	21.88	3.13
1991 (11 March)	111,455,185	24,335,066	27.93	2.79

Sources: ESCAP (1981: 13); and BBS (1984b: 33).

to successive generations. Third, features related to developing "capitalistic modes of production," such as viewing economic roles as a kind of insurance against risks and privileging the productive contribution of male children, induce peasant families to opt for a larger family (see Cain, 1977; 1981; Chowdhury, 1987). The extreme seasonality of farm employment requires additional hands in families dependent upon the sale of their labor. Also, in the absence of any form of social security, children are regarded as the source of security for parents reaching old age or disability. Therefore, until recent years, a deliberate, conscious control of fertility seldom received serious attention from the peasants of rural Bengal.

5.3.2. POPULATION CONCENTRATION IN THE RISK ZONE OF THE RIVERINE ENVIRONMENT

A close scrutiny of the social, economic, and demographic terrain of Bengal suggests some significant relationships between *growth* and *distribution* of the population, especially in the context of the alluvial lands of the region. Arthur and McNicoll succinctly describe these relationships in the context of the Bengal plain. They write:

> When we look more closely at the demographic picture, we see population growth linked tightly to local environmental conditions. Two processes are at work. First, whenever new technology or resources appear, settlement expands to take advantage of them. Population growth, as it were, follows the contours of local economic opportunity. Second, the pressures and momentum of continued population growth sooner or later exhaust whatever "slacks" these opportunities provide; people are forced to seek

> and exploit whatever other local means of subsistence they can find (Arthur and McNicoll, 1978: 31).

Similarly, in the case of the human occupance of the Brahmaputra-Jamuna floodplain, two mechanisms were present: (1) the clearing of forests and jungles along both banks of the river; and (2) the occupance of accreted land. Literary documents indicate that the region between the Barind and Madhupur Tracts were occupied long before the Muslim conquest of Bengal. However, due to the almost stationary population numbers, the density of settlement was extremely low in the swampy and marshy regions. As mentioned above, the gradient of rural population density followed the contours of the changing river courses throughout history. As Nicholas notes:

> Rural Bengal is dependent upon its rivers for economic well-being. It is not merely that rivers carry trade and make possible marketing of local crafts and agricultural products over a wide area; the decline of a river is intimately connected with the decline of agricultural prosperity in its area; the siltation of a stream bed, either by natural processes or because artificial embankments have cut off the natural spill area, is almost always accompanied by water-logging and rise in endemic diseases are intimately connected phenomena (Nicholas, 1962: 28).

An alternate description of the human habitat of rural Bengal by Mukerjee deserves special attention here for its accuracy rather than for its lyrical quality:

> The waters here widen out into endless vistas, and the lands are in a process of perpetual building and unbuilding by a wide and interlacing network of canals and streams, while an endless procession of coconut-clad villages and itinerant hamlets, containing a swarming population, fringes the marshes or the river-banks, which are higher than the surrounding country, and are always accessible by water . . . Man here is essentially a child of rivers. His crops and farming practices are closely adjusted to the timely inundation of red waters [i.e., waters with sediment]; and indeed, if the rivers do not rise in flood . . . he will be a fish out of water indeed . . . His traffic and transport are determined by tides and winds, and he himself lives a more or less amphibious life (Mukerjee, 1938: 19-20).

The intimate relationships over the decades of this century between the riparian environment and the human habitat is clearly reflected in a comparison of population density between (a) the districts directly on the banks of the Brahmaputra-Jamuna, and (b) the adjoining districts located away from the banklines (see Table 5.2). From Table 5.2, it appears that the districts on the bankline have a much higher density of population than the non-riparian districts. This variation in population density between these two broad regions is primarily a product of past population redistribution and the limited influence of natural increase on population growth until the 1950s.

The process of population redistribution in northern Bengal in the early decades of this century involved not only 'pull' factors such as attracting people for trade, commer-

TABLE 5.2. Density of population by bankline and non-bankline districts (new), 1891-1981

Districts	Year									
	1891	1901	1911	1921	1931	1941	1951	1961	1974	1981
A.BANKLINE DISTRICTS										
Bogra	673	688	783	829	797	896	841	1086	1529	1871
Gaibandha	608	683	815	862	834	921	905	1155	1630	1893
Jamalpur	476	546	657	717	821	955	897	1135	1607	1915
Kurigram	548	555	636	768	645	698	732	885	1277	1581
Manikganj	881	919	931	967	1048	1221	1196	1318	1698	1994
Pabna	672	600	670	706	681	767	757	932	1336	1644
Serajganj	796	871	856	937	923	1079	975	1150	1147	1944
Sherpur	451	547	670	662	722	846	840	1161	1512	1791
Tangail	810	914	989	816	853	1033	976	1143	1587	1867
Mean	657	703	780	795	814	935	902	1107	1480	1833
B.NON-BANKLINE DISTRICTS	355	398	435	441	445	508	622	738	1065	1319
Dinajpur	340	415	524	478	680	804	752	949	1438	1727
Gazipur	370	436	533	587	648	705	950	935	1358	1656
Jaipurhat	468	536	651	822	875	1018	966	1180	1649	2003
Mymensingh	347	552	602	627	622	715	659	751	1073	1305
Naogaon	566	577	599	626	635	714	709	1048	1544	1912
Rangpur										
	408	486	557	597	651	744	776	934	1355	1654
Mean										

cial, and agricultural opportunities, but also a 'push' factor from the large-scale expulsion of population from the decaying moribund floodplains. As such, it may be worthwhile to focus on the process of floodplain invasion in the new alluvial regions, where settlement proceeded by forcing the residents of the old floodplain, or of primitive communities in northern India, to clear and settle new floodplain formations. O'Donnell (1893) records, for example, that the *Santhals*[55] were encouraged to clear thick brushwood along the west bank of the Brahmaputra-Jamuna during the late 19th and early 20th centuries with the incentive of rent-free habitation for a few years. These primitive tribes made their livelihood by selling the newly cleared lands to 'resident Bengalis' and then moving on to new plots. The situation along the eastern bank of the river is described by Bourdillon, who records that:

> rents are low, markets good, and the failure of crops unknown; much waste land that ten years ago was jungle and swampy has now been brought under cultivation, while the development of jute industry has induced a gradual prosperity among the cultivating class (Bourdillon, 1883: 45).

Several writers point out that the new waves of floodplain settlers during the early decades of this century were caused by the expansion of the jute industry, along with the need of peasants for cash to pay higher taxes (see Panandikar, 1926; Majumdar, 1943; Kling, 1960). Since production and processing of jute requires annual flushing, the cultivators gradually occupied the natural levees of the rivers closer to the channel flows.

Along with these changes in rural settlement in the doab, river ports and trading centers also evolved along the new river courses. The *Maroary* traders, for instance, began to settle at a location near Belkuchi on the Jenai river in 1809, with the encouragement of the *zamindar* of Belkuchi, who agreed to accept a nominal tax rate. Between 1809 and 1830, the Jenai became a mighty river, and Serajganj evolved into a famous port over the following decades; Beverly (1872: 5) notes that by 1872, Chilmari and Serajganj became places of rising importance as "the great mart of Eastern Bengal." Due to rapid in-migration of people from the surrounding districts, Serajganj town experienced a population growth of 9.8 percent during 1869-1872, which excluded the boat population. By the late 19th century, Gaibandha and Serajganj districts were connected to the railway network linking western regions to the commercial areas on the Brahmaputra-Jamuna. A consequence of this was that many labourers also immigrated to these riparian districts (Gait, 1902: 63).

The other cause of new floodplain habitation was the influx of peasants from neighboring districts or from the moribund floodplains onto the new formations or *chars*. Colebrooke elaborates on how these migrations commonly took place:

> Such . . . islands as are found, on their appearance, to have any soil, are immediately cultivated and water melons, cucumbers, and *surso* or mustard, become produce of the first year. It is not uncommon even to

> see rice growing in those parts where a quantity of mud has been deposited near the water's edge.
>
> Some of these islands, before they have acquired a degree of stability which might enable them to resist the force of the stream, are entirely swept away; but whensoever, by the repeated additions of soil, they appear to be sufficiently firm, the natives then no longer hesitate to take possession of them, and the new lands become an immediate subject of alteration and dispute. The new settlers bring over their families, cattle, and effects; and having selected the highest spots for the sites of their villages, they erect their dwellings with as much confidence as they would do on the mainland (Colebrooke, 1803: 4).

During the early decades of the 20th century, both Rangpur and Bogra districts received a great influx of people, largely consisting of Muslim peasants from Pabna and Mymensingh. They settled on the new mid-channel formations of the Brahmaputra-Jamuna river by a process of *char*land invasion (see Thompson, 1923). The new formations were colonized not only by the people of the adjoining or distant mainland, but also by the people whose land on the other side of the river had been washed away.

While technological progress was very slow over the centuries in Bengal, population growth exhausted the slack that the Brahmaputra channel diversion had offered. Net increase in settlements and population in the areas along the new channel was not only the product of the invasion and the importation of a small but sustained agricultural surplus, but was also due to a natural increase in population produced by a reduction in mortality (see Bourdillon, 1883: 53). As a result of increased dependency and exhausted resource opportunities for settlers, the occupance of floodplain lands soon lost its attraction. Such a situation was observed in the Brahmaputra-Jamuna floodplain as early as the 1930s. In describing the exhaustion of the agricultural frontier based on jute cultivation during the early 1930s, Porter (1933) observes that in many localities of Rangpur district, the decline in jute threw a large number of laborers out of work, forcing them to migrate elsewhere. He adds that in Bogra district a number of villages were washed away causing the inhabitants to move to Mymensingh, Assam, and other parts of West Bengal. The partition of Bengal in 1947 stopped the stream of out-migration from the northern Bengal regions to Assam and West Bengal, both of which became parts of the Indian Union (see Quarishi, 1987). In the following decades, "the relative distribution of the population remained virtually unchanged . . . except for very slight displacements" (ESCAP, 1981: 15).

As a consequence of the above historical processes, involving demographic, social, cultural, and economic factors, the countryside today experiences phenomenal population growth as well as a very high population density. Johnson (1982: 115-7) indicates that the heaviest concentration of population is found today in the southeastern deltaic quadrant centering on the Dhaka-Comilla axis. This region extends "along the Old Brahmaputra and Dhaleshwari distributaries as far as the Jamuna floodplain to the northwest," with an outlier gradually developing in the lower delta area of Faridpur and

Barisal. Besides the social-historical dynamics of the past, the environmental advantages of regular flooding and a sufficiently prolonged rainy season permits double or triple croppings per year, thus significantly influencing the development of high-density settlement along the floodplains. Today, the density of this quadrant surpasses 1,112 persons per square kilometer (see BBS, 1984: 34-6).

While the riparian environment of Bengal has been inhabited for centuries, the benefits of that use, mainly through agriculture, were obtained at a cost of devastating losses from cumulative natural hazards. Today, the massive rural population inhabiting the floodplains is vulnerable to floods, droughts, riverbank erosion, cyclones, tornadoes, tidal surges, among other catastrophic natural events. Also, the population is faced with unemployment or underemployment in agriculture, and with food deficits and a lack of entitlement to social resources. Land and economic losses due to riverbank erosion consequently enhance the prevailing processes of landlessness and rural impoverishment in the country; the size of the population at risk to bank erosion has been estimated to be nearly one-fifth of the entire population of the country (i.e., nearly 19 million; see Currey, 1979).[56]

5.4. State, Land Tenure and Legal Systems: Formation of the Bengal Peasantry

The permanent occupation of the floodplain is not only related to demographic and ecological factors, but also to the land tenure and legal systems that largely determine social relations and entitlement to resources in Bengal society. Land has always been the principal means of production in Bengal, especially since agriculture became the dominant mode of subsistence. The quality of the alluvial land was enhanced by siltation and moisture during normal floods throughout history. However, private ownership of land or the individual's access to land, did not develop in an equitable way. Rather, a complex set of land tenure relations evolved over time through a variety of sociopolitical institutions. In this section, the legal and land tenurial systems are evaluated in light of the role of the state and other institutions. Also, the alluvion and diluvian regulations are reviewed in the context of the *char*lands.

5.4.1. STATE, LEGAL SYSTEMS AND LAND TENURE

Despite disagreement among historians with regard to the individual's relationship to land in ancient India (e.g., Sen, 1962; Baden-Powell, 1882; Mookerjee, 1926), there is a consensus on certain issues. First, the nature of the economy of village communities was 'self-sufficient' but stagnant; and second, with varying forms, land was common property for an entire village. As Desai (1959: 7) elaborates, a 'self-sufficient' village of pre-British Bengal society relied on agriculture practised with the primitive plough using bullock power and manual labor. Village populations were mainly composed of peasants, blacksmiths, carpenters, potters, cobblers, washermen, oilmen, barbers, and other artisan groups. They worked almost exclusively to satisfy the requirements of the village community itself, with very limited motivation for the exchange of commodities with the outside world. Because these feudal social relations ensured an entitlement to land, food, and security for actual producers, famines and other crises took place only when there was an absolute shortage due to extreme natural disasters (Parvin, 1987: 17; also see Sen, 1980, 1981a, 1981b).

A process of differentiation within the peasant society began in Bengal during the late Mughol period (Abdullah, 1980). With the implementation of revenue administration (i.e., 'prebendal domain'; see Hoogvelt, 1978) by the Mughol emperors, inequalities in landholding, as well as other manifestations of commodity production were put in place (see Habib, 1963). The new patterns of revenue-demand from the peasants began to undermine the traditional ties that had been established by the Hindu caste system over the centuries. Instead, an unequal power structure gradually emerged in rural Bengal, controlling land within a 'patron-client' or 'headman-subordinated follower' framework.

As noted above, with the dramatic and large-scale alteration in river systems during the 18th century, Bengal also experienced radical changes in its sociopolitical spheres. These changes influenced virtually all aspects of human habitation during the next two hundred years, but the most far-reaching impact was upon the agrarian and prevailing feudal social relations. The process of change began with the conquest of Bengal by the British East India Company in 1757. Revenue collection staff had been steadily declining in the mid-18th century during the period of decaying Mughol rule. Therefore, a reassessment of land tenure and social relations was imperative for the colonial authorities, especially in order to expropriate surpluses from agriculture and its related economic sectors.

Among the different classes of claimants for the settlement of land revenue, the Company recognized the Hindu *zamindars* (landlords) as the most eligible agents.[57] These *zamindars* were given tenure along with the authority to collect revenue and taxes (see Guha, 1963; Stokes, 1978). During the Mughol period the *zamindars* were only appointed collectors of land revenue for the state or emperor, but the Company insisted that the agents be provided the right to fix their own terms of tenancy. With the apparent assumption that permanent tenure and fixed assessment would induce individuals to reclaim vast areas of potentially cultivable wasteland, the East India Company made agreements with the *zamindars*, along with *independent talukdars*,[58] and other "actual proprietors" of the land. This move, known as the *Permanent Settlement* scheme, was implemented all over Bengal by a proclamation of the Company in 1793.[59]

The impacts of this scheme upon the various socioeconomic spheres appeared soon after its implementation. First, the collection of taxes in monetary form changed agricultural production towards export crops, initially to poppies, indigo, safflower, and later to jute. New divisions of labor and rural-urban linkages began to develop; these had a significant influence upon settlement strategies. Second, land became a saleable commodity for the first time; earlier the transfer of land was only possible as a gift or under hereditary succession (Mamun, 1975). Third, a long chain of intermediary social classes gradually formed in connection with social relations in agriculture and land ownership.

Since the process of rent collection was rigorous, the *zamindars* farmed out portions of their estates to dependent *talukdars*. These intermediaries paid a fixed amount to the *zamindars*, pocketing the difference between their rent collection and the fixed dues. Further, the *talukdars* created their own subordinate tenures. As a result, at least four distinct levels emerged between the state and the cultivators or *raiyats* (who performed the actual cultivation of land): (1) the *zamindars*, (2) the tenure holders (dependent *talukdars*), (3) *raiyats*, and (4) the *under-raiyats*, people who sublet land from *raiyats* (Umar, 1978). *Bargadars* (sharecroppers) and landless laborers were the primary workers on the land, with the *raiyats* "occupying the structural position of aggregating the surplus from sub-tenants and labour and transforming it upward in the form of rent" (Wood, 1981: 8; also see Abdullah, 1976: 69).

The Scheme of 1793 permanently established the revenues to be collected for the state, but the rent paid by tenants was left to the discretion of the *zamindar*. This arrangement, however, failed not only in encouraging *zamindars* to annex new lands or to develop their agricultural lands along the lines of the English feudal lords, but also could not restrict the *zamindars* from imposing excessive taxes which surpassed the established *parganah* (regional) rates (see Panandikar, 1926: 86). Belatedly, the British authorities came to recognize this situation through emerging peasant revolts, especially the revolt of Serajganj in which tenants vigorously challenged the prevailing *zamindari* system (see Umar, 1974: 43-5). Consequently, in 1885 the Bengal Tenancy Act came into force, which gave *de jure* recognition to the rights of others besides *zamindars*. Jannuzi and Peach, in reviewing this revised land legislation, conclude that:

> It is a system in which revenue farmers could be assigned proprietary rights in land within the framework of the Permanent Settlement in the expectation that they would gradually establish a direct interest in improving the productivity of their landholdings. It is a system in which superior rights in land (and greater security of tenure on holdings) were accorded to non-cultivators; it is a system in which actual tillers had inferior (less secure) rights in land (Jannuzi and Peach, 1980: 8).

The overall land tenure system, however, maintained the rights and interests of a landholding elite. The laws concerning relations between *zamindars* and tenants were structured in such a way that the state's right to revenue was guaranteed. The British authorities thus institutionalized in Bengal a dichotomy between rent receivers and actual tillers of land. Some analysts label this group of rent receivers 'mere parasites' because of their absolute dependence upon the surpluses produced by peasant cultivators (Umar, 1974; Saha, 1980). In evaluating the conditions of the peasantry during the colonial period, Abdullah (1980: 18) finds that the Bengal peasantry underwent a double transition: one from 1799 to 1885, when the rural economy was characterized by a pure 'peasant economy,' subjected to surplus appropriation through rent; and the other, from

1885 to 1947, when the peasantry experienced rapid differentiation giving rise to certain exploitative measures through rents, wages, and usury. Thus the process led to the *zamindari* rent gradually losing its dominance; in turn, the rich peasants and petty landlords evolved into a new social class with considerable sociopolitical power.

Following the partition of India in 1947, land legislation aimed at easing the peasant-landlord relationship was enacted both in the Indian Union and in Bangladesh (then East Bengal). The East Bengal State Acquisition and Tenancy Act (EBSATA) of 1950 repealed the Bengal Tenancy Act of 1885 and invalidated the Permanent Settlement Scheme of 1793. The Act was designed to abolish the rights and role of the rent receivers, and brought the state into a direct *de jure* relationship with the rent-paying cultivators of land. However, the Act failed to bring any fundamental changes in the prevailing agrarian structure, or in the relations between people and land since old receivers—the *zamindars* and tenure holders—remained the primary landholding group. The Act also denied sharecroppers their rights to land and their means of acquiring land; the new *malik* (owner) group was given *de jure* rights to continue exercising traditional practices. Likewise, although a ceiling of 375 *bighas* (about 50 hectares) on land ownership was imposed by the East Pakistan Ordinance No. 15 of 1961 (a revised version of the 1950 Act), creating 66,315 hectares of *khas* (government) land for redistribution, the major proportion of this was wasteland (Abdullah, 1976: 85). Although preference for the redistribution of land was to be given to bonafide cultivating families owning less than 1.2 hectares (three acres) of land, as well as to tenants of diluviated land, refugees, and veterans with "long and meritorious service," only a small portion of this land was acquired by "bonafide cultivators." This was because the amount of *salami* (advance payment for the *khas* land) was too high for most of them.[60]

Furthermore, the continued repression of small peasants by local government officials (i.e., the *tahasildars* who were responsible for rent collection) reflected the lack of any effective change in the Act. Responding to this issue, the Land Revenue Administration Enquiry Committee of 1963 reported:

> Everywhere we hear severe condemnation of this system. It was said that the *tahasildars* are generally corrupt and rude to the ordinary classes of tenants, that no rent can be paid in reasonable time without illegal gratification . . . It was further argued that the *tahasildars* treated the Government *khas* lands as their personal property . . . We were told . . . that the *tahasildars* are really *zamindars* in a new garb with all their oppressions and that they are all the more powerful and dangerous as they have the authority of Government behind them (Government of East Pakistan, 1963: 7).

Following the Liberation War and the Independence of Bangladesh in 1971, the land laws of the 1950 Act continued as the valid legal-administrative framework for land administration. Two minor modifications, however, were made by the post-Independence government: first, the landholding of 25 *bighas* (approximately 3.6

hectares or nine acres) or less was exempted from payment of land revenue; and second, a ceiling of landholding per family was fixed at 100 *bighas* (13.5 hectares or 33.3 acres). The prime objectives of such policies were to acquire land through the imposition of a ceiling, and to redistribute this acquired land among the landless. These policies, however, were never implemented, largely because the majority of landholders who exceeded the ceiling disguised the fact by means of *benami,* the fictitious transfer of land title (see Jannuzi and Peach, 1980).

It appears that all these changes in the postcolonial state structure were mainly in the interests of landholders who had been enjoying a permanent occupancy right to land. The changes in the legal system did not bring about any benefits to the actual tillers, or to the landless agricultural laborers. The relationship of the peasantry to land in contemporary Bangladesh thus has remained relatively unchanged, both institutionally and structurally, since the early 19th century, despite the legislation enacted following the Partition and after the separation of Bangladesh from Pakistan.

5.4.2. ALLUVION AND DILUVION REGULATIONS, 1825-1972

The role of the state in administering riverine areas of Bengal became distinctive as early as 1825, when the Bengal Alluvion and Diluvion Regulation was put into place. Before the enactment of this legislation, claims to ownership of alluvion and diluvion land were regulated by local customs and title (Zaman, 1987: 19). According to the 1825 Regulation, in the case of clearly recognizable and established matters, claims and disputes over alluvion land were decided upon by considering the local use of *payasti* (alluvion) and *sikosty* (diluvion; see Gupta, 1940; Kabir, 1969; Ali, 1980). Nonetheless, the Regulation was enacted to protect the mutual interests of *zamindars* and big landlords—a consequence of an alliance between these two dominant groups during the colonial period. That is, land gained by the gradual accretion from receding rivers was annexed to the tenure of the person whose estate it adjoined (Ali, 1980: 296), but a submerged estate that subsequently became attached to an adjoining estate would not fall under the legal control of the latter estate upon reemergence. In these matters, the established property owners prevailed.

The Act of 1950, by abolishing the "Permanent Settlement" with the *zamindars*, imposed two limitations on the restoration to the previous owners of lands lost by diluvion: (1) the restoration was legal if it took place within 20 years following the event of diluvion; and (2) such restoration was legal only if the possession of the land area remained in the hands of rent receivers or cultivating *raiyat* (section 20 of the EBSATA, 1950) and satisfied the limitation of transfer of holdings up to a maximum of 50 hectares (125 acres; section 90 of the EBSATA, 1950; also see Ali, 1980: 197; Zaman, 1987: 20). The Act thus emphasized the *de facto* possession of the land, and reflected the emergence of Muslim landlords as the main sociopolitical and economic power in the riparian areas of Bengal, replacing the absentee Hindu *zamindars*.

In view of the increasing landlessness of the rural masses, the post-Independence government of Bangladesh set goals to recover all alluvion lands from the control of *jotedars* (large landlords), and to redistribute them among the landless and peasant smallholders who owned less than one-seventh of a hectare of land. Specific objectives were also set to encourage cooperatives, collective settlement and farming. The Presidential Orders of 1972 (numbers 72, 35 and 135) asserted that all newly emergent lands previously lost by diluvion should be restored to the government and not to the original owners (Malik, 1983: 27).

However, the provisions of this 1972 legislation were opposed by the locally dominant interest groups of large landlords, and could not be put into operation without their cooperation. The underpinnings of this opposition lay with the potential threat to the prevailing rural power structure posed by the mobilization of the landless as cooperative groups. In 1975, the democratic legislative system of the country was abrogated following the military takeover of government.[61] Along with such abrupt political changes, the political forces supporting large landlords regained their power and rights at the local level (see Haque, 1987a). The renewed policies of the conventional political parties and vested interest groups was reflected in Ordinance 61 of 1975, which established that the owner of land lost by diluvion is the 'first' person eligible for settlement on *khas* land. It thus provided the right to repossess one's old property. In Malik's (1983: 27) words, "therefore, it follows that what has been taken away with one had has been given back with another."

Presently, the settlement of agricultural land in the *char* areas of alluvion land is based upon provisions made by the Memorandum (number 2L-77/70/223-R.L., dated 22nd August) of 1972, though the system has hardly become any more equitable. The lengthy and complicated settlement procedure, which requires surveying, inquiry and verification, and recording of land, still provides benefits only to the powerful landlords or *jotedars* and *talukdars* (see Malik, 1983; Zaman, 1987).

5.5. The Process of Pauperization and Increasing Proneness to Natural Hazards

The contemporary agrarian structure of Bangladesh has evolved into a complex form. For instance, establishing the boundary between landlords and tenants is now difficult because an individual may own land, give out land to others, as well as work on land as an agricultural laborer (see Ali, 1980). Moreover, measuring actual landholdings is problematic because both landowners and tenants are often reluctant to express the actual extent of their holdings. Taking such limitations into consideration, a few surveys have attempted to record the conditions and characteristics of agriculture in the country; these include the Master Survey of Agriculture, 1967-68, the Bureau of Statistical Survey, 1978, and the Land Occupancy Surveys of 1977 and 1978.

Average farm size was reported to be 0.70 hectares (1.73 acres) in the 1978 Land Occupancy Survey, though the distribution of landownership appeared quite uneven when comparisons were made by size. Large landowners, however, accounted for only a small proportion of the total landholders. As is shown in Table 5.3, only 2.67 percent of households owned 4.05 hectares (10 acres) or more land in 1978. However, this group accounted for 19.66 percent of total land area in 1977; this increased to 25.49 percent in the following year. At the other extreme, 14.69 percent of households were absolutely landless in 1978, and more than two-thirds of households owned less than the 1.22 hectares (three acres), the suggested minimum requirement to maintain a bare subsistence for a family (see Saha, 1980).

TABLE 5.3. Distribution of total owned land in rural Bangladesh, 1977-1978 (%)

Area in Acres	% of Total Households		% of Total Area	
	1977	1978	1977	1978
Landless	11.07	14.69	0	0
0.01 - 3.00 (0.02 - 1.22)[a]	72.78	68.58	36.91	33.39
3.01 - 6.00 (1.23 - 2.43)	10.65	10.35	27.03	24.83
6.01 - 10.00 (2.44 - 4.05)	3.56	3.71	16.40	16.30
10.01 - 15.00 (4.06 - 6.08)	1.14	1.51	8.37	10.42
15.01 and Over (6.09 and Over)	0.80	1.16	11.29	15.07

[a] Area in hectares is presented within the parentheses.
Sources: Jannuzi *et al.*, (1977; 1978).

These data suggest that today more than 80 percent of the rural population is below the bare subsistence level. If the definition of a landless household is accepted as being one with no more than 0.2 hectares (0.5 acres) of land, excluding homestead (Jannuzi and Peach, 1980), the proportion of landless then exceeds 50 percent of total rural households (according to the 1978 Land Occupancy Survey). The figures also show the trend of increasing landlessness since the proportion of landless was recorded at 48.08 percent in the preceding year. Evidently, more than half of rural households in the country have virtually no direct access to land, even though the majority are dependent on agriculture.

Both government and non-government statistical sources allow us to assess trends in landownership. As Table 5.4 shows, the average area per household declined from 1.63 to 0.7 hectares (4.02 to 1.73 acres) between 1938 and 1978. A disaggregated examination of landownership by size of holdings for the same period suggests that the proportion of households with less than 1.22 hectares (three acres) of land dropped from 56.89 to 33.39 percent. A remarkable shift also occurred in the larger size group; the percentage of households with more than 4.05 hectares (more than 10 acres) increased from 7.68 to 25.49 during the same period. A similar trend is also evident in the middle farmer category of 2.03-4.03 hectares. Such increases are also marked by the short-term changes in landownership. Table 5.5 shows the trends in landownership by farm size between 1974 and 1979, indicating that the bottom 50 percent of rural households have lost their share in recent years, whereas the top ten and top two percent of owners have increased substantially.[62]

TABLE 5.4. Size distribution of total owned land, 1938-1978

Survey Year	Area Per Family (in Hectares)	% of Families with					
		Less than 0.81 Ha (2 Acres)	0.81 - 1.21 Ha (2-3 Acres)	1.22 - 1.62 Ha (3-4 Acres)	1.63 - 2.02 Ha (4-5 Acres)	2.03 - 4.03 Ha (5-10 Acres)	Above 4.03 Ha (Above 10 Acres)
1938/39[a]	1.63	45.85	11.04	9.29	7.47	16.12	7.68
1967/68[b]	1.27	14.82	13.18	12.00	11.12	27.30	21.41
1977[c]	0.66	23.73	13.18	11.13	9.00	23.30	19.66
1978[c]	0.70	21.11	12.28	10.29	7.93	22.91	25.49

Sources: [a] Abdullah (1976: 102), Table 3.
[b] BBS (1978), Table 2(a) and Summary Table 3.
[c] BBS (1982), Tables 373 and 374.

TABLE 5.5. Trend in landownership pattern by farm size, 1974-1979

Landownership by Farm Size	% of Total Land Owned		
	1974	1977	1979
Bottom 50%	8.9	6.0	4.8
Top 10%	41.5	49.9	53.9
Top 2%	16.1	20.0	25.1

Sources: Alamgir (1977); Jannuzi *et al.*, (1977); and Jahangir (1981).

Nonetheless, these ownership patterns do not provide a complete picture of land relations and dynamics in rural Bangladesh. The tenurial arrangement of land can compensate for this gap because it accounts for the operational patterns in land utilization. Table 5.6 shows that about 13 percent of total land is tenant operated (C + D); slightly over 33 percent is owner operated (A + C), while the remaining 44 percent is operated by hired labor, occasionally in combination with some family laborers. In terms of household distribution, about 39 percent are tenant-cultivators, 23.5 percent are owner-cultivators, and about 38 percent are mostly non-cultivating owners. These tenurial patterns are indicative of the fact that most of the cultivable land in the country is tilled by persons other than the owner. A number of analysts suggest that this prevailig semi-feudal agrarian structure is not only an impediment to the development of agricultural technology and resources, but also prevents the rural masses from participating in development activities (see Jannuzi *et al.*, 1977: 69; Chowdhury, 1987).

The trend in land consolidation is also apparent in changing types of tenure. Table 5.7 shows the changes in the distribution of households by tenure type and by area under possession between 1960 and 1978. Although the proportion of owner-farms did not change significantly between 1968 and 1978, their share of land area increased remarkably from 59 percent in 1968 to 72 percent in 1978 (see Table 5.7). Furthermore, the increase in the proportion of owner-farms is reflected in the reduction of area cultivated by owner-cum-tenant farmers.[63] Abdullah *et al.*, in demonstrating the increasing incidence of pure tenancy between 1960 and 1968, remark: "It would appear that some kind of polarization process was indeed at work during these years, working towards eliminating the hybrid category of owner-cum-tenant, mainly to the benefits of pure tenants" (Abdullah *et al.*, 1976: 211). As explained earlier, this trend is linked to the fact that the marginal peasants cannot escape the burden of indebtedness, and repay their debts primarily by selling their land:

> What is significant to observe here is that while middle and large farmers are either increasing or maintaining their control over the most important of productive assets in rural areas, i.e., land, the small farmers are losing ground (Alamgir, 1978: 91).

TABLE 5.6. Number and area of rural households by type of tenancy, 1977

Type of Tenancy	% of Households	% of Land Owned	% of Land Taken in
A. Owner-Cultivator	23.51	10.47	-
B. Owner-Manager	37.66	43.48	-
C. Owner-cum-Tenant	31.99	23.18	18.48
D. Tenant	6.48	-	4.39
Total	100.00	77.13	22.87

Source: BBS (1978).

Abdullah *et al.*, (1976) and Wood (1981) argue that the internal social structure of rural communities prevents the development of purely capitalistic relations in agriculture. The social relations function as a deterrent to both social and geographical mobility because of strong kinship and quasi-kinship ties operating through "extra-economic and personalized sanctions." Also, the large family size and the prevailing

TABLE 5.7. Households by tenure types and area by possession, 1960-1978

Tenure Type	% of Farm Household				% of Area by Possession			
	1960	1968	1974	1978	1960	1968	1974	1978
Owner Farms	61	66	67	65	54	58	75	72
Owner-cum-Tenant Farms	37	30	27	28	45	38 }	25	23
Tenant Farms	2	4	6	7	1	3 }		5

Sources: Government of Pakistan (1960); BBS (n.d.: 12); Alamgir (1975); and BBS (1982).

laws of equal inheritance among sons fragment landholding and impede the concentrated ownership of land (see Bertocci, 1979; Wood, 1981). In examining the former characteristic, Abdullah *et al.*, (1976) find that the rich, landed families in rural communities gather around themselves persons of lower socioeconomic status, "the nucleus being formed usually by close or distant 'poor relations'." In exchange for their support during village rivalries, local elections, and factional politics, the landlords ensure tenancy and a variety of other benefits to their clients or followers.

Abdullah *et al.*, argue that, "With limited land, expanding rural population, and limited opportunities for urban employment, one would expect rural landlessness to be growing in any case, even without the polarizing effect of market forces" (Abdullah *et al.*, 1976: 215). Wood (1981: 6), in support of this view, asserts that under these demographic and socioeconomic processes, tenants and poor peasants are steadily alienated from their land, "being obliged to sell mortgage arrangements which are impossible to redeem." This pauperized section of the rural population maintains a bare level of subsistence by engaging mainly in agricultural work as laborers. In the case of an extraordinary natural calamity, many of these people are uprooted from their place of origin (see Alamgir, 1978: 31); their options then become severely limited, either to move as squatters to the nearest towns, or to remain in the locality, dependent upon the support of relatives and kin, or as 'dependents' of large landlords (see Zaman, 1987: 14).

The local landlords or *talukdars* regularly extend help because they need these destitute people to provide cheap labor or to work as sharecroppers or bonded employees. However, a sudden and extreme environmental hazard may also cause *talukdars* to limit such assistance, especially when their ability to employ agricultural wage labor is reduced. As a result of such a reduction in 'exchange entitlement' for the destitute, the proneness of the pauperized peasants to natural hazards is increased. No data are available on these people nor has any study attempted to deal comprehensively with these issues. This study is therefore exploratory in its attempt to demonstrate the *proneness* of people to natural hazards in relation to their position in the agrarian social structure.

5.6. Conclusions

In this chapter I have suggested that the human occupance of the Bengal plain, particularly on the active floodplains, should be treated in terms of the ability of individual production units, i.e., a household or family, to command resources through the prevailing institutional and legal means available to them. These include the use of shelter, production possibilities, employment opportunities, social and economic bonds, and entitlements. In essence, the settlement and agricultural occupance of the floodplains of Bangladesh may not be fully explained by analyzing individual decision-making processes that are motivated by profit, as in highly industrialized, capitalistic society (cf. Burton, 1962). In a fully developed, capitalistic society, land-occupancy decisions are generally made from a comparative analysis of a wide range of options with the sole motive of generating rent and revenue for profit maximization (see Burton, 1962). But in a non-industrialized, developing society like Bangladesh, the study of floodplain occupance by individual users requires us to refocus, looking to the *constraints* and *opportunities* offered to individuals through their prevailing natural and social environment. In other words, the analysis of individual decision-making patterns and processes requires an understanding of the relationships between prevailing social, cultural, economic, and political conditions.

Part Three

Riverbank Erosion Hazard in Serajganj District: Impacts and Responses

CHAPTER 6

THE RURAL STUDY DESIGN
The Characteristics of the Samples

The rural component of this study involved two field surveys in Kazipur *thana*—a severely affected area located on the Brahmaputra-Jamuna River. The first field investigation was part of a broader project of the Riverbank Erosion Impact Study (REIS) carried out in the summer months of 1985. This study was sponsored by the International Development Research Centre (IDRC), Ottawa, Canada, and administered jointly by the University of Manitoba, Canada and Jahangirnagar University, Bangladesh, during 1983-1988. I undertook a follow-up study in 1995, sponsored by the Brandon University Research Committee (BURC), under the auspices of the Social Science and Humanities Research Council (SSHRC), Ottawa. In order to allow a direct comparison, the same survey instruments were used in the two rural surveys, though a few items, particularly those concerning flooding issues were added in the 1995 survey.

In this chapter various aspects of the rural study design and the characteristics of the samples are described, chiefly to lay the background for the empirical analysis of riverbank erosion hazard impact and the associated human coping responses. This chapter is divided into two sections. The first section deals with survey and sampling procedures and the second section offers profiles of the samples, with a focus on displacement characteristics.

6.1. Research and Survey Methodology

6.1.1. RESEARCH INSTRUMENTS

The concepts and models concerning natural hazards impacts and associated involuntary migrations are still in a formative stage. Well defined theoretical postulations and models of these phenomena are therefore not readily available; as such, they cannot be represented as a set of ordered hypotheses with specified variables for empirical testing. In consideration of these limitations, our methodology for primary data analysis employs exploratory approaches: first, to trace various conceptual and applied issues in previous studies; second, to formulate potentially testable hypotheses based on both earlier findings and speculative abstraction; and third, to subject these hypotheses to empirical verification.

The design of the survey questionnaire was based on previous local surveys, seminar discussions, and the experience of a pre-test. It consisted of three parts. The first was concerned with demographic and socioeconomic background data, focusing on family size, age, sex, marital status, education, occupation, and birthplace. The second part of the questionnaire dealt with respondents' perceptions of hazards, and the relationship of such cognitive-perceptive processes to the overt responses to hazards. This part also included questions concerning the displacement history and experience of the hazard victims. The third part was directed toward the socioeconomic status of households—information on size of landholdings and land-use patterns, annual income from various sources, expenditures, and values of assets.

All respondents were asked the questions in the first and third parts of the questionnaire, while only the displaced population was asked the questions in the second part. The displaced households were identified from a positive response to the question: "Have you ever been forced to move because your homestead was lost to riverbank erosion?" Translated into and conducted in the local language of *Bangla,* the survey thus provided the necessary data to examine human responses and displacement associated with riverbank erosion from a comparative perspective between non-displacee and displacee households.

The village-level information, which particularly shed light on the dynamics of land possession and violence in *chars*, relied upon Zaman's (1988) survey of Char Chashi during a fourteen-month time period from June 1984 to July 1985. Direct participation and observation in the field was attempted in order to develop an understanding of hazard-resource relationships.

Other necessary data on rural population were obtained from secondary sources, including government publications on the population censuses, agricultural censuses, statistical reports of the Bangladesh Bureau of Statistics, and other published research materials (e.g., archival maps and other materials).

6.1.2. STUDY DESIGN OF RURAL SURVEYS

'Population' of the Surveys

As the term 'population' in a statistical sense denotes the aggregate units to which the survey results are to apply (Moser and Kalton, 1977: 53), the 'population' in the present survey refers to all the households of the Brahmaputra-Jamuna floodplain. A distinction between the 'target population' (i.e., the population for which the results are applicable) and the 'survey population' (i.e., the population actually covered) is worth noting here (see Moser and Kalton, 1977: 53). In the case of the present study, the 'target population' consists all of households occupying geographical space in the Brahmaputra-

Jamuna floodplain, while the 'survey population' is limited only to households located in Kazipur *thana* in the Serajganj district.

Study Area and Areal Stratification

Three reasons justify the selection of Kazipur for the present study. First, this *thana* is cited as one of the areas most severely affected by riverbank erosion in recent years (see Hoque, 1983; Khan and Rashid, 1985; Rogge and Haque, 1987: 2; Haque and Hossain, 1988). Second, this area represents the typical physiographic characteristics of the floodplain. Third, it has the characteristic features of rural underdevelopment, with all the societal syndromes related to bank erosion and flooding (see Zaman, 1987: 7).

As shown in Chapter 4, many studies on the channel characteristics of the Brahmaputra-Jamuna indicate that the west bank of the river is more susceptible to erosion (see Coleman, 1969; Chowdhury, 1973). Further, Currey (1979), Khan and Rashid (1985), and Hoque (1983) clearly show that Kazipur is one of the worst affected reaches of the floodplain (also see Kamaluddin, 1985; Hossain, 1985; Bristow, 1987). Notably, about two-thirds of its areas are *chars*. For example, the Kazipur *thana's* administrative center, as well as eight neighboring villages, were completely eroded between 1979 and 1983 (Haque and Hossain, 1988; also see Hossain, 1984).

The general pattern of rapid encroachment of channels into the west bank is also identified in the examination of sequential Landsat imagery of the area (Hoque, 1983; Hossain, 1984; Bristow, 1987). Khan and Rashid (1985) estimate that the average yearly rate of lateral erosion in Kazipur is as high as 244 meters. Consequently, the *thana* has experienced a large-scale displacement of population, particularly since 1977 (Haque *et al.*, 1984; Haque *et al.*, 1985a). According to a recent report of the *thana* administrator (i.e., *Thana Nirbahi Officer*), land erosion and consequent population displacement are the principal reasons for the overall underdevelopment of the area (Zaman, 1987: 7).

Kazipur is located to the west of the Brahmaputra-Jamuna and covers an area of 143 square kilometers (see Figure 6.1). The population was 213,885 (37,078 households) according to the 1981 census, and increased to 234,804 (46,201 households) by 1991. Thus, the density per square kilometer was 1,495 persons in 1981 and grew to 1,642 by 1991 (BBS, 1984b: 55; 1995). Currently, only one-third of the area of the *thana* is on the mainland; the rest lies in the channels within the river, characterized by *char* land settlements (Figure 6.1). Of the total 11 *unions*, five consist of *chars* and are accessible only by water transport such as country boats and motor-launches. Some 28.88 percent of the total population live in these *char* unions.

The mainland area of the *thana* contains part of the Brahmaputra Right Bank Flood Protection Embankment (BRBFPE)[64] which stretches for 245 kilometers from Phulchari in Rangpur District to Chouhali in Serajganj District. The length of the embankment running through Kazipur is about 19 kilometers (Figure 6.1). It was designed for flood control and protection as well as to increase agricultural production in the area. Construction of the embankment began in 1962 and was completed in 1967. Although there have been a number of breaches since 1967 during high floods, the general pro-

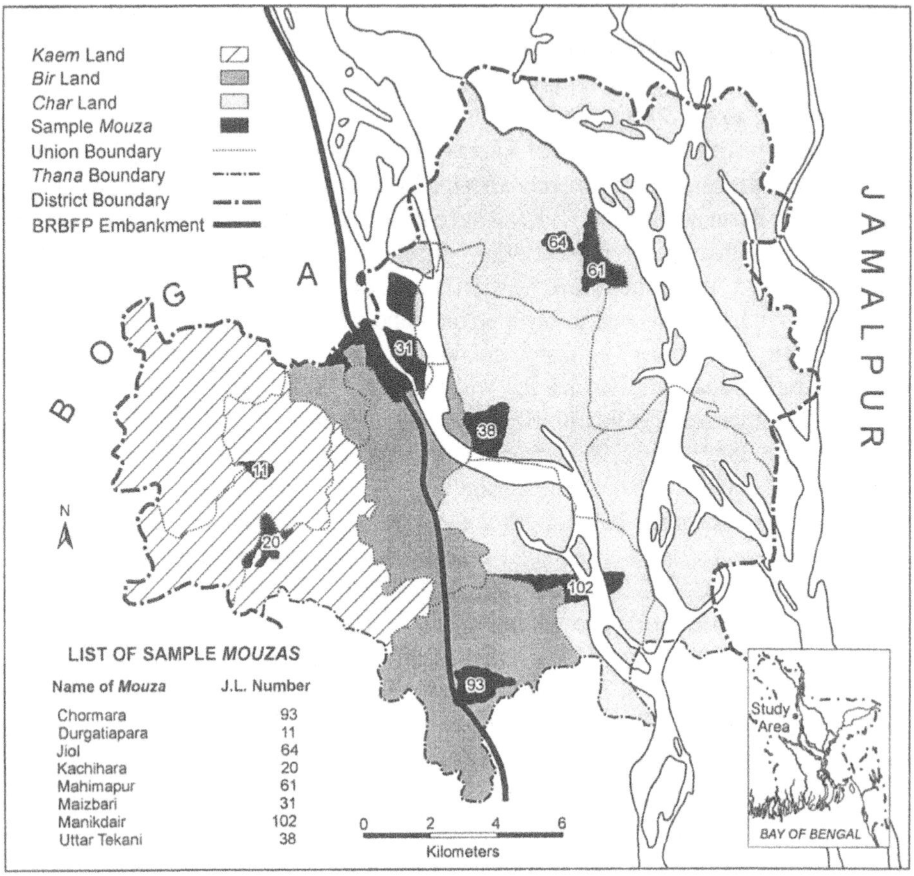

Figure 6.1. Location of sample *mouzas* by stratum, Kazipur, 1985.

tection from floods provided by the embankment has created some distinctive benefits for the areas within the embankment relative to the bankline areas exposed to free flooding.

A consideration of physiographic characteristics together with their social setting reveals variations in livelihood and household characteristics between the following areas: (1) the communities inland from the river; (2) the riverbank communities of the floodplain; and (3) the alluvial islands or *char* communities within the Brahmaputra-Jamuna system (see Hossain, 1984; Zaman, 1988). Hossain's (1984) study shows a noticeable variation in human habitations and responses to hazards in Kazipur, primarily with respect to physiography and distance from the riverbank; similar differences were also observed during several reconnaissance surveys (see Haque *et al.*, 1984: 5). The floodplain of the Brahmaputra-Jamuna is usually composed of a combination of loams on shales and clays, while the *chars* are composed of calcareous loams (Rashid, 1977: 126-7). The relief of *chars* is also relatively lower than that of mainland areas.

These variations in the physiography between mainland and *chars* have resulted in distinctive vegetation and cropping patterns. As Ahmad (1968: 67-8) notes, the *doash*, or clay, silt and sandy loams of mainland areas are of "great fertility, easy to plough and producing a variety of crops." In contrast, the porous sandy soils (i.e., *bali*) of the *chars* are not suited for *aman* paddy, but are suited for watermelons, mustard, sweet potatoes, and pulses. The frequency of settlements on the *char* areas is low and scattered relative to the mainland, partially caused by the non-contiguous location or inaccessibility by land transport. For instance, the density of population on the *chars* is about 515 per square kilometer while it is about 1,586 on the mainland (according to the 1981 census). Boats are the only mode of transport and communication for inhabitants of the *char* zones, and few roads or other transport facilities exist for inter-village contact and communication. The perennial threat of bank erosion makes it necessary for inhabitants to build temporary and moveable dwellings. People on the *chars* therefore equip themselves, both psychologically and materially, to face floods, erosion, and other riverine hazards, provisions most people on the mainland need not make.

Yet variation in the human response characteristics between the communities inland from the river and the riverbank communities shows that not all mainland areas share the same benefits. The BRBFP embankment succeeds in protecting the interior land, its agriculture and settlements from high floods. The distance of the interior areas from the riverbank also reduces the risk of erosion hazard for settlements and other establishments relating to the agricultural economy. In contrast, since part of the river bank communities are outside the embankment, these localities are highly vulnerable to high floods and erosion. Here, the concentration of displacee population from *char* areas and other eroded mainland regions results in a socioeconomic instability relative to protected interior areas.

The areal distinctiveness of the above three zones is reflected in the local popular names for them: (a) *kaem*, meaning land located in interior regions, and relatively elevated compared with the *chars*, (b) *bir*, which is a variation on the word *bhar* land, implying land located between the river and mainland (*kaem*) area, and (c) *char*, meaning mid-channel islands.[65] Given these zonal differences, the Kazipur study area has also been classified into three zones by unions: first, mainland unions without bankline, or the *kaem* zone (KZ); second, mainland unions containing bankline, or the *bir* zone (BZ); and third, unions with mid-channel islands, or the *char* zone (CZ). Table 6.1 summarizes the population and area of the unions by these zones as enumerated in the 1981 census.

TABLE 6.1. Distribution of population and area of the unions in Kazipur by stratum, 1981.

Name of Union	Stratum	Area in Acre	Population
Sonamukhi	KAEM ZONE	6,679	24,106
Chalitadanga	KAEM ZONE	7,042	26,534
Gandhail	KAEM ZONE	6,225	27,105
Total		19,946	77,745
Maizbari	BIR ZONE	4,864	22,324
Kazipur	BIR ZONE	6,783	27,735
Shubhagacha	BIR ZONE	6,776	24,320
Total		18,423	74,379
Khas Rajbari	CHAR ZONE	9,037	7,999
Char Girish	CHAR ZONE	13,489	16,309
Natuarpara	CHAR ZONE	6,737	15,433
Tekani	CHAR ZONE	8,620	13,440
Nischintapur	CHAR ZONE	10,007	8,580
Total		47,890	61,761
GRAND TOTAL		86,259	213,885

Source: Calculated from: BBS (unpublished data).

6.1.3. SAMPLING DESIGN AND PROCEDURE

Explanation of the Key Terms and Units

The *primary sampling unit* (PSU) for both the 1985 and 1995 surveys was the individual household (*khana*). A household or *khana* is defined "as a socio-economic unit, consisting of individuals who live together" (United Nations, 1958: 4). Specifically, a household in Bangladesh consists of a group of persons who share living quarters and their principal meals. The household size thus may vary from a single individual to an extended family. The rationale for the consideration of the household as the PSU lies in its role as the key unit of adjustment decision-making to cope with hazards, as well as in its function as the basic unit of social reproduction of goods and commodities (see Cain, 1977; Haque and Hossain, 1988).

The *respondent* of the survey was the household head (*griha karta*). Representing the unit as a whole, the household *head* is defined as the person who makes the major economic, social, and household decisions as well as the one recognized as the chief of the unit by the members of the household. In case of the long-term absence of a household head, the current proxy who had been maintaining the household on behalf of the head was considered to be the appropriate respondent (i.e., *de facto* head).

The survey design for the study involved four levels of spatial units: (1) *thana*, (2) *union*, (3) *mouza*, and (4) household or *khana*. At the first stage of the survey design, a *thana*-level unit, namely Kazipur,[66] was selected 'purposively'. In general, the *thana* is the basic administrative unit in Bangladesh; its officials are drawn from both government and local public representatives. The average area of a *thana* in Bangladesh is about 181 square kilometers, while the average population size is about 227,460 (BBS, 1995: 24). In comparison, the area of Kazipur is 143 square kilometers, and its population is 234,804 (1991 data). At the second stage of the sampling, the unions—a subregion of a *thana*—were considered for classification into different physiographic zones or areal strata. A *union* consists of a number of *mouzas* (revenue units) and/or villages (social units) and is served by a body of elected public representatives. The average area of a *union* is about 20 square kilometers, containing an average population of around 25,040. The *mouzas* are considered at the third stage of the survey design. As discussed in the preceding chapter, the *mouza* is a revenue unit, usually around 239 hectares, with an average of 333 households or 1,858 people (BBS, 1995: 24-25); each *mouza* may consist of one or more villages. The PSUs, or the household units, made up the fourth level of the sample design—an average household in Bangladesh consists of six persons.

Sampling Design and Procedure of the Rural Main Surveys

Considering the heterogeneity contained in the study area in terms of the magnitude of the natural hazard, the location of settlements, the socioeconomic status and cognitive characteristics of the floodplain inhabitants, a *multi-stage* stratified sample design was deemed appropriate. In addition to the purposive selection of Kazipur as the survey area, the survey design, as noted above, involved three additional levels of spatial units: *union*, *mouza*, and *khana* or household.[67]

Based on findings of earlier investigations, three criteria were considered for zoning the *union*; namely, (i) the geographical location of the *union* relative to the bankline; (ii) the physiographic characteristics; and (iii) the proportion of displaced population relative to total population in that *union*. Three distinct strata were thus constructed based on these criteria and presuming an internal homogeneity within each stratum. Upon the stratification of the 11 unions in Kazipur, the respective number of *unions* in each stratum was three for KZ, three for BZ, and five for CZ (Figure 6.1).

At the second stage, the respective *mouzas* for each *union* were classified by zones, producing 30 *mouzas* for the *Kaem* Zone (KZ), 26 for the *Bir* Zone (BZ), and 61 for the *Char* Zone (CZ). Considering the costs of interviewing respondents, and given the homogeneity in the character of population displacement within the stratum, a sample proportion of five percent of *mouza*-level population was thought to be appropriate. The use of this proportion for *mouza* units resulted in a sample size for the three zones as follows: two *mouzas* from KZ; two *mouzas* from BZ; and four *mouzas* from CZ. The list of *mouzas* is shown in Figure 6.1 as it appeared after the drawing of the sample *mouzas* using a *simple random sampling* (SRS) procedure.[68]

Considerable variation in the size of the *khana* population (or total number of households) was apparent in each stratum considered; the KZ contained 334, BZ contained 1,938, and CZ contained 537 PSUs in 1985. In 1995 the KZ had 90, BZ had 644 and CZ had 110 PSUs. In order to maintain greater control over the sample size for a better representation from each stratum, the selection of the PSUs was made with *probability proportional to size* (PPS).[69]

Heads of households representing the PSUs provided a demographic and socioeconomic profile, and information on their perception of and coping responses to hazards. Data on displacement history were obtained by interviewing only those respondents representing displacee households.

Determination of Sample Size

The present study is primarily concerned with the impacts of riverbank erosion, particularly on the displacement of floodplain inhabitants. This displacement, however, is only characteristic of a certain proportion of the survey population. In such cases, the required calculation of standard deviation of a *mean* is not feasible (see Dixon and Leach, 1978). Instead, a measure of *variability* is employed to calculate the standard error of the proportion—a required element for the determination of the sample size of the survey. The formula used to calculate the standard error of the proportion is as follows:

$$v = \sqrt{p\ (100-p)} \qquad \text{......(6.1)}$$

where p = percentage with the displacement characteristics.

Following Dixon and Leach (1978) and Kish (1965), the sample size required to estimate a given percentage of *confidence level*[70] within a certain degree of *confidence limit*[71] is calculated by applying the following equation:

$$n = \left(\frac{z \times v}{c}\right)^2 \qquad \text{......(6.2)}$$

where c = confidence limit
z = the standard error unit measure for the desired confidence level
v = standard error of the given proportion.

In order to obtain information on the proportion of displacees in each of the three different strata, the census of earlier village studies was employed as a proxy for the required pilot survey. The reconnaissance survey of the KZ *mouzas* indicated that only five percent of all households were displaced by riverbank erosion. Hossain's (1984) survey of three villages in a *union* on the bank of the river, namely Khudbandi, Dhekuria, and Meghai, indicated that 18 percent of households in these villages were displacees. Zaman's (1988) baseline survey of a *char* village, Uttar Tekani, recorded

80 percent of households as displacees; the sample size was determined by using a combination of this *variability* in the three strata within different degrees of the *confidence limit* and with a 95 percent *confidence level*. Given the constraints of time, manpower, and financial resources, the maximum sample size within a five percent *confidence limit* at a 95 percent *confidence level* was thought to be appropriate for the 1985 survey. Incorporating the pertinent data into the equations (1) and (2) for the each of the three strata, the sample sizes were calculated as follows:

$$n = \left(\frac{1.96 \times 21.79}{5}\right)^2 = 73,$$

$$n = \left(\frac{1.96 \times 38.42}{5}\right)^2 = 227,$$

$$n = \left(\frac{1.96 \times 46.00}{5}\right)^2 = 246.$$

These calculations resulted in an aggregate requirement of 546 PSUs to ensure the precision specified above.

Sample size of the 1995 rural survey was determined using somewhat different background information and statistics than in 1985. The Riverbank Erosion Impact Study (REIS) survey of three areas suggested that only five percent of the households in the interior villages were displacees compared to around 65 percent in the villages on the riverbank. The *char* study by ISPAN (1993) revealed that more than 90 percent of households were displaced at least once during the lifetime of the household head. For this study, relevant components such as manpower, time and financial resources, restricted the maximum sample size to within an eight percent *confidence limit* at a 95 percent *confidence level*. The sample sizes for each of the three strata were calculated as follows: 31 for KZ, 125 for BZ and 54 for CZ. Thus, a total of 210 PSUs were covered during the 1995 rural survey.

Supplementary Survey of Squatters

During the reconnaissance surveys in the study area, a sizeable portion of the displaced households were registered as illegal settlers on the BRBFP embankment (Figure 6.2). These displacees are characteristically landless destitutes, and use the embankment as an immediate shelter, primarily because this land is public property and is elevated above the surrounding areas (Figure 6.2). The population on the embankment possesses unique characteristics relative to the rest of the population, variations which suggest a significantly different perspective on displacement. As such, a study of these rural

Figure 6.2. Squatter settlements of displacees on the Brahmaputra Right Bank Flood Protection Embankment. Occupying public land illegally is the only option for most of the erosion-disaster victims.

squatters offers further insights into the dynamics of displacement and the magnitude of hazard impact.

In order to substantiate the findings of the main rural survey of 1985, a supplementary survey was conducted among the displacee-squatter settlements residing on the embankment. The sampling of these displacees represented a departure from the above sampling technique. Since these units were biased, both in terms of their geographical location and their socioeconomic characteristics, they could not be incorporated in the basic frame of the main survey, which was constructed on a *multi-stage* unbiased framework; an alternative procedure was therefore followed. A census enumeration of these squatters conducted in March-April of 1985 recorded a household population of 720 units. Considering the resources available for enumeration, a sample of 10 percent was chosen, which produced a random sample of 72 squatter PSUs.

6.1.4. SOME QUALIFICATIONS

Despite the careful design and implementation of the surveys to minimize sampling and non-sampling errors, a few deficiencies became apparent as the surveys proceeded and the preliminary results were viewed. Attempts were made to resolve some of these deficiencies within the scope of the present study. For example, even after pre-testing and a close scrutiny of the final questionnaire, some concepts and vocabulary appeared difficult to communicate accurately to respondents. Despite efforts to adjust to local dialects, there remained undeniable differences between the thoughts and perceptions of the educated investigators and the largely uneducated and illiterate respondents. Some biases may therefore have been incorporated into the study due to respondents' difficulties in understanding and replying to some questions.

A further possible source of error may have arisen from the influence of the interviewers' own opinions,[72] especially in the case of the annual household labor-input variables. There continues to be a deeply rooted 'cultural bias' towards female labor; consequently, women are readily classified as 'housewives' or 'domestic laborers' by male respondents and are usually excluded from an accounting of productive labor. All interviewer for the survey were male, as were the majority of respondents, but since the survey was designed for a closed questionnaire and was administered by homogenous teams of trained interviewers, it can reasonably be expected that such biases were minimized.

Other limitations of the survey can be summarized as follows. The survey did not include any direct question on the size of land holdings that displacee households possessed at the time of last displacement. As a result, the nature and degree of association between current socioeconomic conditions and pre-displacement landholding status cannot be measured directly from the survey data. Also, questions with regard to the value or extent of material/economic loss due to riverbank erosion, especially with reference to the onslaught of the hazard on homestead land, were not addressed in the survey. This lack impedes any direct measurement of hazard loss. Further, the methodology adopted in the survey was inadequate to examine the extent of hazard loss longitudinally or from a temporal perspective. However, the recording of land-loss during the lifetime of displacee respondents provides some scope for indirectly measuring the relationship between pre- and post-displacement socioeconomic conditions. Similarly, the survey of displacees' perceptions of the hazard impact upon the overall well-being of family members provides some general perspectives on hazard loss. Finally, the limit of the spatial boundary of the survey at the *thana* level further constrains inferences from the survey for the nation as a whole.

The qualifications mentioned above, however, do not appear to pose serious drawbacks to the potential scope and contribution of the study. The survey, for the first time, provides a systematic data set on human responses and population displacement associated with riverbank erosion. It also furnishes adequate data for examining the relationships of the hazard phenomenon to pertinent socioeconomic variables. The results of the survey in general, and the specific application of the data to the objectives set forth in Chapter 1, are presented in the remaining sections of this chapter and in Chapter 7.

6.2. The Profile of the Samples: Displacement Characteristics

For the present study, the basic characteristics of respondents are organized according to their *displacement characteristics* and includes three variables: (i) displacement status, (ii) displacement frequency, and (iii) distance and direction of movement. Generally, the effects of hazards and disasters upon sociodemographic variables are assessed in terms of their geographical location; hence for our purposes, we will consider differences in pertinent variables between the three zones.

6.2.1. DISPLACEMENT STATUS OF THE SAMPLE HOUSEHOLDS

The loss of homestead land and the resulting displacement induced by a riverbank erosion disaster causes serious havoc for households. Both short- and long-term family welfare is jeopardized by such extreme events. It is hypothesized that as a consequence of disaster experience, displacee households develop a very different set of characteristics than those of static, non-displacee native population. In order to test such a hypothesis and to isolate the effects of erosion-induced displacement, a distinction between 'displacee' and 'non-displacee' (i.e. the control group) in the samples is indispensable. As noted earlier, 'displacee' respondents are defined as those who have lost homestead land at least once in their lifetime as a consequence of riverbank erosion. 'Non-displacee' respondents are those who have never experienced displacement by homestead land-loss due to river encroachment, although they might have lost cultivable land due to the hazard.

The high incidence of erosion-induced displacement in the study area is substantiated by the survey data (also see Hossain, 1984; Haque and Hossain, 1988). In the 1985 main survey sample (n = 547), 59 percent of respondents were displacees (Table 6.2). This figure is comparable to a recent study in the southern delta (i.e., Hizla *thana* in the district of Barisal) which identified 63.8 percent of the respondents as displaced at least once in their lifetime (Mamun, 1996: 71). The distribution of displacee households by zones indicates that only one household was displaced by bank erosion among the inhabitants of the KZ (i.e., only 1.4 percent). In the BZ, 34 percent of the sample were displacees, while all of the respondents in the CZ (98.8 percent) lost homestead land and were displaced during their lifetime.

The proportion of displacee households was considerably higher in the 1995 rural survey compared to the 1985 survey (Table 6.2). About three-fourths of the respondents were found to have been displaced at least once in their lifetime in the 1995 sample. Such a high proportion may be attributed, first, to the sample design, and second, to the increasing occurrences of riverine disasters in the region over the last decade. A sharp contrast between the KZ and the CZ was observed since none of the respondents from the KZ were ever displaced while all of the sample households in the CZ lost homesteads due to river encroachment. In the Bankline Zone (BZ), a large majority were also displacees.

TABLE 6.2. Displacement characteristics of respondents by zone, 1985 and 1995

Displacement Characteristics	1985 Survey									
	Zones						Total (Main Survey)		Squatter Settlements	
	KZ		BZ		CZ					
	f	%	f	%	f	%	f	%	f	%
Displacees	1	1.4	77	33.9	244	98.8	322	58.9	72	100.0
Non-Displacees	72	98.6	150	66.1	3	1.2	25	41.1	-	-
Total (n)	73	100.0	227	100.0	247	100.0	547	100.0	72	100.0
	1995 Survey									
Displacees	nil	-	100	80.0	54	100.0	154	73.3		
Non-Displacees	31	100.0	25	20.0	nil	-	56	26.7		
Total (n)	31	100.0	125	100.0	54	100.0	210	100.0		

The above variations in the proportions of displaced population among the different zones primarily reflect the frequency and scale of river encroachment on inhabited land in the floodplain. Each of the three zones possess unique locational characteristics with regard to river channels, landforms, and associated variables such as composition of soil or vegetation cover. The KZ is characterized by old and relatively stable settlements because of the absence of extreme flooding and erosion threats, particularly to agriculture and livelihood. The mainland areas along the bankline of the river contain both permanent settlers and displacees; the latter have migrated from the eroded land of the BZ or from nearby *char* communities. In the *char* areas, people are always on the move from one place to another due to the frequent occurrence of the riverine hazards. Multiple displacement experiences may therefore be assumed. The 1985 supplementary survey confirmed displacee status of all the squatter households residing on the *bandh* (i.e., the BRBFPE).

6.2.2. RECURRENCE OF DISPLACEMENT

Information on displacement history was collected from the displacee household heads. The recurrence of displacement per household in Kazipur in both the 1985 main sample and 1995 sample was high, showing an average displacement of seven and five times respectively (see Figure 6.3). A study of *Hizla* thana in Barisal district (located in the lower Meghna floodplain) records that, on average, the sample households (N = 381) had been displaced three times (Mamun, 1996: 71). One of the reasons for a higher displacement frequency in Kazipur than in Hizla is that most of Kazipur's territory consists of the adynamic Brahmaputra-Jamuna braided river system. In addition, the complex social structure in Kazipur does not generally permit the destitute to move long distances, and thus forces them to live in high-risk locations. This particular situation will be elaborated in Sections 7.1 and 7.2.

The *mean* displacement frequency was remarkably high among the CZ sample compared to that of the BZ—eight and three times in 1985 and ten and three times in 1995 respectively. However, for 1985 the patterns in the variation of this variable within each zone appeared to be relatively similar—these patterns are expressed in the *coefficients of variation* (by converting the *standard deviation* to a percentage of the *mean*; see Table 6.3).

The distribution of displacement frequency by zones is shown in Table 6.3. A bimodal distribution of the displacement frequency is evident; the uppermost displacement frequency group (i.e., 13 or more times) has a higher proportion compared to the group just below it (i.e., 10-12 times). This pattern is more apparent in case of the 1985 CZ, whereas a declining proportion is evident for the BZ. Also, the *mode* of displacement in the CZ is 4-6 times, accounting for 30.3 percent of the respondents in this zone. These patterns together confirm the hypothesis that displacement is more prominent in the CZ than in other zones of the floodplain. It is noticeable that among the 1985 supplementary survey about two-thirds of households

Figure 6.3. This extended family has been displaced twelve times in the past fifteen years. Frequent displacement is mainly influenced by either the lack of affordability to move to a distant place or a willingness to stay in the vicinity of the river to gain benefits from the area.

had displacement experiences from one to three times—this indicates that a majority of squatter households on the embankment were recently displaced. Overall, the distributional patterns in displacement frequency for Kazipur shows that the people affected by erosion do not or cannot usually opt for change in resource use or location during their initial experiences; instead, they are likely to bear the loss or at least attempt to minimize its potential.

6.2.3. DISTANCE AND DIRECTION OF INVOLUNTARY MIGRATION

In studies of migration, one of the most significant concerns for social scientists is the distance moved by migrants. Recently, this aspect has received considerable attention (Johnston, 1976; UNESCO, 1982; Cernea, 1991). One such study, carried out in India by UNESCO (1982) is of particular interest. Designed to determine the characteristics of rural migration, this research provides us with a valuable touchstone since its sociopolitical and cultural environment is comparable to our present study. The UNESCO study reports that more than two-thirds of the total migrants remained within

TABLE 6.3. Distribution of displacement frequency by zone, 1985 and 1995 (%)

1985 Survey

Displacement Frequency Group	Zones BZ n=77	Zones CZ n=244	Total (Main Survey) n=321	Squatter Settle-ments n=72
1 - 3 Times	75.3	21.7	34.6	63.9
4 - 6 Times	14.3	30.3	26.5	19.4
7 - 9 Times	6.5	20.1	16.8	8.3
10 - 12 Times	3.9	11.5	9.6	2.8
13 Times and Over	-	16.4	12.5	5.6
Total	100.0	100.0	100.0	100.0
mean	3 Times	8 Times	7 Times	4 Times
c.v. (%)	87.6	99.4	109.2	86.6

1995 Survey

	n=100	n=54	n=154
1 - 3 Times	86.0	5.6	57.8
4 - 6 Times	9.0	9.3	9.1
7 - 9 Times	5.0	18.5	9.7
10 - 12 Times	nil	24.0	8.4
13 Times and Over	nil	31.5	11.0
Cannot Recall	nil	11.1	4.0
Total	100.0	100.0	100.0
mean	3 Times	10 Times	5 Times

the district boundary; inter-district migration accounted for another one-fifth of total migrants while the rest moved across state boundaries. Libbee and Sopher (1972), supporting these findings, further reveal that the majority of intra-rural migrants in India moved a distance of 12 miles (about 20 kilometers) or less. More, they conclude that the preference for short-distance migration is common except for the following: (1) "those induced by disasters which force people to go to bigger urban centres where relief is expected to be better organized"; (2) those organized by employers; and (3) the voluntary movement of members of affluent families for education and employment. These trends generally parallel those identified by the present study.

The *mode* of the distance category lies in the one- to five-mile range both in 1985 and 1995, confirming the tendency to short-distance moves by the displacees (Table 6.5). This pattern, however, changes when disaggregated by zone.[73] For instance, while the highest proportion of households in the CZ moved between one to five miles (36.1 percent) in 1985 and between a half and one mile in 1995 (49.0 percent), a relatively smaller proportion is reported in these categories among the BZ households (22.1 percent in 1985 and 19.0 percent in 1995). The largest proportion in the BZ moved one quarter to half a mile. It is evident that people in the *char* areas move to relatively more distant locations after displacement compared to inhabitants of the mainland zones. Among the squatters on the embankment, more than 55 percent moved between half a mile and five miles during their last displacement. The people who migrated for a distance of more than a mile to take shelter on government land were mostly displacees from *char* areas. During the survey it was noted that people displaced from mainland areas tended to move only very short distances.

As shown in Table 6.4, less than three percent of total displacees in both 1985 and 1995 migrated a distance of more than five miles. This pattern is typical of the inhabitants of both zones. The respondents indicated that their tendency to remain in the vicinity of their previous residence was usually influenced by two factors. First, a considerable proportion of them could not afford to move greater distances because of their inability to bear the necessary transport costs and the initial expenses at the destination. Second, the majority of displacees had strong belief that their land would reemerge soon, allowing them to return to their place of origin in the near future (Figure 6.3). In most cases, however, such hope soon became despair (see Haque, 1986b: 13).

Although information about the people who moved out of Kazipur due to erosion is not available directly from the survey, it is possible to gather some insights into general migration trends by considering the destination chosen by relatives of displacees. Out of a total of 321 displacee-respondents (excluding the one displacee-respondent of KZ) in the 1985 survey, 199 (i.e., 62 percent) reported that their relatives had also been displaced in recent years and had been forced to move to new locations (see Table 6.4). From Table 6.4, it is evident that the majority of the displacees' relatives (40.5 percent) chose to move to rural areas outside Serajganj district.

Surprisingly, only eight respondents of the 1985 survey reported that their relatives took shelter on the embankment. While 720 households were enumerated on the embankment during the survey, this low frequency may be attributed to under-reporting by respondents. Such under-reporting is likely to be influenced by the possibility that a confession of having destitute relatives might hamper their social status and prestige (i.e., *maan-samman*) and the future social and matrimonial activities of the family.

The 1985 main survey data revealed that about one-sixth of relatives of respondents' remained within the *thana*, but that only a small proportion of them remained within the district of Serajganj. These patterns, in aggregate, indicate that displacees generally tend to remain within close proximity of their place of origin, or alternatively they tend to

TABLE 6.4. Distance moved by displacees during last displacement, 1985 and 1995 (%)

Distance in Miles	1985 Survey: Zones BZ n=77	1985 Survey: Zones CZ n=244	Total (Main Survey) n=321	Squatter Settlements (1985) n=72
Less than 1/4 mile	23.3	13.1	15.5	22.2
1/4 - 1/2 mile	32.5	27.4	28.7	20.8
1/2 - 1 mile	22.1	20.5	20.9	27.8
1 - 5 miles	22.1	36.1	32.7	27.8
More than 5 miles	-	2.9	2.2	1.4
Total	100.0	100.0	100.0	100.0
mean	1,684 yards	2,590 yards	2,410 yards	1,780 yards
c.v. (%)	100.5	124.5	125.4	92.2
1995 Survey	n=100	n=53[a]	n=153	
Less than 1/4 mile	22.0	5.7	16.3	
1/4 - 1/2 mile	41.0	15.1	32.0	
1/2 - 1 mile	19.0	49.0	29.5	
1 - 5 miles	15.0	28.3	19.6	
More than 5 miles	3.0	1.9	2.6	
Total	100.0	100.0	100.0	
mean	1,758 yards	2,503 yards	2,016 yards	

[a] One respondent could not recall the item.

move to rural areas some distance away where, presumably, jobs and higher wages are available. Furthermore, people who moved to urban centers and to destinations outside the country were all from the *char* areas; their frequency is very low, and therefore one cannot infer much from this last trend to larger moves.

In choosing to remain close to their place of origin, displacees attempt to reduce the amount of cognitive restructuring necessary to adapt to a new environment (Wallace, 1957). In case of the Peruvian earthquake victims, Oliver-Smith (1982: 102) finds that survivors tend to be 'conservative' to spatial mobility—"the conservatism is predominantly a defensive stance against the incursion of future stress." The continuity of economic support and sociocultural identity is of vital concern to such hazard victims, particularly in non-industrialized societies.

Zaman (1987), however, rightly points out that this 'cultural-psychological' explanation is not sufficient, and hence the study of such short-distance migration requires investigation into other underlying factors. He notes that "it is not conservatism but the hope to regain access to depositional land that remains a key factor in the migration decision" (Zaman, 1987: 17). The displaced households who do not have their own land to resettle, or who cannot expect any material support from their poor relatives, have their choices limited to one of the two following options: to move as squatters to the embankment, or to remain in the *char* areas 'dependent' on one of the powerful landlords not only for land and other necessary support for resettlement, but also for employment, as low-paid laborers (see Zaman, 1987: 14).

The short-distance migratory behavior of displacees is related to a number of factors; principally: (a) availability of land for resettlement; (b) support from relatives and friends; (c) possibility of maintaining closer ties with their *samaj*[74]; and (d) patronage from landlords and local leaders (see Zaman, 1987; 1991b; Wiest, 1991). Indeed, the interplay of hope and despair with reemerged land, as well as the related sociopolitical factors, are the most crucial influences on migratory behavior in this riparian environment of the Brahmaputra-Jamuna floodplain. Displacees in general have a strong belief that their land will reemerge soon, allowing them to return to their ancestral homes. "My land is right there! I shall return to my land when it reemerges," one respondent commented, drawing my attention to the Brahmaputra-Jamuna River. Such hopes soon disappear, however, in the face of the local power structure and the prevailing landownership systems. Access to newly emerged *char*land by the disadvantaged displacees and by small landowners is limited; they consistently lose, while large landowners gain control through the means of power and violence that govern the economic life of riparian areas (Adnan and Monsoor, 1976; Zaman, 1987; 1989; 1991; Wahed *et al.*, 1988; Ahmed and Jenkins, 1991). An extended discussion of these processes will be presented in Section 7.2. In sum, the short-distance migration behavior in Bangladesh's riparian environment must be viewed in terms of "the social and cultural dynamics of adjustment to river erosion hazard" (Zaman, 1989: 200). When a household is displaced, the decision as to where to migrate is not just one of individual choices for most displacees; rather, it is a choice influenced by a complex set of factors.

CHAPTER 7

IMPACTS OF RIVERBANK EROSION DISASTER
Understanding Differentials in Rural Socioeconomic Characteristics

The three principal resources in Bangladesh are people, land and water. As such, the majority of the population make their living on small land holdings as owner-occupiers, tenants, sharecroppers, or as landless laborers. Although cultivation of at least two crops per year is common in most parts of the country, a rural household receives only limited economic returns from the land. Such a land-welfare imbalance is caused by the fact that the average rural landholding is less than 0.7 hectares per household and the majority of the rural population are functionally landless (Jannuzi and Peach, 1980).

Given the small scale of most landholdings, any loss of land by a household is devastating. While frequent severe flooding creates extreme shocks on household welfare by affecting cropping regimes and damaging or destroying standing crops, river encroachment into settled villages permanently takes away the only source of livelihood—the land—from myriads of households. The changing river courses, either suddenly or through gradual bank erosion, thus generate a process of forced migration among the potential and actual victims of the disaster; involuntary migration is accentuated by frequent abnormal flooding in the region. The socioeconomic impacts of riverine hazards are thus one of the major contributors to the impoverization and marginalization of the rural peasantry.

In this chapter we will consider the impacts of riverbank erosion hazard on the socioeconomic characteristics of floodplain inhabitants. By examining the degree of variations between displacees and non-displacees (i.e., control group), we will attempt to find out whether such differences are statistically significant enough to be able to relate such variations to riverine hazards. The human impacts of riverbank erosion are multidimensional. They may mean simply a separation of individuals from their social groups, such as *gosthi*[75] or *samaj*, or in the extreme, involve the loss of human lives. Needless to say, the threats and the actual erosion disasters begin to affect household socioeconomic conditions when households lose land or are forced to move to a relatively safer area. The loss of land and displacement of floodplain occupants are, consequently, the two principal variables that need special attention.

7.1. Land-Loss and Population Displacement

7.1.1. A NATIONAL PERSPECTIVE

Since the early 1970s, Landsat data have been used to monitor land and water resources in Bangladesh (Hoque, 1983; Bristow, 1987; Haque and Zaman, 1989; ISPAN, 1993). Most of these attempts, however, have concentrated on the broader regional detection of river course and land-use changes, or have been confined to planform changes of water and land. The further use of Landsat data for economic and community development planning has thus remained neglected, despite its capacity to detect changes over time in settlement, population, vegetation and crop cultivation. The nature and extent of river channel changes, in the Bengal delta in general and in the Brahmaputra-Jamuna floodplain in particular, have been discussed in Chapter 4.

As elaborated in Chapter 5, historically, floodplains of the major rivers of Bangladesh have been more densely settled than the other interior zones. More recently, the 1991 national population census has indicated that the density is higher in the central and southeastern districts of the country (BBS, 1995). Thus, every year the encroachment of river channels has put a large proportion of the population of Bangladesh at risk of losing their land and means of livelihood. Currey (1979), employing the 'Delbecq-Delphi Process,'[76] identifies that IN 66 out of the 495 *thanas* of the country, people are vulnerable to riverbank erosion hazard; for comparison, the detected *thanas* are shown with the density of population per square kilometer (Figure 7.1). An estimated 19 million rural people residing on an area of 20,155 square kilometers are at risk from bank erosion in the Brahmaputra-Jamuna and Ganges floodplains alone.

The FAP 3.1 and FAP 16 studies provide relatively precise and reliable statistical figures on forced migration of floodplain inhabitants due to riverbank and island erosion. These recent investigations reveal that:

> The completely unprotected active floodplains of the Jamuna, Ganges, Padma and Meghna are inhabited by 4.3 million people, of whom 2.2 million live on charland. Erosion caused over 728,000 people to be displaced over the period 1981-93, and in the same period 462,000 permanently emigrated from the high erosion risk areas (FPCO, 1995: 6).

The FPCO (1995) reports that the 4.3 million inhabitants of the *chars* in the major river systems comprise the most vulnerable group in Bangladesh. These people frequently face the havoc of abnormal *monsoon* inundation, a threat which often convinces them to abandon their houses and move to safer ground. Then the magnitude of property and resource losses are so severe that most disaster victims cannot recover. For many displacees, a temporary move can turn into a permanent displacement. In addition, "as about 9,000 ha of mainland and 5,000 ha of charland are reworked each year by erosion" (FPCO, 1995: 7), over 60,000 people become landless due to bank erosion problems along the main rivers of the country.

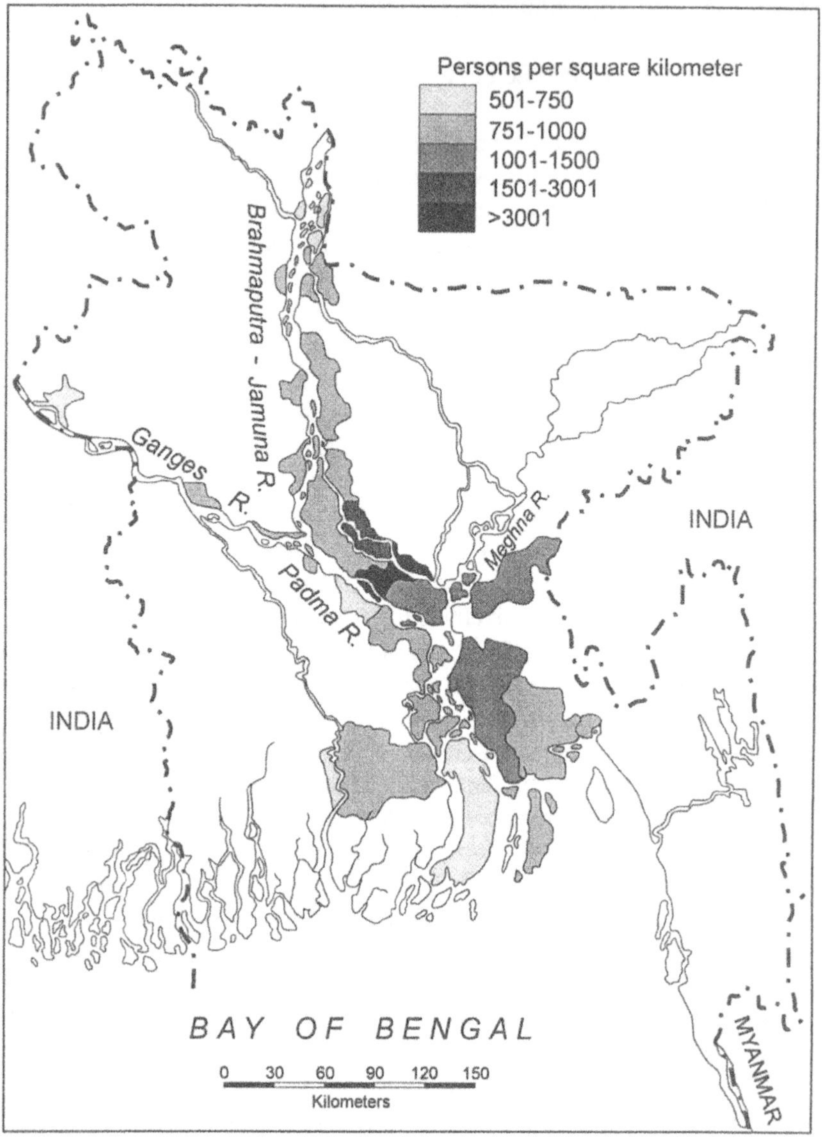

Figure 7.1. Population density in areas (*thanas*) liable to riverbank erosion (after Currey, 1979).

7.1.2. THE LOCAL PERSPECTIVE: KAZIPUR

The above national perspective provides a context for local population displacement and land-loss. The location and amount of eroded and accreted land at the local level is recorded by the Department of Land Revenue of the Government of Bangladesh. However, due to the lack of necessary physical infrastructure, workers and equipment, such surveys are seldom completed; hence, the data on land-loss and accretion are usually unavailable from existing official sources, both at regional and local levels. Remote sensing, nonetheless, offers an opportunity to overcome this problem as it allows us to estimate the amount of land-loss and land accretion over a given period.

Although there are no year-to-year statistics on the amount of erosion and accretion activities in Kazipur *thana*, the Thana Land Revenue Office provides some *mouza*-level information generated by periodic surveys for the years 1977-78, 1983-84, and 1984-85. Table 7.1 presents data on the extent of erosion and accretion in Kazipur during 1977-85. According to the Revenue Office, there are 117 *mouzas* in Kazipur *thana* with a total area of 13,633 hectares, including the existing water bodies. The total amount of land within the *thana* changes almost every year with new erosion and accretion of land. In 1977-78 alone, Kazipur lost 20 entire *mouzas* to the Brahmaputra-Jamuna river channels—a total of 3,986 hectares disappeared, amounting to 30 percent of the total area of the *thana*. In addition, 22 *mouzas* were partially affected during the same period. The amount of land accreted in the same year was 3,707 hectares. This implies that 93 percent of the land lost in 1977-78 was regained through accretion of new land from previous deluvion areas.

TABLE 7.1. Erosion and accretion activities in Kazipur, 1977-1985

Year	Number of *Mouzas* Wiped Out	Land Lost (hectares)	Number of *Mouzas* Partially Affected[a]	Accretion Land (hectares)
1977-78	20	3,986	22	3,707
1983-84	8	1,629	27	3,454
1984-85	12	1,933	3	1,771
Total	40	7,548	52	8,932

[a] Data on land lost by the partially affected *mouzas* were not available.

Source: Haque and Zaman (1989: 305).

One must, however, view these statistics with some caution—the process of erosion and accretion is a very complex phenomenon with a random character. In such a situation, a closely monitored Landsat data series provides a better scope for depicting the loss and gain of land. By using Landsat data of Multi-Spectral Scanning (MSS) bank 7, the present study attempts to estimate the amount of land-loss and the volume of migrating population in Kazipur. The selected Landsat images were enhanced to the scale of 1:50,000, then superimposed by the use of transparent overlays; estimates of land eroded and accreted were made by the application of the Lasico digital planimeter.[77] The required technical assistance was provided by the Space Research and Remote Sensing Organization (SPARRSO), based in Dhaka.

The interpretation of the superimposed images reveals that, out of the eleven unions of Kazipur, seven were affected by bankline shifting between 1973 and 1980 (Figure 7.2). Not surprisingly, among these affected unions, five were from the *char* zones (see Table 7.2). Calculations show that about 26.4 percent of the total area (3,935 hectares) of these affected unions was eroded during the study period, while only 13.6 percent of the total area of these affected unions (2,032 hectares) was accreted between 1973 and 1980. Furthermore, some six percent (904 hectares) of the area contained water in both 1973 and 1980. The images also show that the proportion of accreted land was nominal in the case of the unions located on the west bank of the river (i.e., Maizbari and Kazipur) (Figure 7.2). The detection of eroded and accreted land, in aggregate, shows that 10.6 percent of the total area of Kazipur was eroded while only 5.5 percent reemerged from the riverbed between 1973 and 1980.

The impact of such land-loss in the floodplain areas of Bangladesh involves primarily the loss of homestead land, housing structures, crops, cattle, trees, and household utensils. Thus, in contrast to floods and other pervasive types of riverine hazards, bank erosion generally leaves little scope for immediate recovery for marginal households. Without question, the loss of a homestead forces people to move to a new location. Such sudden disruption to their usual livelihood causes enormous economic, social, and psychological costs for displacees, as well as for non-displacees in the recipient society (Haque, 1986a; Haque and Zâman, 1989).

For the present study, attempts were made to assess these costs from displacees' own experience, both in the 1985 main survey and the 1995 rural study. In both surveys, a significant proportion of respondents (i.e., more than 40 percent) listed three common effects of erosion-induced displacement which influenced the well-being of family members: (i) a severe cut in their standard of living, (ii) the occurrence of mental illness in the family, and (iii) the loss of valuable assets (Table 7.3). Also, more than a quarter of the total respondents in both surveys indicated the perpetual deterioration of physical health and consequent illness in their family in association with displacement. A comparison of 1985 and 1995 survey results further reveals that in the latter, a considerably higher proportion of displacees experienced the loss of economic assets and social position (Table 7.3). Overall, the deterioration in economic conditions and the increase of psychological stress, both due to erosion disaster-induced displacement, cause severe degradation of both the short- and long-term well-being of affected families.

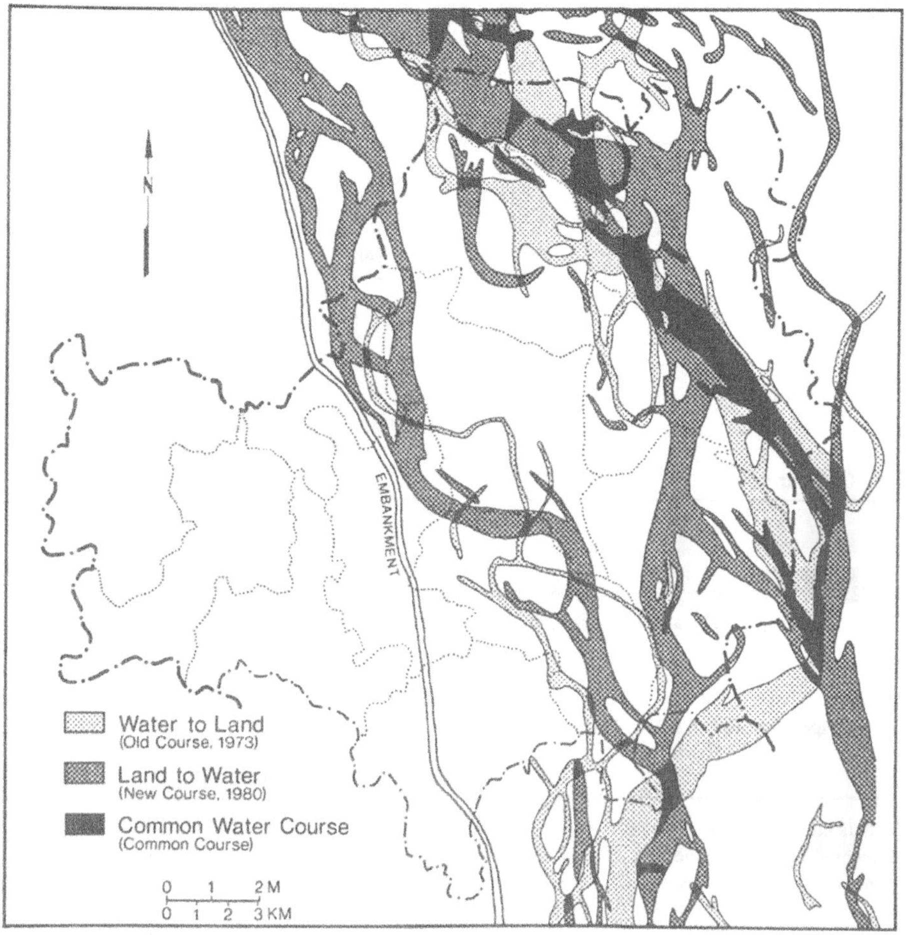

Figure 7.2. Change detection map of the Brahmaputra-Jamuna River in Kazipur *thana*, 1973-1980 (based on Landsat images; Haque and Zaman, 1989).

As noted above, due to the lack of any systematic efforts by governmental and non-governmental agencies to collect vital statistics and other data on migration at the regional or local levels, it is impossible to estimate the total volume of natural hazards-induced migration in the study area. However, using an indirect method of estimation,[78] we can calculate that a total of 18,032 persons moved from their usual place of residence in Kazipur during 1974-1981 (see Table 7.4). Of these, 11,759 moved out from their own *mouza* but stayed within the same union, while 6,273 moved to other *thanas*.

TABLE 7.2. Detection of land eroded and accreted in Kazipur between 1973 and 1980

Name of Union	Zone	Total Area (hectares)	Cumulative Area Eroded (hectares)	Cumulative Area Accreted (hectares)	Area With Water Both in 1973 and 1980
Maizbari	BZ	1,256	412	95	40
Kazipur	BZ	1,358	308	40	-
Khas Rajbari	CZ	1,785	665	339	58
Tekani	CZ	3,304	948	292	26
Nischintapur	CZ	2,578	603	245	87
Char Girish	CZ	3,448	684	864	680
Natuarpara	CZ	1,184	316	157	13
Total of Affected Unions		14,918	3,936	2,032	904
Total (Kazipur *thana*)		37,066			

TABLE 7.3. Distribution of perception of impacts of displacement (excluding land-loss) on family well-being, 1985 and 1995 (%) (multiple response possible)

Impacts on Family Well-Being	1985 Main Survey n=322		1995 Survey n=154	
	Frequency	%	Frequency	%
Serious cut in standard of living	171	53.1	62	40.3
Hampered children's education	17	5.3	25	16.2
Increased illness in family	82	25.5	49	31.8
Mental illness in family	161	50.0	67	43.5
Loss of valuable assets	145	45.0	147	95.5
Loss of social position	36	11.1	83	53.9
Fragmented *bangsha*	22	6.8	9	5.8
Others	9	2.8	6	3.9

The above figures suggest that about two-thirds (65.2 percent) of all migrants moved from their usual place of residence but stayed within the same union; the rest (34.8 percent) moved out of the *thana* during the study period. We can infer that the majority of these migrants are natural hazards-induced displacees, since the unions under consideration all experienced severe river encroachment and land-loss during the study period (see Table 7.2). The majority of them moved only a short distance, but about one-third chose to move out of this hazardous *thana*, presumably to a relatively safer place. This observation substantiates the earlier findings (presented in Section 6.2) on the distance of the last move by displacees. The effects of land-loss due to river encroachment and the consequent population displacement among floodplain communi-

TABLE 7.4. Migration of population in Kazipur by union, 1974-1981

Name of Union	Total Migration	Net Inter-Union Migration	Net Intra-Union (i.e., Inter-*Mouza*) Migration
Maizbari	8,362	-7,925	437
Kazipur	1,715	3,519	5,234
Khas Rajbari	1,334	-636	698
Tekani	1,539	884	2,423
Nischintapur	1,430	-963	467
Char Girish	2,247	-752	1,495
Natuarpara	1,406	-400	1,005
Kazipur (Total)	18,032	-6,273	11,759

Source: Haque and Zaman (1989: 307).

ties can be seen further in the varying socioeconomic status of displacees compared to non-displacees. Of the numerous socioeconomic variables which are affected by land-loss and forced migration induced by riverbank and island erosion, only the most pertinent characteristics are selected for the study; these include: a) landownership status, b) aggregate income, c) value of assets, d) educational attainment, e) type of labor-force activity, f) type of occupation, and g) household size.

7.2. Accretion, Power and Violence: The Dynamics of Impoverization

The term *lathiyal* originated from *lathi*, the bamboo stave which is the peasant weapon in Bengal (Figure 7.3). *Lathiyals*, historically, has referred to the 'private army' employed by the *zamindars* and *talukdars*. The *lathiyal*, in contemporary usage, is a 'clubman' or 'hired thug' of the *talukdar* who assists him in violent fights in order to gain control over new *char*land (Ali, 1980). In this section we will examine the nature and influence of the existing sociopolitical systems with focus given to the *lathiyali* system, in the context of *char* or accretion land settlement in the Brahmaputra-Jamuna floodplain. The analysis draws from Zaman's (1988) anthropological field study, conducted as part of the REIS project in the village of Char Chashi—an island village of peasants—between June 1984 and July 1985; we will also look to non-ethnographic data from press and published reports. Village Char Chashi lies within the extreme western channel of the Brahmaputra-Jamuna along with many other *char* villages. This village is one of the ten *mouza* villages of Natuarpara *union* under Kazipur *thana*. The 1920 Cadastral Survey of Bengal recorded Char Chashi with three map sheets (Jurisdictional Land-JL number 38, Sheets 1, 2, and 3), a detail that indicates the large size of the island. Population settlement in this area began around 1940, reaching total household numbers of 85 by 1974 and 255 by 1981. During the field survey of 1985,

Figure 7.3. Villagers in Kazipur celebrating a ritual with weapons such as machetes, harpoons, and bamboo batons. *Jotedars* and their followers frequently use such weapons in violent clashes with other groups.

however, only 205 households were registered, mainly due to land-loss during 1981-85. In the past forty years the village site had to be shifted several times: in 1951 massive erosion devoured nearly half of the *mouza* land; in 1983 another one-third was lost to the river; and in 1995, only a parcel of the village settled by 110 households remained unencroached. All of the 205 households were surveyed by the REIS socioeconomic questionnaire in 1984-85.

7.2.1. THE *LATHIYALS*, THE LANDED ELITES AND THE STATE

As noted earlier, there are 2.2 million inhabitants of the *char*lands in the major river systems of the country (FPCO, 1995). As more than 5,000 hectares of *char*land are eroded each year, tens of thousands more people become landless and homeless. But shifting river channels of the braided system cause not only erosion and loss of settlement, they also open up frontiers of settlement in new depositional land. This dynamic riparian environment historically has promoted the development of two institutions. The first is a set of land laws regulating ownership and access to emergent *char*land and the conflicts arising over it; the second refers to a social institution known as the *lathiyali* system—a hierarchical peasant-lord structure which controls accretion land through the means of power and violence. By this system, the locally powerful *talukdars* and *jotedars*[79] use their patron-tied dependents as *lathiyals*. In the past, the combined force of ownership rights and violence among competing local *talukdars* to control newly emerged land has played a pivotal role in the resettlement and adjustment process, and still remains an enduring force in the social structure of the riparian villages (Mukerjee, 1938; Nicholas, 1962; Ray, 1979).

Settlement and ownership rights over accretion land in the riverine areas have always been complicated by the difficulty in ascertaining ownership of new land, whether it should belong to the state or to some other riparian proprietor upstream or even to somebody on the other side of the river. Consequently, inhabitants on both banks of the river have become very quickly involved in conflicts, which in most cases end in battles and bloodshed over the disputed land (see Figure 7.3).

Zaman (1987; 1988) assesses 40 cases relating to land conflict and organized violence in *char* villages, stories which appeared in the daily newspaper *Ittefaq* over a period of five years, 1978-1982. Of these 40 cases, 11 cases occurred in 1981 alone, while 22 (56 percent) were reported from the southern active deltaic districts of Barisal and Noakhali. The new land being formed led to conflicting claims by "land grabbers." In the violent clashes over the five-year period (1978-82), a total of 733 persons reportedly died; the numbers of injured and 'missing' were 96 and 123 respectively. In the eight cases to which the use of *lathiyals* was linked, Zaman notes that a total of 5,600 *lathiyals* were used and 1,308 houses were burnt to ashes. He writes:

> The extent of brutality and terror cannot possibly be described unless someone has lived and experienced it in char life. It is a different and difficult life out there. What appeared in the daily *Ittefaq* should be considered only some "glimpses" of the horror of everyday life in char villages. It is said that "violence is the name of the game in char life (Zaman, 1987: 4).

The grip of local landed elites on *char*lands is sustained with the assistance of networks of political 'agents' and their dependent *lathiyals*. Since such private armies of *lathiyals* are illegal in the country, the *talukdars* need to use strong measures to protect them from arrest and conviction by the state. Already, several studies have confirmed the association of land ownership with political power at the village and state levels (Rahman 1981; Blair, 1978; Jones, 1979; Ahmad and Jenkins, 1991; Jahan, 1980). Jahan (1980) reports that in the two national parliamentary elections held in Bangladesh in 1970 and 1973, more than a quarter of the members elected were large landowners in the countryside.

The regime in the capital, Blair (1978: 70) claims, has been absolutely dependent for its survival on the 'kulak class' of farmers to generate unrest and master votes in election times. At the village level, the landed elites provide a network of support for their urban counterparts made up of dependent sharecroppers, clients, and armies of retainers. In return, public policies designed in urban centers are tailored to assist and protect the interests of the large landowners. The evolution of the legislative measures meant to administer alluvion and diluvian land clearly demonstrates this reciprocal relationship.

Traditionally, ownership and disputes over alluvion and diluvian land were regulated by local custom and titles whereby previous owners had access to newly depositional land (Gupta, 1940; Kabir, 1969). From the first legislative enactment for the regulation of such disputes in 1825, and continuing until 1972, the rights of previous owners to *char*land were recognized. The Presidential Order of 1972 significantly changed the rights of tenants on accretion land, proclaiming that newly emerged *char*land previously lost by diluvion would now be taken over by the state (i.e., Government of Bangladesh) as *khas*land. The stated objective was to redistribute newly emerged land among landless peasants for cooperative or collective farming and settlement.

The content of the new legislation was not well received, particularly by the powerful landed elites, as it introduced a potential threat from landless groups. Under immense political pressure, a new amendment was enacted in 1975 (Ordinance LXI of 1975) which defined the owner of land lost by diluvion as one who is eligible 'first' to the settlement of *khas*land, thus reopening access to land lost by erosion. This move by the state virtually reinstituted the bands of local 'chiefs,' providing them with opportunities to manipulate and deprive smallholders and landless peasants of access to depositional land.

7.2.2. THE CASE OF *CHAR* NATIPARA, KAZIPUR

The Char Natipara is a well-known case in Kazipur *thana*. It is a complex and long-standing dispute in which the ownership and control of the *char* has changed hands several times in the last four decades. The events all commenced in the 1950s following the abolition of the *zamindari* system. The original owner, Rani Dinamoni of Tangail, made settlement of the *char mahal* (estate) with some of her local agents who later began disputing each other for larger shares. Bahar Haji of Rehai Suriber took advantage of the situation and recorded a large proportion of the estate in his and his followers' names during the 1962 Land Records Survey. Rafiq Talukdar of Char Girish made attempts to regain the land on behalf of the previous owners, a dispute which eventually cost five lives in Char Natipara. The *char*, however, remained under the control of Bahar Haji and his supporters.

In 1974 Hashem Sikdar, a *jotedar* and veteran Awami Leaguer of Bhetua Jagannathpur attacked Char Natipara with 500 *lathiyals*, looting the crops and slaughtering three of Bahar Haji's agents to gain control. After Bahar Haji took Hashem Sikdar to court, Sikdar was arrested, but then released after a couple of hearings for "lack of evidence"; his political connections with the ruling party helped him get away with the murders. Meanwhile, he maintained his grip over the *char* and became the *defacto* owner of Char Natipara. Within a few years Sikdar began intimidating the peasants to extract more surplus from them, building up pressure to the extent that they found him unbearable.

In 1982, when the Serajganj District Survey was being conducted in Kazipur, Rafiq Talukdar, in collusion with settlement surveyors, recorded almost one-half of the total 390 hectares of land in Char Natipara. Through these shady dealings and renewed violence he eventually established his control over a large area. Local sources revealed that in 1984 Talukdar arranged the murder of the young pregnant wife of Barkat Ali—his own dependent in the *char*—and then filed the murder case against Hashem Sikdar in a bid to take over the rest of the *char*land. This incident, however, did not frighten Sikdar and his supporters. Sikdar was in police custody for three weeks in 1985, but the case was yet undecided in Serajganj Court during the time of the REIS field survey in 1985.

7.2.3. VILLAGE SOCIAL STRUCTURE, POLITICS AND PATRONAGE

Landownership in *char* villages is more polarized than in mainland communities; about 45 percent of the households in Char Chashi have absolutely no agricultural land of their own. This is almost double the Kazipur mainland figures (24 percent). In addition, over half of the 205 households (the 1985 survey) in the village are smallholding peasants who own between one to three acres of land. During our fieldwork it was extremely difficult to collect landownership data on *char* villages, especially when dealing with those who were allegedly involved in the local land conflict.

The case of Rahim Chakladar illustrates this issue clearly; as a typical *jotedar*, he is one of the most powerful agents in violent local and regional land grabbing. On different occasions Chakladar provided to the field investigator conflicting figures for the amount of land he owns. Chakladar stated:

> You may hear many different stories about me; I do not own very much land, maybe about 30 acres. But you see, you cannot guarantee your ownership in *char*land. It changes almost every year. You may lose all your land tomorrow! You even may not have control over your land if you don't have enough power (Zaman, 1988: 83).

Local informants reveal that Chakladar may 'officially' own 13 hectares but unofficially controls another 60 to 80 hectares of arable land in several *char* villages around Char Chashi and in mainland Kazipur. Chakladar, a violent and effective political 'chief,' maintains his own *lathiyali* troops to enforce and extend his domain over new *chars*. The percentage of landlessness in *char* villages is thus not only induced by environmental factors, it is also a product of the social and political processes at work in the riparian villages.

How then are these political alignments formed and how do they lead to group cohesion? The socioeconomic context of hazards-induced displacement needs attention in order to understand these issues. Displaced households with no land of their own to resettle, or who cannot expect any material assistance from their equally poor relatives, have very limited choices. As most displacees tend to remain within the close vicinity of their original settlements, a large majority of them remain in *chars* as "patron-tied" dependents of *talukdars, chakladars* and other *jotedars*, who provide them free land and necessary support for resettlement. The 1985 survey data indicated that over 50 percent of households in *char* villages live on land provided by their land-rich patrons and relatives in the form of *kot*, or 'free-use'—the 'free-users' are locally known as *uthuli*, meaning helpless and destitute people. As might be expected, the free-use status is itself a mechanism of exploitation and bondage based on so-called generosity as members of the those households constitute a cheap labor pool for the patrons (Haque and Zaman, 1989: 310-311; Zaman, 1991a: 680-681).

The social and economic dependence on the local landed elite is almost absolute, both on the part of those made destitute by erosion, and also by the marginal landholders. The adaptation process following a dislocation in the *char* social environment is entirely controlled by landed elites through the following means: (i) provision of 'free-use' land for homesteads; (ii) regular and often yearly contracted agricultural wage labor (*kamla*); (iii) sharecropping (*bargadari*); and (iv) tenant farming (*chukani*). In return, the *talukdars* and *jotedars* extract considerable surplus wealth annually from 'tribute' and dependent 'clients,' and more importantly, receive a loyalty expressed in terms of physical and emotional participation in violent fights over disputed *char*land. Furthermore, the network of relationships among the village faction leaders extends to many different neighboring villages through a system of alliance with, or opposition to, one or another prominent *talukdar*. These village leaders often work as political 'agents' of *talukdars* to organize the land conflicts are endemic in the riparian villages in the floodplain.

In sum, the use of violence, dispossession, murder, rape and confiscation of crops and animals have become almost established practices of *char* life in Bangladesh (Adnan and Monsoor, 1976; Ahmad and Jenkins, 1991; Wahed *et al.*, 1983; Zaman, 1987; 1988; Haque and Zaman, 1989). The selective use of violence by local landed elites and their dependent *lathiyals* works as the ultimate arbitrator of all disputes over new *char*land. When a *talukdar* succeeds in a violent fight, resettlement takes place, usually by his own support groups and possibly himself and his dependents; in this way the *talukdars* remains the virtual owner of the entire *char*. These aspects of local-level (a village or multiple villages) factionalism and patronage relationships represent systematic sociopolitical adaptations to the entire spectrum of erosion and displacement phenomena.

7.3. Effects of Displacement on Socioeconomic Characteristics

In this section we will examine the hypothesis that riverbank erosion-induced displacement causes significant degradation in socioeconomic conditions. It may be formally stated that *the 'disadvantaged' socioeconomic characteristics of floodplain users are related to displacement experience caused by riverbank erosion occurrences.* The testing of this postulation requires that it be disaggregated into subsets, taking into consideration that: (i) the socioeconomic characteristics of displacees are significantly different from non-displacees (i.e., native population), and (ii) the socioeconomic characteristics of displacees in the squatter settlements (i.e. the most disadvantaged displacees) are significantly different from those of other displacees. In both cases variation is associated with riverbank erosion-induced displacement. As indicated in Section 7.1, survey data on seven selected socioeconomic and sociodemographic variables are employed in testing the above hypotheses: (i) landownership status, (ii) aggregate income, (iii) value of assets, (iv) educational attainment, (v) type of labor-force activity, (vi) type of occupation, and (vii) household size.[80]

A disaggregated analysis of the two subsets of floodplain users (i.e., displacees and non-displacees) is required since each of these groups represents an aggregate perspective. A comparison of 'disadvantaged' sections of displacees with other floodplain users (i.e., relatively 'privileged' displacees and non-displacees as a whole) is emphasized in this disaggregated analysis. The supplementary sample of the present survey then provides a scope for testing the pertinent hypotheses.

7.3.1. LANDOWNERSHIP CHARACTERISTICS

The discussion in Chapter 5 revealed the complex structure of contemporary agrarian relations in Bangladesh. Land is the most important means of survival in an economy predominantly characterized by subsistence agriculture, thus landownership patterns largely determine the economic status of rural households. A distinction between *landownership* and *landholding status* is necessary to isolate the real economic base in terms of property assets and labor input. The former refers to the legal proprietorship, whereas the latter relates to use of land, such as through sharecropping, *kot* or mortgage, *bundok*, and *khas khalasi*.

In the scope of the present study, landownership is considered an appropriate variable because of its direct relevance to riverbank erosion. However, the dynamics involved with the use of land is analyzed through a consideration of aggregate income. This distinction is explained in the following subsection (7.3.2). As Amin (1982: 130) elaborates, very low per capita land use (less than 0.25 acres) along with limited non-agricultural employment opportunities in rural areas makes landownership a highly significant determinant of access to other social entitlements.[81] Floodplain users with direct access to land are usually characterized by better education and occupation, and often send their offspring to the cities and towns for education, jobs, and better living conditions (see Choudhury, 1978; Islam, 1976).

In contrast, displacees who have limited or no access to land, or have little education and few economic assets, are the ones who migrate to the embankment or to the nearby rural shelters for survival as *uthulis* (see Hossain, 1984). Meanwhile, the households with moderately sized farms often become destitute as a result of their direct loss of land due to river encroachment. These middle and small landowners are generally the targets of 'land grabbers' or local *talukdars* (see Wahed *et al.*, 1983; Zaman, 1987). Zaman explains:

> The local talukdars bought land records from . . . people . . . after being displaced and paid very little. The displacees were willing to sell for whatever money they could get, because these records were of no value to them. When asked about such incidents, one talukdar said: How can you pay more? We just buy water with the hope that the land will show up one day. It may take ten, fifteen or more years, who knows! Even I may not get possession. You know, papers don't matter very much. It is power that matters (Zaman, 1987: 16).

In the 1985 data it can be seen that while the landless[82] in rural Bangladesh were 14.7 percent in 1978, their proportion in Kazipur was 24.1 (see Table 7.5). This suggests that landlessness in the study area is much more acute than that shown by the rural average. When the squatters were included in the sample, the proportion of landlessness increased to 30.9.

To examine if there exists any significant variation between the displacees and non-displacees in terms of direct access to land, the distribution of landownership by displacement status is presented in Table 7.6. The results suggest that a disproportionately high incidence of landlessness is caused by riverbank erosion-induced displacements. More than 35 percent of displacees are landless while only eight percent of non-displacees are landless. Ironically, the proportion of large landowners is also higher among displacees than among non-displacees. In landownership status among the supplementary sample, it appears that more than 83 percent of households have no direct access to land; the rest own only very small parcels. Those findings thus support the hypothesis that while a certain section of the displacees become completely landless due to erosion and experience a gradual pauperization, a small but powerful section within the displacee category remains a relatively 'privileged' social class.

Besides the above cross-sectional analysis of the impact of bank erosion-induced displacement on landownership, we also need to examine the magnitude of land-loss at the household level. Such an attempt provides a temporal dimension to the analysis of hazards impact on land. According to the 1985 main survey results, out of the total of 322 displacee households, 55 reported that they had never possessed homestead land (17.1 percent). This suggests that at the time of displacement they were residing on either government land or on private land owned by others. As explained earlier, these *uthuli* usually live with close relatives, friends, or other villagers. In some cases, they take up leases for homestead land on a short-term basis, known locally as *kot*. Further, more than ten percent of all displacee households reported that they had never had any cultivable land (n=33 households). Table 7.7 shows the loss of cultivable and homestead land by displacees during their lifetime. It appears that in the case of both cultivable and homestead land the majority of holdings are small (0.02 - 2.50 acres), which implies that the majority of floodplain users lost small land parcels at the individual level.

TABLE 7.5. Distribution of landownership groups of sample households, Kazipur, 1985 and rural Bangladesh, 1978 (%)

Landownership Groups (in Acres)	Landownership Broad Groups	Kazipur (Main Survey) n=547		Rural Bangladesh* n=12,031,272	
0	Landless	24.1 }	24.1	14.7 }	14.7
0.01 - 1.00		40.8		44.7	
1.01 - 2.00	Small	16.1	66.0	15.2	68.6
2.01 - 3.00		9.1		8.7	
3.01 - 5.00	Middle	6.1	7.9	8.2	11.8
5.01 - 7.00		1.8		3.6	
7.01 - 12.00	Large	1.3	2.0	3.1	4.9
12.01 and over		0.7		1.8	

* According to 1978 Land Occupancy Survey (Source: BBS, 1982: 783).

TABLE 7.6. Distribution of landownership groups of households by displacement status, 1985 (%)

Displacement Status	Landownership Groups (in Acres)			
	Landless (0)	Small (0.01 - 2.50)	Middle (2.51 - 7.50)	Large (7.51 and over)
Displacees (n=322)	35.4	51.9	9.9	2.8
Non-Displacees (n=225)	8.0	78.2	12.9	0.9
Total(Main Survey) (n=547)	24.2	62.7	11.1	2.0
Squatters (n=72)	83.3	16.7	-	-

Row 1 and Row 2: $X^2(p < 0.001; df=3) = 56.30$ (Significant)
Row 1 and Row 4: $X^2(p < 0.001; df=3) = 54.82$ (Significant)
Row 2 and Row 4: $X^2(p < 0.001; df=3) = 159.85$ (Significant)

In contrast to many other hazard phenomena, the process of river channel shifting also produces some outputs; specifically, land reemerges from the riverbed to form *chars*. Repossession of such reemerged land is a complex process in Bangladesh which involves local government agencies, local factional politics (i.e., *doladoli*), and conflicts of social and economic power. It is generally found that the powerful middle and large landowners are the ones able to acquire these reemerged lands. In the sample of the 1985 survey, while less than a quarter of the group losing small amounts of land

TABLE 7.7. Distribution of displacee households by amount of land lost and acquired, 1985

Type of Land	Amount of Land (in Acres)			
	Did Not Own Any Land (*Kot*, Mortgage, etc.)	Small (0.01 - 2.50)	Middle (2.51 - 7.50)	Large (7.51 and Over)
A. LAND LOST (% to total displacee households)				
Cultivable n=322	10.3 (33)	61.5 (198)	20.3 (65)	8.1 (26)
Homestead n=322	17.1 (55)	82.0 (264)	0.9 (3)	-
B. LAND ACQUIRED (% to land-loss size group)				
Cultivable	n.a.	23.2 (46)	49.2 (32)	50.0 (13)
Homestead	n.a.	20.1 (53)	33.1 (1)	-

Figures in parentheses are showing the actual numbers of households.
n.a. = not applicable

acquired cultivable land in their lifetime, about 50 percent of the group losing middle and large size holdings repossessed their land (Table 7.7). Similarly, in the case of homestead land, the proportion of the group acquiring mid-size plots appears to be higher (33 percent) than that for small plots (less than 20 percent; see Table 7.7).

Although the Department of Land Revenue keeps a chronological account of eroded household land (i.e., *Viti Jomi*), in reality, all reemerged land falls into the category of arable rather than homestead land. Repossession of cultivable land is therefore considered by the government only for cross-tabulation of current landownership and land lost (see Table 7.8). About three-quarters of small landowners and the currently landless have lost between 0.01 - 2.50 acres of land. In contrast, 77.8 percent of the currently large landowners have lost more than 7.5 acres of land. Thus, although large landowners have lost larger acreages, they have maintained their large landownership status through their relatively privileged socioeconomic positions. The number of these large landowners is, however, small—only nine households in the main survey sample. Overall, riverbank erosion has been a major factor in the marginalization of the small landowners.

TABLE 7.8. Cross-tabulation of current landownership and land lost by displacee households, 1985 (%)

Current Landownership Categories (in Acres)	Amount of Land-Lost (Cultivable; in Acres)		
	Small (0.01 - 2.50)	Middle (2.51 - 7.50)	Large (7.51 and Over)
Landless (n=93) (0)	76.3	19.4	4.3
Small (n=156) (0.01 - 2.50)	72.4	21.8	5.8
Middle (n=31) (2.51 - 7.50)	41.9	38.7	19.4
Large (n=9) (7.51 and Over)	11.1	11.1	77.8

7.3.2. AGGREGATE HOUSEHOLD INCOME

The analysis of landownership characteristics in the preceding subsection focused only on legally owned land and did not take into account other income sources. Yet personal income of all members of a household can be used as an effective proxy for actual land use, considering that personal income is:

> the flow of income that accrues to the individual or household. The income may be a result of the sale of labour and/or ownership of assets, or it may take the form of a transfer income [income which cannot be regarded as payment for services, e.g., retirement pensions, gift, *zaqat*] . . . There are also some sources of household income which do not arise from directly productive activity, such as government and business *transfer payments* and personal interest income (Pearce, 1986: 324-5; also see Peterson, 1984: 71).

Furthermore, by analyzing income we are able to determine the aggregate economic earnings of household members from all non-agricultural sources and thus provide a comprehensive picture of the economic status of a household unit. It is difficult to obtain total annual income of a PSU in rural communities because of the considerable role of the non-monetized sector. As such, the data on income were therefore collected by item for the following categories: cultivable land, livestock, pisciculture, and other non-agricultural sources. Gifts, loans, and relief assistances were excluded from the accounting aggregate annual incomes.

In general, the average annual income of a household in Kazipur is low, as reflected in the 1985 main survey figure of Tk. 13,347 (approximately US $370; Table 6.16). If the average household size of six persons is taken into consideration, per capita annual income is only Tk. 2,225 (US $62). In order to facilitate a direct comparison of the 1995 data with the 1985 information, the 'current income' figures of 1995 are adjusted according to the consumer price index (CPI) for rural families (with base: 1984-85 = 100; BBS, 1996: 271). From this, the surveys of 1985 and 1995 evidently show that rural household income has deteriorated considerably over the last decade; this is a general trend affecting both displacee and non-displacee categories. Alam

(1992) identifies the principal factors for the perpetual decline of rural income as: (i) an annual population growth rate over 2.6 percent, (ii) a declining role of the agricultural sector in the total GDP, (iii) a skewed landownership pattern, and (iv) an urban bias in development plans. In Alam's words:

> The strategy of rural development in the country at all times did not spell out coherent measures of eliminating the threads that web to trap the rural poor in deprivation. Rural impoverishment in the country is the product of a system with interlocked political, social and economic components (Alam 1992: 86).

To examine the disaggregated patterns in income distributions of the sample households, four broad groups are used: marginal (Tk. 1-6,000), lower (Tk. 6,001-12,000), middle (Tk. 12,001-24,000) and upper (Tk. 24,001 and above). In Kazipur, more than 61 percent of the 1985 sample households—or the marginal and lower groups combined—earn less than the average annual income. Among the rest, only 11 percent belong to the upper income group. In comparison, two lowest income groups account for 82.3 percent of the 1995 total sample households (Table 7.9), figures which demonstrate the gradual pauperization of the rural masses. Both the 1985 and 1995 survey data reveal that the income of displacee households in Kazipur is significantly lower than those of the native households.

These large variations in income distribution between displacee and non-displacee households is seen in the corresponding coefficient of variation (c.v.) of 119.1 and 81.7 percent respectively in 1985 and 152.2 and 113.1 percent in 1995. Differences in the distribution of income between displacees and non-displacees are further substantiated by the fact that in the 1985 main survey the proportion of households in the below-average income groups totals about 65 percent for displacees and 57 percent for non-displacees. Also, only eight percent of displacees have income levels in the upper group, compared to 16 percent for non-displacees. Among the squatters, 93 percent earn less than the average income of the 1985 main sample, while none belong to the upper income group. The 1995 survey data conform to the patterns set in 1985, with the proportion of households in lower and marginal income groups at 86.4 percent for the displacees and 71.4 percent for non-displacees (Table 7.9). That the differences in income status between displacees and non-displacees are not more pronounced can be attributed to the overall decline in rural income during the 1980s and 1990s. As a result of this minimizing effect, the variation in the 1995 income status between the displacee and non-displacee households is found not to be statistically significant.

It is worth employing here the conventional tools for examining inequalities in income distribution—the computation of an *index of concentration* (I).[83] The degree of inequality in the distributions of income within the 1985 sample is indicated by the values I=0.4547 for displacees and I=0.4527 for non-displacees. The inequality in income distribution among floodplain users as a whole, as well as the differentials in income levels by displacement status, are more easily observed when the shares of income are examined by fractile groups. The bottom group, in terms of share of income, consists

TABLE 7.9. Distribution of household income by displacement status (%)

Displacement Status	Household Income (in Taka per annum) Marginal (6,000 or less)	Lower (6,001 - 12,000)	Middle (12,001 - 24,000)	Upper (24,001 and Over)	Average Annual	c.v. (%)
A. 1985 Survey						
Displacees n=322	28.3	36.3	27.6	7.8	12,575	119.1
Non-Displacees n=225	17.8	39.6	27.1	15.5	14,452	81.7
Total (Main Survey) n=547	23.9	37.7	27.4	11.0	13,347	103.3
Squatters n=72	52.8	40.2	7.0	-	6,360	53.6
B. 1995 Survey [based on real (adjusted) income]						
Displacees n=154	68.2	18.2	7.8	5.8	7,900	152.2
Non-Displacees n=56	57.1	14.3	19.6	8.9	9,325	113.1
Total n=210	65.2	17.1	11.0	6.7	8,280	140.6

Row 1 and Row 2: X^2)p <0.01; df=3) = 13.79 (Significant)
Row 1 and Row 4: X^2(p <0.001; df=3) = 25.50 (Significant)
Row 2 and Row 4: X^2 (p <0.001; df=3) = 46.63 (Significant)
Row 5 and Row 6: X^2)p <0.05; df=3) = 6.98 (Not Significant)

of 86 percent of the 1985 sample households and accounts for only 60 percent of total income (see Table 7.10). On the other hand, the upper 12 percent of households share 30 percent of total income. Similarly, only two percent of sample households share ten percent of total income. The degree of variation among displacees and non-displacees is minimal. The overall analysis of income distribution nonetheless supports the correlation between disadvantaged socioeconomic characteristics and displacement experience.

TABLE 7.10. Extent of inequalities in household income distribution by displacement status, 1985

Displacement Status	Share of Income (%) Bottom 60%	Middle 30%	Top 10%	Index of Concentration (I)
Displacees n=322	85.8	13.0	1.2	0.4547
Non-Displacees n=225	83.9	13.4	2.7	0.4527
Total (Main Survey) n=547	85.8	12.4	1.8	0.4547
Squatters n=72	47.2	48.6	4.2	0.2808

7.3.3. ASSETS OWNED BY HOUSEHOLD MEMBERS

An asset is defined by economists as "an entity possessing market exchange value, and forming part of the wealth or property of the owner" (Pearce, 1986: 20). However, for our purposes, and given the nature of the agrarian economy of the study area, only 'real' assets—tangible resources (like rickshaws, bullock carts, boats, tubewells, radios, plows) rather than financial assets (such as money, bonds, and equities)—are taken into account. As mentioned in Chapter 5, the rural economy of Bangladesh is largely based on subsistence agriculture and is often characterized by non-monetized exchange (also see Chowdhury, 1987). The real asset holdings in such an economy not only play an economic role in terms of mortgage, *kot*, loan, and other exchanges, but also provide socioeconomic status through their 'prestige value.' Although the latter phenomenon is a cultural one, it plays a major sociopolitical role in rural Bangladesh. So in order to transform the value of assets into a measurable form, the average market value at current prices (i.e., 1985) for each asset is assessed.

Overall, about half the sample did not have any economic assets while the majority of the remaining households possessed assets valued at less than Tk. 1,000 (US $28). Thus, less than one fifth of households had assets worth more than Tk. 1,000. A comparison of displacee and non-displacee households by a chi-square test indicates that the difference in assets between these two categories is not significant (Table 7.11).[84]

The percentage distribution in assets in 1985, however, reveals that more than 50 percent of displacees possessed no assets, compared to 43 percent of non-displacees. Also, the proportion of assets with a value below Tk. 5,000 (US $140) was higher among the non-displacees (49.8 percent) compared to the main survey displacees (38.9 percent). In comparison, more than 80 percent of squatter settlements did not have any economic assets while only one squatter household was found to have assets valued in the range of Tk. 5,000-10,000 (US $140-278); the balance possessed assets of less than Tk. 5,000. A comparison of assets among squatter settlements with the 1985 main survey households indicates that differences are statistically significant at the 0.001 level (see Table 7.11).

TABLE 7.11. Distribution of household assets (at current price) by displacement status, 1985 (%)

Displacement Status	Value of Assets (in Taka)				
	Have No Asset	Less Than 1,000	1,001-5,000	5,001-10,000	10,001 and Over
Displacees n=322	50.6	30.5	8.4	9.6	0.9
Non-Displacees n=225	43.1	37.8	12.0	6.7	0.4
Total (Main Survey) n=547	47.5	33.5	9.9	8.4	0.7
Squatters n=72	80.6	9.7	8.3	1.4	-

Row 1 and Row 2: $X^2(p < 0.05; df=4) = 7.27$ (Not Significant)
Row 1 and Row 4: $X^2(p < 0.001; df=4) = 24.46$ (Significant)
Row 2 and Row 4: $X^2(p < 0.001; df=4) = 32.31$ (Significant)

7.3.4. EDUCATIONAL CHARACTERISTICS

Literacy in rural Bangladesh is poor, as is reflected in both the 1985 and the 1995 samples. For these studies, educational attainment is grouped into four categories based on years of schooling. These categories are: (1) illiterate (i.e., no schooling), (2) primary and *madrasha*[85] (i.e., 1-5 years of schooling), (3) secondary (i.e., 6-12 years of schooling), and (4) postsecondary (more than 12 years of schooling). About two-thirds of household heads in both the sample-years never attended any school in their lifetime (Table 7.12), while the number of respondents with postsecondary background was minimal (1.5 percent in 1985 and 1.4 percent in 1995). With the exception of the higher proportion of household heads with secondary education in 1995, all other patterns remained almost the same as they were in 1985. The increase in the secondary education of sample respondents can be associated with the commendable success of infrastructural development and a shift in public policy in recent decades towards the promotion of elementary and secondary education (Rahman, 1992: 82-87; Rahman, 1988).

TABLE 7.12. Distribution of educational attainment of household heads by displacement status, 1985 and 1995 (%)

Displacement Status	Educational Attainment (Years of Schooling)			
	No Education (0)	Primary and *Madrasha* (1 - 5)	Secondary (6 - 12)	Post-Secondary (More than 12)
A. 1985 Survey				
Displacees n=225	59.6	24.9	13.8	1.8
Non-Displacees n=322	68.6	17.4	12.7	1.2
Total (Main Survey) n=547	64.9	20.5	13.2	1.5
Squatters n=72	73.6	18.1	8.3	-
B. 1995 Survey				
Displacees n=154	65.6	16.2	16.0	1.2
Non-Displacees n=56	44.6	16.1	37.4	1.9
Total n=210	60.0	16.2	22.4	1.4

Row 1 and Row 2: $X^2(p < 0.05; df=4) = 8.32$ (Not Significant)
Row 1 and Row 4: $X^2(p < 0.05; df=2) = 5.64$ (Not Significant)
Row 2 and Row 4: $X^2(p < 0.05; df=2) = 1.54$ (Not Significant)
Row 5 and Row 6: $X^2(p < 0.05; df=2) = 10.68$ (Significant)

Much of the homogeneity in 1985 can first be attributed to the relatively higher weight given to the proportion of illiterate persons in the overall distribution, and second, to the educational status of households heads might have achieved before displacement took place. In comparison, the benefit of locational stability is clearly mirrored in the 1995 survey data, especially in the relatively low proportion of illiterate persons and the relatively higher share in secondary education among non-displacee household heads as compared to the displacee sample (Table 7.12).

In the context of educational achievement, the 1985 supplementary sample shows distinct contrasts with the above data. The percentage of respondents with no educational background was highest in the squatter settlements (73.6 percent) as compared to both the 'other displacees'[86] and the non-displacee respondents of the 1985 sample (59.6 and 68.6 percent respectively). Similarly, the proportion of squatters in the other educational categories were generally lower than among the rest of the sample, and none attained postsecondary levels of education (see Table 7.12).

In the 1985 survey data, levels of educational attainment among main survey displacee households appear to be higher than that of the non-displacees. This anomaly can be explained by the spatial distribution of educational institutions in Kazipur—most of the primary and secondary schools are located near the bankline, although some were relocated inland after the encroachment of the main river channel on the old *thana* headquarters in the late 1970s.

Yet the examination of educational attainment for household heads provides only a partial picture of the whole population. A study of variations in the complete sample population is likely to give a more comprehensive picture, especially to indicate whether displacement affects children's education. To determine this, the following analysis focuses on all members of the sample households, using only the 1985 survey data, considering its higher reliability. Looking at the example of Kazipur, as in other rural areas of Bangladesh, more than two-thirds of the total population eligible for education in formal institutions (i.e., population five years and over) have never attended school. Furthermore, while 13.1 percent of the male population have attained education above the primary level, only four percent of the female population have attained that same level of schooling. These differences in education by sex are significant at the 0.001 level using a chi-square test and do not appear to be significantly different from the national norm (see Table 7.13).

Although displacement does not significantly affect the educational attainment of household heads (see Table 7.13), significant impacts occur for the displacee population as a whole. Tests for the distribution in educational attainment of displacees and non-displacees reveal that the variations among them are statistically significant at the 0.001 level (see Table 7.14). The rate of literacy among the non-displacee population appears much higher (37.9 percent) than among either the displacee population (30.1 percent) or the squatters (26.8 percent). Also, the percentages for all categories of educational attainment are higher among the non-displacee population than for other floodplain users. Generally, the differences in educational characteristics appear to be statistically significant, which strongly supports the hypothesis that bank erosion-induced displacement significantly affects the educational opportunities of floodplain users.

7.3.5. LABOR-FORCE ACTIVITY

For the analysis of occupational characteristics the household heads and the population of the PSUs (primary sampling units) are first placed into manpower[87] categories. In the main surveys—for 1985 and 1995 respectively—about 83 and 87 percent of total respondents were employed while another 14.1 and 7.1 percent did not earn wage-income while still contributing to the household economy (Table 7.15). As might be expected from traditional bias, this latter section of respondents was comprised mostly of females. With further examination of the survey data, the distribution of manpower

TABLE 7.13. Distribution of educational attainment of population by displacement status, Kazipur, 1985 and Bangladesh, 1981 (%)

Displacement Status	Sex	Educational Attainment (Years of Schooling) of Population 5 Years and Over			
		No Education (0)	Primary & Madrasha (1 - 5)	Secondary (6 - 12)	Post-Secondary (More than 12)
Kazipur (1985 Main Survey)	Male n=1,351	57.0	28.9	13.1	1.0
	Female n=1,279	77.1	19.3	3.6	-
	Both Sexes n=2,630	66.8	24.2	8.5	0.5
Bangladesh (Rural)*	Male n=31.0 million	62.2	24.1	13.2	0.5
	Female n-29.9 million	78.7	17.1	4.1	0.1
	Both Sexes n=60.9 million	70.2	20.7	8.7	0.4

Row 1 and Row 2: $X^2(p < 0.001; df=2) = 145.52$ (Significant)
* According to 1981 Census (Source: BBS, 1984a: 192-208).

categories among household heads by displacement status provides strong evidence of significant variations between paired groups: displacees versus non-displacees; displacees of the 1985 main survey versus squatter displacees; and non-displacees versus squatter displacees (see Table 7.15). These data suggest that displacement affects manpower distribution quite significantly.

For a valid chi-square test of the 1985 survey data the 'economically inactive' persons are grouped together. This results in two manpower categories: 'economically active' (i.e., employed) and 'economically inactive' (i.e., unemployed, students, housewives, and unpaid household workers). Comparisons of the paired groups show significant variations in the distribution at the 0.01 level, which confirms an adverse effect of displacement upon employment status. These data further show that in 1985 the proportion of employed persons was higher among the non-displacee household heads (88.9 percent) than among both main survey displacees (78.3 percent) and squatter displacees (only 62.5 percent; see Table 7.15).

A comparison between the 1995 displacee and non-displacee samples reaffirms the significant difference in the labor-force activity of household heads. Evidently, more of the displaced household heads are active in income generating activities than their non-displacee counterparts, a situation reflected in the significantly higher proportion of students among non-displacees than displacees (14.3 and 2.6 percent respectively). The

TABLE 7.14. Distribution of educational attainment of population (5 years and over) by displacement status, 1985 (%)

Displacement Status	Educational Attainment (Years of Schooling)			
	No Education (0)	Primary and *Madrasha* (1 - 5)	Secondary (6 - 12)	Post-Secondary (More than 12)
Displacees n=1, 578	69.9	21.5	8.2	0.4
Non-Displacees n=1, 052	62.1	28.4	8.8	0.7
Total (Main Survey) n=2, 630	66.8	24.2	8.5	0.5
Squatters n=299	73.2	19.1	7.7	-

Row 1 and Row 2: $X^2(p < 0.001; df=2) = 19.04$ (Significant)
Row 1 and Row 4: $X^2(p < 0.05; df=2) = 1.35$ (Not Significant)
Row 2 and Row 4: $X^2(p < 0.001; df=2) = 13.08$ (Significant)

Source: BBS (1984a: 192-208; 1984b: 645-649).

most severe effects of erosion disaster upon work status are demonstrated by the highest rate of unemployment among the squatters (8.3 percent), among whom also no respondent claimed student status. In addition to the economic hardship of living in a flood-prone area, the loss of self-esteem is a major reason for non-attendance in schools for individuals from squatter communities.

7.3.6. OCCUPATIONAL CHARACTERISTICS

The occupations of individuals in the production system of a society reflect the organization of and the relations within the labor process. Through an understanding these patterns, occupational categories can be used effectively to examine differences between the economic activities of displacees and non-displacees.[88] Table 7.16 shows the distributions of economically active respondents in 1985 and 1995 according to major occupational categories by displacement status.

Significant differences in occupational status by displacement category may be attributed to downward occupational mobility caused by bank erosion-induced displacement (Table 7.16). For instance, more than 60 percent of employed non-displacee household heads were engaged in farming and other primary-production activities both in 1985 as well as in 1995. The corresponding figure for the 1985 main survey displacees was 49 percent, which dropped to 22 percent for the 1995 survey

TABLE 7.15. Distribution of manpower categories of household heads by displacement status, 1985 and 1995 (%)

Displacement Status	Manpower Categories			
	Employed	Unemployed	Student	Unpaid Household Workers
A. 1985 Survey				
Displacees n=322	78.3	2.8	0.3	18.6
Non-Displacees n=225	88.9	3.1	0.4	7.6
Total (Main Survey) n=547	82.6	2.9	0.4	14.1
Squatters n=72	62.5	8.3	-	29.2
B. 1995 Survey				
Displacees n=154	89.6	n.a.[a]	2.6	7.8
Non-Displacees n=56	80.4	n.a.	14.3	5.4
Total n=210	87.1	n.a.	5.7	7.1

[a] n.a. = not available

Row 1 and Row 2: $X^2(p < 0.01; df=1) = 10.42$ (Significant)
Row 1 and Row 4: $X^2(p < 0.01; df=1) = 7.88$ (Significant)
Row 2 and Row 4: $X^2(p < 0.001; df=1) = 26.30$ (Significant)
Row 5 and Row 6: $X^2(p < 0.005; df=2) = 10.56$ (Significant)

displacees, and to only nine percent for displacees on the embankment (1985 supplementary sample). This observation indicates that as a result of bank erosion, a considerable proportion of floodplain users are uprooted and separated from their land and the associated agricultural activity; these displacees thus have limited scope for self-employment. This assertion is substantiated by the finding that in 1985 a considerably higher proportion of casual and wage labor[89] was reported by squatter household heads (68.9 percent of the total employed) as compared to both "other displacees" (37.3 percent) and non-displacees (20.5 percent). Similarly, the proportion of casual and wage labor was more than double among the displacee households than those of non-displacees, according to the 1995 survey data (Table 7.16).

TABLE 7.16. Distribution of occupation (primary) of household heads (employed) by displacement status, 1985 and 1995 (%)

Displacement Status	Major Occupational Categories: Farmers (Self-Employed) and Other Primary-Production Workers	Manufacturing Workers	Commerce/ Banking and Service Sector Workers	Laborers (Including Unclassified Workers)
A. 1985 Survey				
Displacees n=252	46.8	3.6	12.3	37.3
Non-Displacees n=200	60.5	7.5	11.5	20.5
Total (Main Survey) n=452	52.9	5.3	12.0	29.8
Squatters n=45	8.9	6.7	15.6	68.9
B. 1995 Survey				
Displacees n=138	21.7	29.0[a]		49.3
Non-Displacees n=46	67.4	13.0[a]		19.6
Total n=184	33.2	25.0[a]		41.8

[a] Manufacturing and commerce/business and service workers were combined as non-agricultural workers in the 1995 survey.

Row 1 and Row 2: $X^2(p < 0.001; df=3) = 17.78$ (Significant)
Row 1 and Row 4: $X^2(p < 0.001; df=1) = 15.63$ (Significant)
Row 2 and Row 4: $X^2(p < 0.001; df=1) = 41.45$ (Significant)
Row 5 and row 6: $X^2(p < 0.001; df=3) = 35.42$ (Significant)

Comparatively, in the 1985 survey data only 12 percent of household heads were engaged in the commerce/business and the service sectors, and only 5.3 percent in the manufacturing sector. Like other rural areas of the country, the lack of capital and physical infrastructure limits industrial entrepreneurship in Kazipur. Commercial activities are essentially confined to small-scale enterprises like retailing, petty trading, hawking, or boat renting (see Haque and Hossain, 1988).

In terms of overall involvement in economically gainful occupations in Kazipur, it is important to examine the gender dimension. The higher percentage engaged in manufacturing in Kazipur (17.8 percent) relative to rural Bangladesh as a whole (7.2 percent according to 1983-84 National Labour Force Survey) can be attributed mainly to the involvement of males in this sector (+11.9; see Table 7.17). In comparison, the calculated deviations reveal a higher rate of female involvement in primary-production activities in Kazipur (41.7 percent) relative to other rural areas of the country (14.6 percent). The deviations further show that in the mid-1980s a lower proportion of females were employed in the service sector in the study area (only 8.3 percent) compared to rural Bangladesh (36.5 percent). However, a caution is required in drawing inferences for Kazipur females since in the sample only 12 women reported to be employed. Considering the low share of female household heads, the entire employed population 10 years and over was included in the 1985 main survey.

TABLE 7.17. Distribution of occupation (primary) of employed population by sex, Kazipur, (1985) and Bangladesh, (1984) (%)

Area	Sex	Major Occupational Categories[a] of Employed Population (10 Years and Over)			
		Agriculture	Manufacturing	Trade and Commerce	Service
Kazipur (the 1985 Main Survey)	Male n=661	66.0	17.5	7.7	8.8
	Female n=12	41.7	33.3	16.7	8.3
	Both Sexes n=673	65.5	17.8	7.9	8.8
Rural Bangladesh[b]	Male n=22.8 million	72.8	5.6	10.7	10.9
	Female n=1.5 million	14.6	30.0	18.9	36.5
	Both Sexes n=23.3 million	69.1	7.2	11.2	12.5
Deviation	Row 1 from Row 4	-6.8	+11.9	-3.0	-2.1
	Row 2 from Row 5	+27.1	+3.3	-2.2	-28.2
	Row 3 from Row 6	-3.6	+10.6	-3.3	-3.7

[a] Data for wage laborers were not available for rural Bangladesh, hence this category was excluded from occupation categories.

[b] Calculated from BBS (1984b: 143,162).

The relatively skewed distribution of local resources, particularly of land, brings a higher proportion of female employment to agriculture. Because of marginal socioeconomic conditions among the 'disadvantaged households,' females in these families are often forced to work in the field. As noted above, opportunities for work in the service sector are very limited in the study area. However, the difference in the distribution of female employment in Kazipur compared to national rural conditions may be partly influenced by variations in the sampling techniques of the two surveys; the former covers only a small fraction of the population of Kazipur *thana* (i.e., less than two percent) while the latter draws a 20-percent sample from the entire population of the country.

In any case, it is important here to cross-examine the above results through temporal analysis. Changes in occupation resulting from displacement involve both temporal and spatial dimensions, and have important bearings upon current levels of household income and other entitlements. Obviously, while settlement in a new location involves many different adjustments for a household, the question of occupational recovery is one of the most vital factors in coping with hazard loss. Cross-tabulation of the respondents' past (i.e., prior to last displacement) and present primary occupations can provide a precise picture in this respect, as shown in Table 7.18 from the 1985 survey results. Evidently, cultivation and other primary production, including animal husbandry and fishing, dominate the past occupations of most displacee household heads (61.5 percent). This is followed by unpaid household work (14.9 percent), and then by general labor (13.3 percent). Non-agricultural occupations, such as manufacturing, commerce, and service, account for minimal proportions (only 5.3 percent), while the proportion of those 'not economically active' is also less than five percent.

Comparing the distributions of past occupation with present occupation, some interesting patterns can be seen. With the effects of displacement and other social changes, the proportion of people in cultivation and primary production activities declined considerably: from 61.5 percent to 37.3 percent between the time of the displacees' last move and the time of the survey in 1985. Persons who had been engaged in primary production and had direct ownership of land became uprooted. Riverbank erosion was therefore a prime determinant in such changes. Further, the rapid marginalization of peasants can be seen through changes in other occupational categories among household heads. While the laborer category accounts only for 13.3 percent of total household heads and ranks third in past occupation, it ranks second as the present primary occupation (28.9 percent); Table 7.18 shows that a significant proportion of former cultivators (26.3 percent), students or unemployed persons became laborers following their displacement. As a result of this ongoing marginalization process, real income and other sociocultural entitlements among affected households have deteriorated considerably.

7.3.7. HOUSEHOLD SIZE

The general household size in Kazipur has already been introduced in Section 6.2. Earlier research suggests that hazard-prone inhabitants and hazard victims in developing societies usually maintain an extended family or group coherence to share their losses

TABLE 7.18. Cross-tabulation of the past and present occupation of household heads (primary), 1985 (%)

Past Occupation Categories	Present Occupation Categories					
	Self-Employed Farmers	Non-Agricultural Production Workers	Laborers (Including Unclassified Workers	Unpaid Household Workers (Including Housewives)	Students, Unemployed and Others	Total
Self-Employed Farmers (n=198)	55.6	8.6	26.3	5.5	4.0	100.0 (61.5)
Non-Agricultural Production Workers (n=17)	17.6	76.5	5.9	-	-	100.0 (5.3)
Laborers (Including Unclassified Laborers; n=43)	9.3	4.6	79.1	7.0	-	100.0 (13.3)
Unpaid Household Workers (Including Housewives; n=48)	2.1	2.1	6.2	85.4	4.2	100.0 (14.9)
Students, Unemployed, and Others (n=16)	12.5	37.5	28.9	18.3	3.4	100.0 (5.0)
Total (n=322)	37.3	12.1	28.8	18.3	3.4	100.0

Figures in parentheses show the percentages for the last column.

to meet their labor needs during emergencies (see Kates *et al.*, 1973). Such conditions among bank erosion-induced displacees are reflected in the data on household size.

As shown in Table 7.19, in both 1985 and 1995 more than 85 percent of displacee households had more than four members; the respective proportion for non-displacees was 76.5 percent in 1985 and 85.7 percent in 1995. This preference for large families among people in the *char* zone is reflected by the 11 percent of CZ households with 10 or more persons in each unit in 1985 (Table 7.20)—almost all of these households (98 percent) were displacees. Obviously, the necessary economic support for such people can only be provided by the wealthy land proprietors. The 1995 survey data also reveals that the proportion of households with more than seven members was higher in the CZ than those in the KZ and BZ (Table 7.20). A comparison between the 1985 and the 1995 survey data indicates that the variation between displacee and non-displacee household size is significant at the 0.02 level for 1985, whereas this difference levels off by 1995. One of the reasons for this change is the decline in 'smaller' household size (1-3 members) among the native, non-displacee population, which dropped from 23.5 percent in 1985 to 14.3 percent in 1995.

Resulting from bank erosion, fragmentation of relatively 'disadvantaged' households leads to a smaller household size among squatter settlements. The proportion of households with three or less persons is highest (27.8 percent) among these groups compared to either displacee or non-displacee households in the 1985 main survey (14.9 and 23.5 percent respectively). This fragmentation of households is further substantiated by the very small proportion of large households (with more than 10 persons) among the squatters. A chi-square test for the distribution of households by size between squatter settlements and the main survey displacees confirms a significant variation between them at the 0.05 level (see Table 7.20). This finding is indicative of a wide range of variation in certain demographic variables, especially those related to the socioeconomic context of displacees.

7.4. The Model for Regressing Displacement Status on Household Income

The variations analyzed through the cross-tabulation of data in the preceding section are 'zero-order' differences computed without controlling other variables. Although the zero-order differences in income and landownership status between displacees and non-displacees are found to be statistically significant, the patterns may not appear the same once all pertinent variables are controlled. In the following section we will use a multivariate regression analysis for the simultaneous evaluation of different possible explanations.

TABLE 7.19. Distribution of household size by displacement status, 1985 and 1995 (%)

Displacement Status	Household Size (Number of Persons)			
	(1 - 3)	(4 - 6)	(7 - 9)	(10 or More)
A. 1985 Survey				
Displacees n=322	14.9	50.6	23.6	10.9
Non-Displacees n=225	23.5	42.3	27.6	6.6
Total (Main Survey) n=547	18.4	47.2	25.3	9.1
Squatters n=72	27.8	43.0	27.8	1.4
B. 1995 Survey				
Displacees n=154	14.9	55.2	26.0	3.9
Non-Displacees n=56	14.3	48.2	28.6	8.9
Total n=210	14.8	53.3	26.7	5.2

Row 1 and Row 2: $X^2(p < 0.02; df=3) = 10.73$ (Significant)
Row 1 and Row 4: $X^2(p < 0.05; df=2) = 6.83$ (Significant)
Row 2 and Row 4: $X^2(p < 0.05; df=2) = 0.83$ (Not Significant)
Row 5 and Row 6: $X^2(p < 0.05; df=3) = 2.48$ (Not Significant)

7.4.1. THE MODEL

The model assumes that aggregate household income is a function of the following factors: displacement status, household size, age of the household head, amount of land inherited, number of household housing structures, distance between homestead and river bankline, value of assets, amount of land owned, level of educational attainment by the household head, and labor-force status of the household head. Some other similar variables may also contribute to variations in aggregate household income. Nonetheless, the objective of the present investigation is to concentrate only on the principal variables available in the data set. Thus the model is not an exhaustive one; rather it is limited to determining the significant factors influencing the dependent variable (i.e., income).

TABLE 7.20. Distribution of household size by zone, 1985 and 1995 (%)

Zone	Household Size (Number of Persons)			
	(1 - 3)	(4 - 6)	(7 - 9)	(10 or More)
A. 1985 Survey				
KZ n=73	23.3	34.2	34.3	8.2
BZ n=227	20.3	47.5	24.7	7.4
CZ n=247	15.4	50.6	23.0	10.9
Total (Main Survey) n=547	18.4	47.2	25.3	9.1
B. 1995 Survey	16.1	54.8	22.6	6.5
KZ	15.2	54.4	25.6	4.8
BZ	13.0	50.0	31.5	5.6
CZ				
Total	14.8	53.3	26.7	5.2

The estimated relationship can be expressed in the following form:

$$I_j = f(M_j, HS_j, A_{ij}, LI_j, S_j, D_j, AH_j, LO_j, E_{ij}, L_{ij}, \text{random errors}) \qquad ......(7.1)$$

I_j denotes the aggregate annual income of all members of household j; M_j the displacement status of household j; HS_j the size of household j; A_{ij} the age of household head i in household j; LI_j the amount of land inherited by all members of households j; S_j the number of housing structures in homestead j; D_j the distance between homestead j and river bankline; AH_j the value asset owned by all members in household j; LO_j the amount of land owned by all members of household j (excluding inherited land); E_{ij} the educational status of household head i in household j; and L_{ij} the labor status of household head i in household j.

As can be seen among the explanatory variables in the equation (1), displacement status of household (i.e., floodplain user unit) is the main focus of the model. This variable has two dimensions: households are either displacee or non-displacee. As the multiple regression analysis requires variables with interval or ratio-level measurements, a dummy variable is created to represent a nominal form of the variable (viz., displacement status, education status, and labor status). Since dummy variables are represented with arbitrary metric values of 1 and 0, they are treated as interval variables in the regression procedure. The inclusion of all the dummies constructed from a given variable is likely to render the normal equations unsolvable,[90] requiring the exclusion of one of the categories of dummy variables. Indeed, in that case, the excluded category becomes a reference by which the effects of the other categories of dummy variables are evaluated.

In the present research, the non-displacee category is excluded from the equation to represent the reference category. Similarly, persons with less than five years of schooling are used as the reference category in the educational status variable, and persons who are not engaged in economic activities as laborers are the reference category in the labor status variable. These adjustments are made in consideration of the limitations of the primary data of the study.

7.4.2. THE EMPIRICAL RESULTS

The Ordinary Least Square (OLS) results are presented in Table 7.21. As shown in Table 7.21 the Coefficient of Multiple Determination (R^2) is reasonably high (0.63) when all independent variables are included in the estimated relationship. The model is also fully supported by the F-ratio which is significant at the 0.001 level (see Table 7.21). After controlling other explanatory variables, the individual effects of the independent variables are in the expected direction, as represented by the corresponding slopes upon the dependent variable. Furthermore, the amount of land owned by the members of a household is seen to be the most significant predictor of household income, an observation substantiated by the level of significance (0.0001) and the corresponding standardized slope of 0.4117, which is the highest among all the standardized coefficients. As indicated by the standardized coefficient (0.2292) for the 'value of assets of household,' this variable appears to be the next important predictor of household income. In rural Bangladesh assets provide a scope for mortgage, *kot, khas khalashi, bundok,* and other economic exchanges; thus, in turn, assets become positively correlated with income.

Landownership is gained by floodplain users through a number of means: inheritance, personal purchase, lease from the government, and illegal occupation. Among these means, inheritance has a special significance because it changes landownership status directly without any changes in income or other household characteristics. Thus, in order to evaluate this special case the amount of land inheritance and the amount of landownership are included in the model as two separate categories. The significant status of inherited land suggests the prevalence of the feudal relationships by highlighting the magnitude of intergenerational property transfer in the community. The significance of the land inheritance variable is at the 0.03 level and the standardized estimate of slope is 0.016; these figures imply that income significantly depends upon the amount of land inherited from older generations, while remaining independent of other means of gaining landownership.

Household income also depends on household size, one of the two demographic variables included in the model. In contrast to conventional findings, the relationship

TABLE 7.21. Regression estimates of displacement status and socioeconomic variables on household income, 1985

Independent Variables	Slope	Level of Significance for Slope	Standardized Estimate	Percent of Variance Explained in Dependent Variable	F-Statistics	Level of Significance for F-Statistics
M_j (Displacement status)	-3826.43	0.0005*	-0.1249	63%	65.56	0.0001
HS_j (Household size)	781.46	0.0021*	0.1366			
A_{ij} (Age of the head household)	-54.93	0.1275	-0.0530			
LI_j (Land inherited)	12.99	0.0262*	0.1554			
S_j (Housing structures)	202.19	0.6787	0.0195			
D_j (Distance between homestead and bank-line)	-0.09	0.6395	-0.0164			
AH_j (Assets)	1.44	0.0001*	0.2292			
LO_j (Landownership status)	23.61	0.0001*	0.4117			
E_j (Education)	-1399.69	0.1824	-0.0445			
L_{ij} ("Labor" status)	576.67	0.6893	0.0139			
Intercept	7279.42	0.0002				

* Significant at the 0.05 or lower level.

between household size and income in Kazipur is positive. This actually reflects the livelihood of the locality: able bodies are needed in these communities as a survival strategy for disadvantaged households, and alternatively as a means of power and authority for the landed *talukdars*.

Meanwhile, other independent variables did not factor significantly in the state of household economy. These include: distance between the homestead and the bankline of the river, age of the household head, number of housing structures on the homestead, education exceeding five years of schooling, and labor status. Already, a recent study has indicated that education is not a significant factor in explaining the control of resources and manpower in rural Bangladesh (Amin, 1982). Also, because of limited occupational diversity in rural areas of the country, particularly in non-agricultural sectors, education may not help to enhance one's income. In effect, educational status becomes an insignificant factor for explaining variations in income.

In the above analysis, when economic variables (landownership, land inherited, occupation, assets), sociodemographic variables (education, household size, age of the household head) and geographic variables (distance and hazardousness of an area) are controlled or held constant, the displacement status *per se* is found to be linked to the lower income of floodplain users. How then can this relationship be explained? Consider that displacement hampers the ability to earn by damaging the normal personal and social life of the household. As shown elsewhere (Haque, 1986b; 1992), increased psychological stress, serious health problems, social inhibition, and fragmentation from social groups such as *samaj* or *gosthi*, severely affect families displaced from their birthplace or usual residence. Though not generally quantifiable, the sociocultural and psychological effects of involuntary or forced migration are thus embodied in the OLS results for displacement effect.

Based on the above observations, a refined model will eliminate the insignificant variables and strengthen our previous findings. The estimated relationships of equation (1) can be modified to read:

$$I_j = f(M_j, HS_j, LI_j, AH_j, LO_j, \text{random errors}) \qquad \ldots\ldots(7.2)$$

The results of this modified regression are presented in Table 7.22. The Coefficient of Multiple Determination (R^2) of this modified model remains the same (0.62), even after the exclusion of five insignificant explanatory variables. In other words, compared to the earlier model, the present five explanatory variables account for the same amount of variance in the dependent variable. Also, the regression coefficients for the independent variables do not change to any considerable degree after the exclusion of insignificant variables from the regression equation. Notably, the level of significance for the slope of the displacement status shifts from 0.0005 to 0.0001 when the insignificant variables are removed from the model (see Tables 7.21 and 7.22).

Overall, the OLS results confirm that displacement induced by riverbank erosion considerably reduces the level of household income, and this inference is statistically

TABLE 7.22. Modified regression estimates of displacement status and socioeconomic variables on income

Independent Variables	Slope	Level of Significance for Slope	Standardized Estimate	Percent of Variance Explained in Dependent Variable	F-Statistics	Level of Significance for F-Statistics
M_j (Displacement status)	-3838.22	0.0001*	-0.1253	62%	128.19	0.0001
HS_j (Household size)	740.02	0.0004*	0.1293			
LI_j (Land inherited)	12.93	0.0255*	0.1547			
AH_j (Assets)	1.49	0.0001*	0.2367			
LO_j (Landownership status)	23.97	0.0001*	0.4780			
Intercept	4761.02	0.0003				

* Significant at the 0.05 or lower level.

significant. It implies that the welfare of floodplain users is significantly dependent upon their displacement status when all other independent variables are controlled.

7.5. Conclusions

The main thrust of this study has been to determine the impact of natural hazards on household income in rural Bangladesh. In particular, we have examined the case of riverbank erosion hazard in Kazipur *thana* of the Brahmaputra-Jamuna floodplain. Based on the results of three pertinent hypotheses concerning the relationships between household income and displacement induced by natural hazards, the findings may be summarized in five major points. First, the average annual income of households living in areas prone to natural hazards is lower than the national average. Next, the income status of natural hazards victims (in this case, riverbank erosion-induced displacees) is significantly different from those of non-victims; this implies that natural hazards experience has an important bearing upon the socioeconomic conditions of floodplain households. Third, the OLS results confirm that displacement induced by natural hazards does cause a significant reduction in income among floodplain users, independent of other socioeconomic and demographic variables. Fourth, the disadvantaged hazard victims (i.e., the lower socioeconomic groups) were more severely affected by disaster and experienced a steadier decline in their income than the privileged hazard victims; if this is taken into account, we can conclude that the degree of inequality has widened further in the hazard-prone floodplain and coastal areas than in other regions of the country. Finally, inequality in the distribution of landownership and the extent of landlessness are both higher among hazard victim groups than among non-victims.

The above conclusions add new considerations for planners and policy makers—specifically, that without addressing the natural hazard problems in the floodplain zones of Bangladesh, success in development efforts can only be minimal. Nonetheless, these conclusions are based, first, on a particular area of the country and on a limited number of sample units, and second, on a cross-sectional data set. Moreover, only riverbank erosion hazard is analyzed. To surpass the limitations of these findings would require an extension of the research, including comparable areas and different types of geophysical and biological hazard effects.

CHAPTER 8

COPING RESPONSES OF FLOODPLAIN USERS IN RURAL KAZIPUR

Human coping responses to natural hazards consist of two primary components: perception and overt behavior (Chapter 3). Furthermore, we have seen that variations in these responses depend on socioeconomic-structural, experiential, and locational factors—in other words, their social, individual and geographical dimensions. This study postulates that perception and adjustment behavior of floodplain users with varying degrees relate to all of these three factors simultaneously. The following analysis thus focuses on both individual and social-collective responses to riverine hazards and involves three variables: (i) socioeconomic structure; (ii) previous experience of displacement due to river encroachment; and (iii) hazardousness of current residence by zone.

The hypotheses to be tested state:

(i) That (a) perception of and (b) measures to cope with riverbank erosion hazard are associated with an individual's *position in the socioeconomic structure*. The variables relating to socioeconomic structure are: (i) educational status; (ii) income level; (iii) landownership size; and (iv) occupational status.

(ii) That (a) perception of and (b) measures to cope with bank erosion hazard are related to an individual's *previous experience* of disasters. Experience of erosion disaster specifically refers to displacement by loss of homestead and is measured in terms of displacement frequency.

(iii) That (a) perception of and (b) measures to cope with bank erosion hazard are related to *potential hazardousness of an area*. The three zones are assumed to reflect different levels of hazardousness with respect to risk of river encroachment on land.

We first look to the characteristics of perception as they will later create the basis for formulating coping strategy and behavior. The test results are presented in Section 8.3.

8.1. Patterns in Perception of Hazards

8.1.1. PERCEPTION OF HAZARDS AMONG FLOODPLAIN OCCUPANTS

The perception of extreme natural events, albeit subjective in nature, plays a profound role in adjustment decision-making by people living in hazard-prone areas. The term 'perception' implies the "individual organization of stimuli relating to an extreme event or a human adjustment" (White, 1974a: 4). An examination of hazard perception allows one to determine how people view occurrences or threats, how such attitudes are influenced, and how such views relate to the available options for coping with the hazard effects. For our purposes, the concept of perception takes on a broader sense, meaning the subjective evaluation of the 'phenomenal environment' which surrounds a human subject, and reveals his or her consciousness of it and intentionality towards it. Thus perception is treated as a phenomenal experience as well as a directive for action (see Billinge, 1986: 135).

To generate data on the perception of the riverbank erosion hazard three interrelated questions were asked of respondents: (1) "Can you tell us what type of natural hazards (*durjog*) are likely to occur and adversely affect you and your household?" (2) "Why do you consider rapid river encroachment to be a serious hazard?" (3) "What do you think causes unusual or rapid river encroachment?" In total, nine hazard *concepts*[91] were described by respondents to the questions in the samples. These concepts fall into three broad categories according to their principal causal agents: meteorological, geomorphic, and biological (see Burton *et al.*, 1978: 20). Out of the nine events that are recognized as hazards (*durjog*) by respondents, five are meteorological: cyclones, nor'westers, hailstorms, droughts, and floods. The three biological events include famines, epidemics, and pests, while only riverbank erosion falls under the geomorphic category.

Most of the floodplain occupants (93 percent in both 1985 and 1995) consider flood as a threat to them. A considerable proportion of the respondents also perceive riverbank erosion, nor'westers and drought as hazards. The remaining five events are seen as hazards by less than one-quarter of respondents. It is interesting to note that in the 1985 survey, while only half of the sample (53.4 percent) named river encroachment as a threat (i.e., without prompting), when asked specifically about erosion problems about 99.3 percent of the sample revealed that they consider the phenomenon as a serious threat to their resources. However, in the follow-up survey of 1995, the majority of respondents cited erosion as a hazard without prompting by the interviewers. The above patterns are indicative of the fact that perception of hazard is at least partially related to previous experience of the events in the floodplain environment. Notably, in Bangladesh's floodplains the flood frequency is relatively high; Kayastha and Yadava (1985: 21) record four to five moderate floods and three to four low floods in the region every ten years.

In order to determine the perceived threats of bank erosion, respondents were asked the following question: "Why do you consider rapid riverbank erosion to be a serious hazard?" Three broad types of threats were identified from the answers: (1) damage to tangible commodities; (2) damage to community resources; and (3) adverse effects to socioeconomic opportunities. The majority of the 1985 and 1995 respondents perceived the threat of erosion hazard in relation to probable damage to tangible commodities or personal properties such as land, crops, trees, and houses. In the 1995 survey sample more than two-thirds of respondents also associated reduction in income with erosion disaster. In the 1985 survey sample, more than 86 percent of respondents indicated that erosion may damage crops and property, while only 55.4 percent suggested that the phenomenon was responsible for loss of land. This somewhat moderate proportion may be attributed to three features of the floodplain: first, the majority of floodplain inhabitants perceive flood and erosion as closely interrelated phenomena and view their impacts as inseparable variables; second, people residing in the interior lack any direct exposure to land-loss problems; and third, a considerable proportion of respondents are landless. The above patterns in the perceived threats to resources indicate that respondents' cognition of the hazard phenomena are principally confined to immediate surroundings and personal experience of the event. The identification of other explanatory variables requires further investigation and will be addressed in the following section.

TABLE 8.1. Distribution of perception of factors that causes riverbank erosion to be a hazard (%)

Perceived Causes of Riverbank Erosion (Multiple Response Possible)	Number		Percentage of Total Respondents	
	1985 (n=547)	1995 (n=210)	1985	1995
"Will of Allah"	185	57	33.8	27.1
High Floods	110	152	20.1	72.4
Breaking of Embankment	22	31	4.0	14.8
High Precipitation in the Region	45	90	8.2	42.9
High Discharge in River	114	27	20.8	12.9
"Too Many Chars"	30	48	5.5	22.9
Shallowness in River	69	70	12.6	33.3
Unconsolidated Soil	61	30	11.2	14.3
Storm Surge and Waves	39	3	7.1	1.4
Fast Current	320	104	58.5	49.5
Nature of the River	44	21	8.0	10.0

In describing their perception of causes of riverbank erosion, the majority of 1985 survey respondents suggested 'fast currents' as a major cause of bank erosion; this cause was also given by almost half of the 1995 sample. While floodplain occupants were aware of the risks they face in inhabiting floodplain areas, about one-third in 1985 and one-quarter in 1995 perceived the occurrences of erosion events as the 'will of Allah.' Among the many other factors related to bank erosion, hydrological variables were perceived by only one-fifth of respondents as a major cause of the event in 1985, although 72.4 percent of the 1995 sample saw 'high floods' as the cause of bank erosion (Table 8.1).

Two major patterns appear in the perception of erosion hazard. First, people's view that hydrological phenomena such as fast currents, high floods, or high discharge are the principal causes of erosion is influenced by both their familiarity with such stimuli (see Kates, 1962; Tuan, 1974) and the level of information available to them. Second, a simplistic and fatalistic view concedes that erosion is not related to physical surroundings but to the 'will of Allah.' Such believers perceive hazards to be out of their hands, or deny the existence of hazards to justify their own reasoning (see Saarinen, 1976; Burton and Kates, 1964).

Several Western writers attempt to explain this feature by adopting the concepts of 'locus of control' and variability in personality (see Saarinen, 1966; Simpson-Housley and Bradshaw, 1978; Parker and Harding, 1979). Rotter (1966) and Simpson-Housley and Bradshaw (1978) suggest the use of 'locus of control,' a measure that evaluates individuals according to their *personality* by considering whether they view events as contingent upon their own behavior or as a result of forces beyond their control. Festinger's (1957) 'theory of cognitive dissonance' also puts emphasis on personality characteristics to explain fatalistic attitudes towards the environment. The theory suggests that individuals reject information and suppress facts that might be damaging to their prevailing value system. Nonetheless, as Stokols (1978) points out there is a limitation in approaches that invoke personality factors to explain individuals' attitudes towards the environment: personality characteristics are wrongly considered as *en tout* independent variables; thus these approaches ignore the complex interaction between individual personality and social processes.

In a developing society like Bangladesh a variety of factors complicate people's responses to hazards. The traditional social norms, limited technological progress, lack of social, economic, legal, and political entitlements for large numbers of peasants—these characteristics leave the majority of rural people virtually without any means of material and effective support against extreme natural events. Under such conditions of absolute 'helplessness,' people view the events beyond their control. They accept the "inevitable as inevitable" (see Ittelson *et al.*, 1974: 316), and hence tend to favor reparative action once an extreme event has occurred.

8.1.2. DESCRIPTION OF FLOODS

Regular annual flooding in Bangladesh depends on the 'normal' hydro-climatic conditions which cause precipitation in the region from June to October (see Section 4.1). Such normal annual flooding is desirable since it starts at the right time for cultivation, lasts for an appropriate duration for paddy rice and other major crops, and is of a limited severity which the local conditions of settlement and housing, agriculture, and physical infrastructure can normally withstand (Haque and Zaman, 1993: 95). When this regular inundation (known locally as *barsha)* occurs, crop damage is minimal and farmers are able to harvest all of their *aus* and *aman*[92] rice. For centuries, *barsha* has played a vital role in maintaining the fertility of cultivable land, thereby supporting an expanding sedentary population in this region. Over time, farmers of the region, especially those who live in the floodplains, have developed numerous adaptive strategies to reap the benefits of this regular event. Floodplain users' concerns hence focus only on unusual floods which surpass their coping ability and threaten to devastate both their lives and their resources. Such high floods, which cause widespread damage to standing crops and properties, are known locally as *bonna* (see Paul 1984: 9-14).

The 1995 rural survey specifically examined the issue of unusually high flooding. Almost half of the sample respondents (n=210; nearly 49 percent) suggested that *barsha* is characteristically different than *bonna* (Table 8.2). However, when asked about the benefits of the normal floods, an overwhelming majority of respondents (92.7 percent) indicated at least one beneficial item. In this list the following items received most prominence: increase in crop productivity (64 percent), increase in freshwater fishery (39 percent), increase in local job opportunities (28 percent), and facilitation of water transport (23 percent).

Notably, respondents' awareness was found to be clearer regarding *bonna*—unusual events—than *barsha*. A vast majority of respondents stated that a *barsha*-related disaster occurs in terms of timing (i.e., if it occurs either earlier or later than the usual time of June-August), duration, and magnitude (i.e., whether the water level overtops the broadcast *aman* paddy and the homestead land). Alternately, respondents cited the following as major adverse effects of *bonna:* damage of crops (86.7 percent), significant reduction in income (73.2 percent), and loss of properties (49.3 percent; Table 8.2). The majority of the sample described *bonna* as those floods in which floodwater overtops the broadcast *aman* paddy as well as the homestead land. In addition, several respondents added timing and duration along with the height of floodwater as criteria for defining an abnormally high flood.

During the later half of the 20th century, the inundation of literature on floods has been remarkable. Notably, most of the studies, especially those in the West, view floods solely in terms of their damaging role (e.g., White, 1945; 1961; 1964; Burton and Kates, 1964; Smith and Tobin, 1979; Smith and Handmer, 1986; Carper, 1990). As we have seen, flooding in the Ganges-Brahmaputra basin is a constituent part of rural village-life and is deeply imbedded in the culture. Only the *bonna* threatens riparian settlements and hence is considered a hazard in rural communities; *barsha*, the more frequent phenomenon, is a necessary and desired occurrence. Studies from other floodplains in Bangladesh also record a similar perception among the inhabitants. For instance, Paul (1984) in the Jamuna floodplain, and Ralph (1975) and Islam (1980) in the Meghna floodplain observe that the selection of crop variety and other pre- and post-flood adjustments are attempted by the farmers based on a distinction between *bonna* and *barsha*.[93]

TABLE 8.2. Distribution of perceived threats of high floods (*bonna*) by types of resources, 1995 (%) (multiple response possible)

Threats of *Bonna* to Resources	Number N=203	% of Total Respondents
THREATENING TO LIFE	85	41.9
DAMAGING TO TANGIBLE COMMODITIES		
Damages Crops	176	86.7
Damages Other Properties	100	49.3
DAMAGING TO COMMUNITY RESOURCES		
Creates Food Shortage	79	38.9
Creates Health Crisis	83	40.9
Creates Transport Problem	85	41.9
AFFECTS SOCIOECONOMIC ENTITLEMENT		
Reduces Jobs	89	43.8
Reduces Income	150	73.9

A total of 96.2 percent of respondents in the study anticipate future loss due to the impact of unusually high floods on homestead land, while 90.9 percent anticipate future damages because of the vulnerability of their cropping lands. In rural Kazipur an overwhelming majority of floodplain inhabitants expect that their properties are subject to future *bonna* events. In general, a high level of awareness about flood magnitude and frequency is reflected in the 1995 survey data.

8.2. Patterns in Coping Responses to Riverine Hazards

Besides exploring variations in hazard perception and in the threats of these hazards to lives and property, we will now evaluate differences in the overt responses among resource users. While some people buffer themselves against the devastating effects of nature through extensive precautionary efforts, others accept the danger and loss as the proceeds of fate or the 'will of Allah.' As a general principle, the range of potential adjustment measures depends on the awareness of possible options, the available technology, the economic feasibility of available strategies, and the social context in which such responses arise (Burton *et al.*, 1968; Brooks, 1972; Blaikie *et al.*, 1994; Cannon, 1994).

By evaluating the adjustment options versus passive or no action, and then determining the relative costs of these alternatives, one decides whether or not to adopt a positive measure to satisfy one's goals (see Brooks, 1973: 116). Such a proposition, as follows from the neoclassical economic theories, is predominant in advanced capitalist society where individuals are considered relatively free to make decisions (see Burton, 1962: 1-2). But as is readily observed in a developing society like Bangladesh, the rural masses are largely governed by the constraints of the prevailing social structure and institutions (see Blaikie *et al.*, 1994: 46-71). In most cases individuals have limited or no choice but to abide by the existing social customs and local sociopolitical forces. In the light of the above constraints, the theoretical range in coping measures are outlined in the following subsection.

8.2.1. RANGE AND LIMITS TO COPING MEASURES: A CONCEPTUAL CONSIDERATION

Human coping responses to natural hazards reflect human perceptions and efforts to minimize the negative consequences of hazards. The attitudes and actions involved in adjustment processes to any particular hazard are usually put into effect within a short span of time, and in most cases they are temporary relative to the cultural adaptation process.[94] As Burton *et al.*, (1978) suggest, coping adjustments are to be separated into those that are *purposefully* adopted and those that often are not primarily related to the hazard. The latter actions, called *incidental* adjustments, have the effect of reducing potential losses. For example, instead of using brick, steel, concrete and other quality (*pucca*) housing materials, floodplain occupants in Bangladesh primarily use thatch, bamboo, wood, and corrugated iron sheeting. Houses made of these materials are readily moveable in emergencies and thus are less susceptible to loss by erosion than are permanent brick structures (Figure 8.1)

As shown in Table 8.3, 58 and 77 percent of sample households in Kazipur used corrugated iron sheets as roof material in 1985 and 1995 respectively; the rest used thatch, bamboo, and other locally available materials. Incidentally, none of the sample households in Kazipur had brick or concrete structures. Nonetheless, the general poverty of rural communities is also of one of the main reasons for using local materials, a fact reflected in the relative homogeneity of building materials in 1995 among different floodplain zones [$X^2(p < 0.05; df=2) = 3.23$].

Since *incidental* adjustments are not distinctly related to the hazard phenomenon they are often difficult to identify. For example, in the 1985 survey significant variation in the use of roof material is visible between geographical zones.[95] Yet to put this in context, moveable materials for housing have been in use throughout the region for centuries and represent a cultural adaption over time. However, a positive correlation of the use of cheaper and moveable roof materials with the hazard potential of the zones suggests that such coping adjustments are characterized by both incidental and purposeful adaptations.

Agricultural adjustments also reflect both incidental and purposeful characteristics of coping with bank erosion hazard. Farmers in this region commonly respond to recur-

TABLE 8.3. Distribution of households according to roof materials by zone, 1985 and 1995 (%)

Roof Materials	Zones 1985			
	KZ N=73	BZ N=227	CZ N=247	Total N=547
Corrugated Iron Sheet	75.3	63.0	48.2	58.0
Thatch, Bamboo and Other Local Materials	24.7	37.0	51.8	42.0
	1995			
Corrugated Iron Sheet	77.4	80.8	68.5	77.1
Thatch, Bamboo and Other Local Materials	22.6	19.2	35.4	22.9

1985 $X^2(p<0.001; df=2) = 21.11$ (Significant)
1995 $X^2(p<0.005; df=2) = 3.23$ (Not Significant)

Figure 8.1. Dismantled houses to be moved to a safer location in a *char* village in Kazipur. Salvaging roofing made of corrugated iron sheets is a frequent loss-reduction strategy—an example of incidental adjustment.

rent flood problems through elaborate adaptations of different crops to varied flood depths (see Paul, 1984; Rasid and Paul, 1987). Similar cropping adjustments are also noticed in erosion-prone areas. For example, the floodplains along the Brahmaputra-Jamuna are composed of a combination of loams on swales and clays, while the *char*lands, with relatively lower relief, are composed of calcareous loams (Rashid, 1977). Ahmad (1968: 67-68) points out that the *doash* clay, silt, and sandy loams of the mainland areas have "great fertility, [are] easy to plough and produc[e] a variety of crops." In contrast, the porous sandy soils (i.e., *bali*) of the *diars* and *chars* do not suit *boro* paddy, but work well for pulses and other winter crops.

As shown in Table 8.4, most households in the KZ grow *aman* and *boro* paddy, while the majority of *char* households cultivate *aus* and *aman* paddy.[96] The *char*land in Kazipur also seems to be more suited to the cultivation of pulses than other winter crops. The significant interzone variations are reflected in the multinominal chi-square estimate of homogeneity for the overall distribution which is 203.02 ($p < 0.001$ level). The paired comparisons between the zones also indicates significant differences at the 0.001 level. The cropping patterns of the different strata show that people living in hazardous areas, especially in the *char* and *bir* lands, tend to reduce the magnitude of their crop loss from erosion by cultivating low-cost varieties. For instance, among all the crops, *boro* paddy and potatoes involve relatively higher costs and intensive care in terms of water requirements. As such, the proportion of households cultivating these two crops is lower in the BZ than in the KZ, and lower in the CZ than in either the BZ or the KZ.

TABLE 8.4. Cropping patterns by zone, 1985 (%) (multiple response possible)

Type of Crop	Zones			Total
	KZ n=73	BZ n=227	CZ n=247	(Main Survey) n=547
Aus (Paddy)	35.5	68.8	93.0	74.0
Aman (Paddy)	90.3	64.7	76.6	73.7
Boro (Paddy)	80.6	43.4	15.1	37.1
Cane	-	13.3	1.2	6.1
Potato	48.4	29.5	23.3	29.7
Pulses	25.8	19.7	60.5	37.8
Oilseeds	67.7	48.0	34.4	45.2
Spices	64.5	67.1	58.7	63.1

For overall distribution $X^2(p < 0.001; df=14) = 203.20$ (Significant)
Col. 1 and Col. 2 $X^2(p < 0.001; df=7) = 35.05$ (Significant)
Col. 1 and Col. 3 $X^2(p < 0.001; df=7) = 148.40$ (Significant)
Col. 2 and Col. 3 $X^2(p < 0.001; df=7) = 127.03$ (Significant)

In examining the types and characteristics of purposeful coping measures, we must modify the model of 'modes of adjustments' offered by Burton *et al.*, (1978) for application to riverbank erosion problems. Human coping behavior can be grouped into the following categories: (a) acceptance of loss; (b) reduction of loss; and (c) change in location or resource use. These responses can be further distinguished according to their corresponding threshold levels: (a) awareness and ability; (b) voluntary action; and (c) intolerance and/or involuntary action. The acceptance of loss relates to the *awareness and ability* threshold because such responses are taken within the knowledge of the members of a given society and in correspondence with their economic, social, legal, and cultural viabilities. Society in general, or individuals, arrange to bear loss as a passive strategy, either by sharing it within the broader social sphere or by bearing it personally.

The mode of loss reduction results in some active or positive measures to reduce loss either collectively or individually. Such attempts to modify loss potential may be realized by technological responses to control the cause of hazards or by social measures aimed to minimize hazard impacts or modify damage potential and losses. These actions essentially fall under the category of *voluntary* measures. In the final threshold hazard loss surpasses the *tolerable* range of resource users. Possible causes may include social manipulation of abnormal situations created by extreme events—namely, a sudden change in either location or resource-use patterns. Involuntary or forced migration is a common behavior at this *intolerable* threshold.

In the following section we will analyze these three modes in the context of bank erosion hazard. As well, we will examine the relationships of coping strategies to the pertinent explanatory variables. In the process, a distinction between *individual* and *collective* responses to hazards must be maintained in order to acount for varied objectives and coping modes (see Burton, *et al.*, 1978; Rowntree, 1974; Blaikie *et al.*, 1994).

8.2.2. FLOODPLAIN USERS' COPING STRATEGIES

Perceived Potential Coping Measures

The high frequency and recency of flooding, and the seriousness of bank erosion, coupled with a lack of structural protection in *char* and *bir* areas, suggest that high levels of erosion hazard awareness should prevail for people in these regions. It follows that a more widespread awareness should be found in the *char* areas than in the *bir* areas, and similarly more in the *bir* areas than in the *kaem* areas.

The findings of both the 1985 and 1995 surveys support these assertions. Almost all respondents (96 percent) anticipate the loss of their homesteads to future bank erosion, while more than half of the respondents (58.9 and 66.2 percent in 1985 and 1995 respectively) expect to lose their homesteads within a year, 22 percent within 2-4 years, and the rest within the next five years or thereafter (Table 8.5). Likewise, almost all households having cultivable land anticipate future land-loss due to river encroachment.

The relationship of future hazard expectation with spatially distinctive risk zones shows up in the significant chi-square tests for both homestead and cultivable land (see Tables 8.5). Table 8.5 displays an overall high level of perception among floodplain inhabitants and a positive correlation with hazardousness of the zones.

Table 8.6 presents the percentage distribution of potential coping strategies aimed at reducing bank erosion disaster impact. Evidently, among the passive strategies at the awareness and ability threshold, two measures are common. Of the 1985 sample, 42.8 percent favor the abandonment of land while 42.4 percent of the 1995 sample intend to remain and wait in the homestead land until affected directly by river encroachment. In contrast, 18 percent of the 1985 survey respondents and 32 percent of the 1995 survey respondents do not intend to take any active response other than 'pray to God' when affected by erosion. This relatively high degree of acceptance of loss is directly related to their perceived causes and awareness of the hazards. In the 1985 survey, among all the potential coping strategies, the salvaging of housing accounts for the highest proportion of respondents (69.3 percent); this is one of the most common loss-reduction strategies undertaken by hazard victims (Figure 8.1). As a long-term loss-minimizing strategy, 23 percent intend to retain their ownership of submerged land by continuing to pay taxes to local revenue offices. One third of respondents state that they would leave the village in search of a safer place. But only a few (11.5 percent) indicate no other option but to move to the embankment to illegally occupy government land (see Figure 6.2; Chapter 6). As we can see, the general trend among floodplain inhabitants is to adopt multiple coping strategies.

TABLE 8.5. Distribution of perception of whether land will be affected due to river encroachment by zone, 1985 and 1995

Anticipating a Loss Due to Future Erosion Hazard	Zones[a] KZ n=73	BZ n=227	CZ n=247	Total (Main Survey) 1985 n=547	All Areas 1995 n=210
i. HOMESTEAD					
Yes	83.6	98.7	98.4	96.5	96.2
No	1.4	0.9	1.2	1.1	2.4
Do Not Know	15.1	0.4	0.4	2.4	1.4
ii. CULTIVABLE LAND					
Yes	74.0	69.6	66.8	68.9	51.4
No	6.9	6.2	2.8	4.8	5.3
Do Not Know	12.2	0.4	1.2	2.3	4.3
Landless	6.9	23.8	29.2	24.0	39.0

X^2 for homestead ($p < 0.001$; df=4) = 58.74 (Significant)
X^2 for cultivable land ($p < 0.001$; df=6) = 51.10 (Significant)
[a] Zonal data were obtained from the 1985 survey.

The 1995 survey data show some noticeable changes in coping strategies with potential bank erosion hazard compared to the patterns in 1985. Compared to the 1985 sample, more floodplain users in 1995 tended to stay in the homestead until the onset of the erosion disaster (Table 8.6; Figure 8.2). Such a shift in response pattern is indicative of the desperation of floodplain inhabitants in holding to their survival means in recent years.

It appears that potential individual coping strategies, as perceived by the respondents, fall under the *corrective* rather than *preventive* type. In 1985 the majority of respondents tended to reduce hazard loss by salvaging housing structures and/or trying to bear the loss by abandoning their land. Similarly, most respondents in 1995 tried to minimize hazard loss by moving their families to safer areas and/or by salvaging housing structures.

TABLE 8.6. Distribution of potential measures to cope with riverbank erosion hazard (%) (multiple response possible)

Types of Coping Strategy	Number		Percentage of Total Respondents	
	1985	1995	1985 n=547	1995 n=210
BEARING THE LOSS				
Remain in the Homestead Land Until Affected by Erosion	13	89	2.4	42.4
Abandon land	234	1	42.8	0.5
Pray to *Allah* (God)	97	67	17.7	31.9
REDUCTION OF LOSS				
Protect Land	24	3	4.4	1.4
Dismantle Housing Structure	379	98	69.3	46.7
Keep Ownership for Re-emergence of Land	126	5	23.0	2.4
CHANGE USE/LOCATION				
Leave Village	188	93	34.4	44.3
Move Family to Embankment	63	159	11.5	75.5

Overt Behavior in Coping Measures: Riverbank Erosion Prevention and Mitigation

The overt behavior in dealing with hazards is a reflection of perceived adjustment strategies. The perceived and actual measures, however, may not follow an identical path because the former reveals the possible alternatives, whereas the latter shows the actual action. The following discussion thus focuses on actual measures undertaken to cope with bank erosion hazards in the light of individuals' perceived responses to potential adjustments.

Figure 8.2. A village community facing riverbank erosion hazard in Serajganj district. Most people tend not to move until after their homestead land is eroded, resulting in increased hazard potential.

In the present study the actual adjustment measures taken to cope with erosion could only be obtained from displacee respondents. It appears from the 1985 survey that the most common coping strategies among displacees are the acceptance of loss through abandoning land (63 percent), salvaging houses (primarily the roof; 52.2 percent; Figure 8.1), moving the family from disaster-affected areas (43.5 percent), and selling livestock (23 percent; Table 8.7). The survey conducted a decade later shows that moving family and belongings to a safer area is the most frequent measure among the displacees (see Figure 6.2). This implies that floodplain users are taking more pragmatic measures than before as they now tend to accept losses only when actual disasters occur; they also tend to save lives and properties by changing the location and use of household goods and tangible properties.

By comparing Tables 8.6 and 8.7, we can see little deviation in actual behavior from perceived adjustments. For instance, both the perceived and actual adjustment strategies suggest that people undertake more than one measure to minimize hazard loss. In both cases, abandoning land, salvaging housing structures, and transferring the family and belongings from erosion-affected areas are characteristic of a high proportion of respondents. Notably, the above three strategies fall into three different modes of ad-

TABLE 8.7. Distribution of coping measures taken by displacees before last move (%) (multiple response possible)

Type of Coping Measure Taken Before Move	Number		Percentage of Total Displacees	
	1985	1995	1985 n=322	1995 n=154
BEARING THE LOSS				
Abandoned Land	203	34	63.0	21.9
REDUCTION OF LOSS				
Protected Land	9	nil	2.8	-
Sold Livestock	74	45	23.0	29.2
Sold Cultivable Land	6	nil	1.9	-
Sold Homestead Land	11	1	3.4	0.6
Sold Belongings	10	1	3.1	0.6
Dismantled Housing Structure	168	46	52.2	30.0
Cut Trees	18	37	5.6	24.0
CHANGE USE/LOCATION				
Moved Family to Other Area	140	146	43.5	96.1
Moved Livestock to Other Area	30	12	9.3	7.8
Moved Belongings to Other Area	50	80	15.5	51.9

justment: abandoning land is an unequivocal acceptance of hazard loss; salvaging housing structures is a measure to reduce loss potential; and moving the family from disaster-affected zones is a change in the use of a resource unit.

A much higher proportion of displacees in the 1985 sample (63 percent; n=225) actually abandoned their land compared to the proportion of respondents who perceived abandonment as a potential strategy (42.8 percent, n=547). A similar pattern can be seen in the 1995 sample; 21.9 percent abandoned their homestead land while only 0.5 percent of respondents considered abandonment as a potential coping method (Tables 8.6 and 8.7). Furthermore, while 69.3 percent of respondents of the 1985 sample considered the salvaging of housing as a potential measure, only 52.2 percent of displacees actually undertook this measure. In the 1995 sample, likewise, 47 percent of the sample suggested this strategy as a potential mitigation measure and only 30 percent of displacees dismantled their houses to save them from 'grabbing' by the river. Evidently, displacees opted for a number of responses aimed at reducing hazard loss through purposeful measures— these measures fall into the *corrective* type of responses to hazards.

For an individual to protect the land from the physical impact of a large braided river like the Brahmaputra-Jamuna is obviously impossible. Yet to buy time to move belongings, a nominal proportion of the respondents attempted to protect their homestead land from slumping and river erosion. This kind of measure consists

primarily of building protective bamboo structures, locally called *chegar*. As a further loss-reduction strategy, about one-quarter of displacees sold their livestock. In rural Bangladesh livestock often constitute the only tangible asset that peasants can readily convert to cash. In contrast, only about one-tenth of displacees moved their livestock along with their family. Aside from these two principle means, other loss-reduction strategies appeared to be insignificant.

While conducting the field survey in 1985 we observed that a considerable number of marginal peasants sold the title to their submerged land (already affected) or to their potentially affected land. In the survey responses *per se* this strategy did not appear to be significant, particularly in the 1995 survey data. Since these peasants are generally fearful of public officials they show a reluctance to provide information relating to land and income. The under-reporting of such sales is obviously due to fear of legal complexities, not to mention the fears of local 'land grabbers' and other local patrons. Usually the large landowners (the *talukdars* and *chakladars*) buy titles to submerged land for two reasons. First, because they are economically more capable of paying. For them the price is modest and prospects of eventually repossessing the land upon reemergence is good; they can afford to wait. Second, such action also serves to enhance their social power in the community; they appear to be doing the marginal peasants a favor (see Zaman, 1987; 1991a; Zaman and Wiest, 1985; 1991; Haque and Hossain, 1987).

Perceived coping strategies and actual mitigation measures are closely correlated, but the actual coping measures appear to be more diversified than the perceived range of coping options. These patterns suggest that resource users look for a wider range of alternatives while actually performing an action relative to their perception of potential strategies. Meanwhile, the lack of preventive measures in both perceived and actual coping strategies indicates a high level of awareness of the nature of the hazards and related problems. Floodplain users recognize that technological or engineering-structural controls of a braided river like the Brahmaputra-Jamuna are not feasible because of the geomorphological and hydrological characteristics of the river, as well as because of the sheer scale of the problem (see Coleman, 1969; Khan and Rashid, 1985; Khan, 1988; 1995; also see Chapter 4). As such, in both perceived and overt responses, floodplain inhabitants place less importance on *preventive* actions. While the *ability* to reduce hazard loss through *corrective* measures are not homogeneous due to variations in socioeconomic characteristics, it is not surprising that a considerable proportion of respondents feel absolutely 'helpless.' The responses of these people thus become an unequivocal acceptance of their losses.

Overt Behavior in Coping Measures: Flood Prevention and Mitigation

Coping strategies with flood hazard in Bangladesh at the household or individual level primarily consist of *corrective* measures. These chiefly involve the non-structural measures taken before and during floods for mitigating loss, but do not include major structural-engineering work. Most of the measures originate indigenously and relate to

social organizations and relationships, while others deviate from material responses at an individual level. Several scholars have explored the role of resource sharing and income redistribution in ameliorating famine effects in a predominantly agrarian economy (Desai, 1959; Alamgir, 1980; Sen, 1981a), but attempts at studying the role of indigenous social organizations and other cultural elements in mitigating hazard effects have been lacking.

I conducted a questionnaire survey, in collaboration with Mohammed Zaman, in December 1988 in the Sreenagar *thana* of Munshiganj district to explore indigenous coping strategies. A sample of 280 households was selected (Haque and Zaman, 1993; 1994). In sum, the survey showed that the majority of respondents took some corrective measures to minimize flood loss. More than 71 percent of the sample attempted to reduce their loss to floods by selling their land, livestock, or belongings, while some moved housing structures, livestock and family members to other safer places.

Most respondents welcomed outside assistance as they dealt with flood loss. Although a nominal proportion of flood-affected households received support from local or national government sources (less than 12 percent), most of the victims received some assistance in coping with the disaster from other local, community-level sources. Assistance received from relatives (79 percent of flood affected households) and other community members (33 percent) consisted of moral support, free shelter and accommodation, free help with moving family or livestock, and loans of cash or food. Notably, a majority of flood victims received food and clothes, housing materials, seeds and loans from non-government organizations (NGOs). More than half of the sample considered it important to remain within or to maintain close ties with their corresponding *samaj* while coping with disasters. This is due to the fact that *samaj* participants are obliged to help with difficult situations confronting other members.

At the household level some small-scale structural measures are commonly employed to mitigate flood loss. One of the most common strategies is the construction of homesteads on *natural levees*. The plinth of houses, locally called *bhiti*, are further elevated by digging earth from local depressions (Currey, 1979). In this way, non-metalled rural roads (*katcha*), flood protection embankments, and the courtyards of local schools and mosques are often raised above the level of abnormal floods and often used as flood shelters. People also build platforms or *machans* using bamboo, straw, water hyacinth and banana stalks during abnormal flooding years (see Figure 8.3). Use of banana stalks to construct floating platforms, locally known as *vela*, during the monsoon period is common in many parts of Bangladesh.

In order to obtain realistic perspectives on coping measures for unusually high floods, the 1995 survey questionnaire included non-prompted, unstructured items. The coping measures of resource users were categorized as *pre-flood* and *during-flood* actions. A total of 190 respondents of the 1995 sample (90.5 percent) experienced temporary displacement from their usual residence due to previous floods. However, only one-third of the sample attempted pre-flood loss-mitigation measures. Of these, two strategies received priority: construction of *machans* (23.9 percent of total respondents who under-

Figure 8.3. Submerged houses in rural Serajganj during the 1984 flood. People often remain in house *mechans* (platforms made of bamboo) until the flood level reaches the roof. A bamboo bridge is a common coping measure to overcome transport disruption in rural communities.

took pre-flood actions), and moving belongings to safer areas (35.2 percent). The survey data thus reveal that pre-disaster mitigation measures at the individual level are limited, yet whatever measures are taken, they are also of a corrective nature. One would expect disaster victims to undertake coping measures in a 'pro-active manner' during the high flooding. Nevertheless, the data obtained from the sample show that only 45.7 percent (n=91) took some kind of actual coping measures to ameliorate flood loss (Table 8.8). The strategies were also mainly confined to only a few measures, among these: moving families to higher ground (67.7 percent to total respondents who took coping measures during abnormal flooding), construction of *machans* (52.1 percent), and moving belongings to safer places (29.2 percent).

A low degree of coping response to mitigate hazard loss at the individual level, as revealed in the 1995 survey, may be attributed to the unprompted form of the survey instrument. Additionally, residents of Kazipur are so familiar with flood events that they commonly do not react until unbearable conditions emerge. In many cases disaster victims do not get enough time or they do not possess the required resources to under-

TABLE 8.8. Distribution of coping measures taken by respondents during high flooding years, 1995 (%) (multiple response possible)

Type of Coping Measures	Frequency n=96	% of Total Respondents Who Took Measures During Flooding
STRUCTURAL		
Constructed *machan*	50	52.1
Built *vela* (floating platform)	3	3.1
Constructed or strengthened *bandh* (embankment)	2	2.1
NON-STRUCTURAL		
Moved family to higher ground	65	67.7
Moved belongings to safer place	28	29.2

take any corrective measures (see Haque and Zaman, 1989; 1993; Paul, 1984; Rasid and Paul, 1987).

Of indigenous responses to flooding in Bangladesh, a significant number are agricultural coping measures. As Rasid and Paul (1987: 168-169) observe, farmers of this wet monsoon region have made a careful selection of the most adaptive varieties of rice over the centuries to enable them to face high floods. For example, certain varieties of *aus* rice are cultivated only in elevated areas (*taans*). Since some broadcast and transplanted *aman* varieties grow during the flood and are harvested following the recession of floodwater, they are grown in low-lying areas (doabs) such as backswamps, channel scars and shallow beds of oxbow lakes. *Chamara*, a broadcast variety of *aman*, can grow by more than 15 centimeters a day with rising flood levels.

Overall, it is evident that flooding is vital to the rural habitat and agriculture of Bangladesh. Although some extreme events turn out to be serious threats to local resources and human lives, their frequency is measured in the 50- to 100-year range. The floodplain inhabitants are well adapted to usual floodings and any attempt to prevent annual inundation could pose a serious threat to local ecosystems and cultural ecology.

8.2.3. COMMUNITY-LEVEL COPING MEASURES

Several analysts elaborate on the importance of collective decisions and subsequent action; these include Kates (1971), Kates and Burton (1986b), White (1974a), Burton *et al.*, (1993), Mileti *et al.*, (1975), Nagpaul (1984), and Palm (1990). Moreover, sociological studies place greater emphasis on community or organizational response systems than do other disciplines (see Dynes, 1970; Quarantelli, 1970; 1977; Mileti *et al.*, 1975; Drabek, 1986). The complexity of collective response processes make them difficult to comprehend compared to individual responses; they involve the roles of individuals in multifaceted forms (e.g., user, manager, guide, dispenser of services, and agents) at various levels of social institutions and interest groups.

The precarious socioeconomic conditions and low level of technological know-how in rural Bangladesh limit community-level resources and responses. Local government and other formal institutions are not only deficient in capital, technical, and planning resources, but they also have limited contact and communication with the disadvantaged rural masses. Under these circumstances, the traditional social organization based on norms, social customs, and laws plays a crucial role in linking local people to public and formal institutions.

The Brahmaputra-Jamuna floodplain, with its recent geological origin, non-cohesive soil conditions, and highly variable river discharge produces frequent and dramatic changes in river channels which result in the loss of homes and land, particularly in the *char* area (see Chapters 4 and 5). The inhabitants in the mainland areas usually enjoy a relatively stable environment. Yet in Kazipur, channel changes have been a serious threat since the 1974 flood. At that time, the *thana* townsite was located on a minor chute-channel of the Brahmaputra-Jamuna system known locally as *Daokoba*. This western-most channel of the Brahmaputra-Jamuna system began vigorously to erode areas in close proximity to the Kazipur *thana* headquarters in 1977; by 1981-1982, Daokoba had become one of the main channels. The magnitude of land-loss (3,935 hectares) and consequent population displacement (17,623 persons) due to erosion-disaster in the *thana* during 1974-1981 has been described in Section 7.1 (see Haque *et al.*, 1985a).

Furthermore, the rapid river encroachment during 1974-1977 threatened government and private property in the area. In order to minimize potential loss, local leaders formed a committee to contact regional and national officials to alert them to this hazardous situation. This Riverbank Erosion Protection Committee estimated the value of public property at the *thana* headquarters site at about Taka 253.62 million (approximately US $16.1 million).

Following a mass political and popular mobilization in Kazipur during 1977-1979, the government assigned the Water Development Board as the planning and implementing authority for the protection of the Kazipur headquarters against river encroachment. After an initial failure to arrest river encroachment by placing bamboo baskets filled with bricks at the bank slumping sites, the Board developed an alternative scheme to construct a cross-dam in collaboration with local leaders which cost Taka 150 million (US $10 million). The project, however, was abandoned after only partial implementation because its execution was after the fact, and because of conflicting opinions among local leaders and public officials.

Among the few other community-level measures, the development of a flood-shelter at a location some five kilometers from the then river bankline was undertaken by the Ministry of Relief and Rehabilitation in conjunction with the World Bank. Local displacees, however, never used this facility because the local government offices were transferred to the flood shelter in 1980 following the river's encroachment on the local headquarters. Also, local displacees gave two other reasons for not using the shelter: first, they were fearful of exploitation by government officials who were alleged to impose undocumented 'rents,' and second, their relocation to this shelter would separate them from their *samaj*, on whom they are especially dependent for post-displacement needs.

As noted earlier, relief and rehabilitation programs are seldom undertaken by either non-governmental agencies or local government offices in this community. The lack of any effective support from local government officials, community leaders, and relief agencies is clearly reflected in the available sources of assistance to displacee households at the time of their last displacement (see Table 8.9). About one-fifth of displacee households received some kind of support; the majority were left without any assistance. Among the limited assistance received by displacees, most was provided by relatives and friends. Other forms of assistance to hazard victims were almost non-existent in the study area.

8.3. Differentials and Explanatory Factors: A Test of Hypotheses

The 1985 rural survey achieved a higher reliability, a larger sample size, and a more comprehensive survey instrument than did the 1995 survey. As such, data for testing pertinent hypotheses were obtained from the former field investigation. It may be worthwhile to reiterate that this research postulates that resource users' perceptions and coping behaviors are related simultaneously to the socioeconomic-structural, experiential and locational dimensions. In general, the research covers individual, social and collective coping responses to riverine hazards.

8.3.1. DIFFERENTIALS AND EXPLANATORY FACTORS IN PERCEPTION OF *DURJOG* (HAZARDS)

Studies on floods, cyclones, and storm surges in Bangladesh suggest that the patterns of adjustment to natural hazards relate to differing perceptions. If this is so, varying assessment of bank erosion hazards should evoke particular sets of coping measures. Should an individual deny the existence of a threat of bank erosion, he or she would behave according to an illusion of safety. Should an individual recognize the threat of extreme events in terms of their damages to resources, he or she would undertake coping measures according to their perception of that threat. The analysis of explanatory factors for the differences in perceptions of hazard is therefore important to understand response strategies and coping behaviors. A total of 16 chi-square tests were undertaken to test the pertinent hypotheses. Table 8.10 presents the summary of relationships between (a) human perceptions of hazards and potential coping measures and (b) explanatory variables.

TABLE 8.9. Sources of assistance received by displacees during last displacement, 1985

	Zones		
	BZ (n = 77)	CZ (n = 244)	Total (n = 321)[a]
Assistance received during last displacement (%)			
Yes	23.4	20.5	21.2
No	76.6	79.5	78.8
Source of assistance (multiple response possible)	n = 18	n = 50	n = 68
Relatives	88.8	64.0	70.6
Friends	50.0	36.0	39.7
Other villagers	5.6	26.0	20.6
Community leaders	-	6.0	4.4
Thana officials	-	4.0	2.9
National government	-	4.0	2.9
Relief agencies	5.6	2.0	1.5

[a] Excluding one displacee household in the KZ.

Socioeconomic-Structural Dimensions

As we reviewed in Chapter 3, the existing literature on the relationships between human responses to natural hazards and socioeconomic variables clearly indicates two important features. First, findings based on the stratification of socioeconomic variables such as education or income suggest contradicting conclusions; some researchers find no relationship between them (Roder, 1961; Kates, 1962; M. Islam, 1974), while others find a positive relationship (Baker and Patton, 1974; Leigh and Sim, 1983; Varley, 1994; Blaikie *et al.*, 1994). A disagreement also arises over which socioeconomic indicators should be used. The conventional views follow the social stratification approach, and use discrete groupings of relevant variables; the other critical approaches deny the appropriateness of using the cultural indices of social stratification as true indicators of socioeconomic relationships. The latter approaches promote the use of social class analysis to explain variations in human responses to hazards (O'Keefe, 1975; Cannon, 1977; Watts, 1983; Blaikie *et al.*, 1994) and thus place emphasis on determining the individual's entitlements to ownership of land, machinery, and other socially available resources (see Wood, 1981; Sen, 1980, 1981a, 1981b; 1982). Considering the integrative approach of this research, we will deal with variables related to both of the above approaches. However, I do view the social-structural approach more appropriate to analyze the socioeconomic dimensions of hazard response.

In this study, four socioeconomic-structural variables are tested as explanatory factors for the perception of bank erosion hazard in Kazipur. These are: (i) educational status

TABLE 8.10. Summary of relationships between (a) human perceptions and potential coping responses to riverbank erosion and (b) explanatory variables, 1985

Components of Human Responses	Significance Level of Chi-Square Test Explanatory Variables							
	Experiential Dimensions		Spatial Dimensions		Socioeconomic-Structural Dimensions			
	Displacement Status	Displacement Frequency	Potential Hazardousness of an Area	Distance Between Homestead & Bankline	Educational Attainment	Household Income	Landownership	Occupation
A. PERCEPTION OF HAZARDS								
Perceived Hazard Concepts	0.001*	0.025*	0.001*	0.001*	NS	0.05*	0.01*	0.01*
Perceived Threats of Erosion	0.001*	0.01*	0.001*	0.001*	NS	NS	0.01*	NS
Perceived Causes	0.001*	NS	0.001*	0.001*	0.001*	0.05*	0.01*	0.01*
B. POTENTIAL COPING STRATEGIES	0.001*	0.01*	0.001*	0.001*	NS	0.05*	0.001*	0.001*

*Significant; NS = Not Significant

of the household head; (ii) aggregate household income; (iii) landownership of the household; and (iv) occupation of the household head. Education and income variables are primarily considered here in order to compare the existing relationships with international findings. Land ownership is the most important variable for social class analysis in Bangladesh society; occupational status is used as a proxy for types of production and labor.

The chi-square test for 'perceived hazard concepts' according to educational attainment of the household head reveals that the null hypotheses cannot be rejected at the 0.05 level. This suggests that the perception of hazards among floodplain users is not dependent upon their educational attainment. The statistical test results for the distribution of 'perceived threats of erosion' also conform with the above assertion (Table 8.10).

Not surprisingly the perception of causes of erosion does relate to educational attainment. The educated, with relatively better accessibility to the media and to the other sources of technical knowledge, can be expected to demonstrate better perception of the causes of hazards. Education also relates inversely to the response that erosion was an act of God; about 38 percent of respondents with 'no education' stated that erosion was the 'will of Allah,' whereas only 17 percent with secondary and post-secondary education expressed the same view. Finally, the examination of the overall relationship between hazard perception and educational attainment negates the hypothesis that they are correlated; this agrees with the conclusions of Roder (1961), Kates (1962), and M. Islam (1974).

The relation of household income to hazard perception has not been widely tested. Household income is an aggregate index of assets, education, and occupation, and thus can be expected to have some influence upon an individual's responses. In Table 8.10, the chi-square tests for the 'perceived hazard concepts' and 'perceived causes of erosion' show that these variables are significantly related to household income; the level of significance for each of these is 0.05. A noticeable variation in the 'perceived hazard concepts' is seen for the biological events (e.g., pests); the relatively higher income groups are more aware of these hazards compared to the marginal ones. The finding of a significant relationship between the 'perceived causes of erosion' and household income conforms with the expected result; objective and technical knowledge gathering relates directly to educational attainment, which in turn depends largely upon the income status of households in rural Bangladesh.

The data also indicate that the threat of erosion is not related to household income; the null hypothesis cannot be rejected at the 0.05 level (Table 8.10). Although the data show that people with lower incomes are less aware of the threat of erosion than are the higher income groups, the overall difference in the responses does not appear to be statistically significant. Thus, the influence of income appears to relate primarily to the other hazard perception components such as 'hazard concepts,' and 'causes of hazards.' Since income in rural Bangladesh primarily correlates with landownership, it follows that hazard perception relates to the landownership status of households.

In contrast, the landless (59.8 percent) appear to recognize bank erosion as a hazard to a higher degree than marginal (51.9 percent) and middle and large landowners (47.9 percent). Yet the lower percentage among landless respondents who recognize that erosion is a threat to land (i.e., 40.9 compared to 51.4 percent for middle and large landowners) evidently reflects their lack of concern about land—cultivable land is largely inaccessible to poor households. Furthermore, only a nominal proportion of landless respondents (1.5 percent) regard hailstorms as a hazard, while 11 percent do in the other landowner categories. Hailstorms are mainly harmful to crops and therefore cause more of a concern to those who own cultivable land. Such data substantiate the postulation that the perception of hazards is related to the control of resources.

In perceived 'threats of erosion,' large landowners (having more than 1.01 hectares) feel more of a threat of erosion to personal properties (such as land, crops, food grain, and other tangible properties) than do the landless. In contrast, a higher proportion of the landless feel the threat of erosion to jobs and income opportunities compared to the landowner groups. Remarkably, none of the middle and large landowners suggest that erosion threatens employment opportunities. The fatalistic belief that erosion is due to divine will is most common among the marginal landowner group (38.2 percent), the people most vulnerable to erosion-hazard. A combination of the immediacy of the threat and their helplessness to act is likely to convince them that catastrophic events are caused by forces beyond their control.

The surveys assessed occupational status in terms of types of produce, production processes, and divisions between material and mental labor.[97] The responses to 'perceived hazard concepts' are found to be significantly related to occupational status. Noticeable differences are apparent in responses to perception of hazards between laborers (which include wage laborers, boatmen, rickshaw peddlers and unclassified workers) and other occupational groups. For instance, the proportion of laborers responding to 'perceived hazard concepts' is generally lower than 'farmers and other workers engaged in primary-production' or 'persons engaged in non-agricultural production.' In the responses to 'threat of erosion,' no significant variation is shown by occupational categories. In contrast, the responses to "causes of erosion" do show some significant differences by occupational categories. More than 65 percent of respondents in non-agricultural occupations stated that "fast currents" were one of the major causes of erosion; the respective proportion for the unpaid household workers (which also includes unemployed persons and students) is 53.7 percent. Also, the proportion of respondents viewing bank erosion as an 'act of God' appears to be lowest among non-agricultural workers (25.6 percent; see Table 8.10).

Experiential Dimensions

As noted above, perceptions of hazards among the floodplain users are examined in three components: hazard concepts, threats of erosion to resources, and causes of erosion. The perceptions of hazards were not homogeneous among respondents—how then might this variance be accounted for? The first hypothesis proposes that the perception of bank erosion hazard varies with experience of the event *per se* and with the frequency of displacement.

The results for all three components of hazard perception among respondents confirm that the perception of bank erosion varies significantly (at the 0.001 level) according to displacement status. Compared to displacees, a higher proportion of non-displacee respondents perceive meteorological and biological events as hazards. In the 'threats of erosion to resources' component some similar differential patterns are also evident. A higher proportion of displacees, for instance, indicated that they consider erosion as a serious hazard because it damages crops (88.2 percent) and creates food shortages (30.4 percent). The respective figures for non-displacees are 83.3 and 14.7 percent. Compared to displacee respondents (44.7 percent), a considerably higher proportion of non-displacees (70.7 percent) expressed their concern about erosion hazard because it destroys land. This suggests that the displacees feel more of a threat in terms of ultimate effects, whereas non-displacees are more aware of immediate and visible impacts. A remarkable variation in the 'perceived causes of bank erosion' by displacement status is noticed for the following natural events: storm surges and waves, high floods, high discharge, and high precipitation.

The number of displacement experiences provides further insights into the experiential differentials for the perception of hazards. The chi-square test results for the perceived hazard concepts by displacement status show that the former variable is not independent of the latter; the level of significance is 0.025 (see Table 8.10). Noticeably, the proportion of those who consider bank erosion as a hazard in the upper displacement frequency category (i.e., more than 12 times) varies considerably with the other categories.

Among floodplain users, the perceived threats of erosion to resources are also found to be related to displacement frequency. The chi-square test results for the overall distribution show that these variables are not independent of each other; the level of significance is 0.01 (df=24). Again, the patterns in perceived threats of erosion for those with more than 12 displacement experiences differ considerably from other displacement frequency categories.

The distribution of the 'causes of erosion' component of perception according to displacement frequency offers some interesting findings. Most displacee respondents (61.6 percent) with less than three displacements perceive "fast current" as a major cause of bank erosion, while one-third (32.1 percent) believe that it occurs according to the 'will of Allah.' In contrast, among all the displacement frequency categories, respondents with more than 12 displacements account for the maximum proportion (45.0 percent) of those expressing the fatalistic view the erosion is dependent upon God's will.

Following Burton (1972), we may presume that persons with displacement experience are relatively aware of the scientific causes of erosion, but they attempt 'dissonance reduction'[98] to eliminate hazard threat. The chi-square test results for perceived causes of erosion indicate that these variables do not vary significantly according to displacement frequency categories. While all other tests for experiential differentials of hazard perception are found to be statistically significant, it is more likely that the responses for this component are characterized by a Type II error.[99] The findings of the chi-square tests reinforce Saarinen's (1966) and Brooks's (1973) observations that hazard perceptions are related to previous experience of the event.

Locational Dimensions

The physiographic or landform variety, biotic diversity, and uneven spatial distribution of riverine events in floodplain areas like Kazipur, provide sufficient justification for classifying the study area into the three zones of *kaem*, *bir*, and *char* (see Chapter 6). This stratification of the study area represents the low, moderate and high erosion-hazard potential of these zones. But is this stratification reflected in the perceptions of hazards among the people of these zones? Hypothetically, if perception varies in accordance with the potential hazardousness of an area, persons in the zones of high erosion potential will show the greatest concern of erosion-threat, whereas those residing in zones of low hazard potential will demonstrate less awareness of the event.

In the present study two subsets of the spatial dimension, one by zone and the other by distance between riverbank and homestead location are analyzed to reveal their relationships to perceptions of erosion hazard. People residing in the highly hazard-prone areas (i.e., the *char* areas) demonstrate an awareness of more riverine hazard concepts relative to those in the interior zone (i.e., the KZ). Compared to CZ respondents, a higher proportion of KZ respondents listed meteorological and biological extreme events as threats to them. As pointed out earlier, the population in the intermediate zones tends to fluctuate between two extremes in perception of hazards (Burton *et al.*, 1968). In the case of Kazipur, this assertion appears to be quite apparent.

Similar to the 'perceived hazard concept' component, the 'threats of erosion to resources,' as well as the 'causes of erosion' components take part in determining the hazardousness of an area. A relatively higher percentage of respondents in the CZ compared to the KZ consider erosion a hazard because it threatens crops, food, income and job opportunities. The threat of erosion to land is also more prominent among mainland zone respondents those than in the CZ.

Respondents in the KZ also emphasize high floods as the cause of erosion (54.8 percent) compared to either the BZ or the CZ samples (less than 15 percent). Among the three zones, the proportion of CZ respondents is highest for those who perceive erosion events as the 'will of Allah.' The above findings confirm the hypothesis linking differential perception to hazard zone; people in zones with high hazard potential perceive the hazard components more objectively than do the people of the low hazard potential zones. Also, people in the high hazard potential zone (i.e., the *char* areas) are more capable of perceiving the long-term effects of hazards; the people in the low hazard potential zone (i.e., the KZ) tend to identify more the immediate impacts.

For a further understanding of spatial differentials in hazard perceptions, the geographical distance between riverbank and present location of homestead is also considered as an explanatory variable. More than 60 percent of respondents in each distance category up to half a mile view erosion as a hazard. In contrast, less than 25 percent of respondents in each of the distance categories above half a mile recognize erosion to be a threat. Also, people residing at locations distant from the bankline are more concerned with biological events (i.e., famines and epidemics). Chi-square tests show that both hazardousness of the zones and distance between homestead and river bankline are significant explanatory variables at the 0.001 level for differentials in all three of the hazard perception components (Table 8.10). Thus, this subset of the hypothesis relating to spatial differentials in hazard perception substantiates that the perception of a hazard is related to the distance between floodplain users' usual place of residence and the source of the natural event.

8.3.2. DIFFERENTIALS AND EXPLANATORY FACTORS IN POTENTIAL COPING MEASURES

The 'individual choice model' of Slovic *et al.*, (1974) and Burton *et al.*, (1974) suggests that humans are rational beings capable of evaluating the consequences of selected adjustments, and then selecting one or a combination of actions. This presumption may not be valid in developing societies because, first, the range of alternatives is socially constrained, and second, the scope of voluntary choice in adjustments is primarily dependent upon economic ability and the local sociopolitical power structure. In other words, an individual's socioeconomic position and ability may not permit them to perform rationally selected coping measures. Rather, individuals are often compelled to react and behave according to the limited options offered them by the prevailing social structure. Our understanding of potential hazard coping strategies should then include entitlements to the socioeconomic opportunities, the extent of hazard encounters, and the potential hazardousness of their usual place of residence.

Socioeconomic-Structural Dimensions

Kates (1971) and other researchers (Brooks, 1972, 1973; McDonald, 1979; Irish and Falconer, 1979) assert that the "initial set of known adjustments is . . . a function of an individual manager's attributes, specifically his casual and specialized access to communication networks" (Kates, 1971: 446), particularly those in which education plays a significant role as one of the widely accepted socioeconomic indicators. Yet the survey data show that potential coping strategies have no significant relationship to the educational status of resource users ($X^2=14.97$; df=14; see Table 8.11). The chi-square tests, in contrast, reveal that the individual's status in terms of income, landownership, and occupation are significant variables explaining variation in potential adjustment strategies.

Income is the aggregate product of landownership, non-agricultural activities, and other formal and informal economic activities in rural Bangladesh. The proportions of middle (51.3 percent) and upper (58.3 percent) income groups are much higher than the lower (37.9 percent) and marginal (33.6 percent) groups who opt to abandon their land.

TABLE 8.11. Relationships between potential coping strategies and socioeconomic-structural dimensions, 1985

Socioeconomic Dimensions (Explanatory Variables)	Types of Potential Coping Strategies (Dependent Variables)			
	Calculated Chi-Square	df	Critical Chi-Square	Significance Level
Educational Attainment of Household Head	14.97	14	23.68	NS
Household Income	33.55	21	32.67	0.05*
Landownership of Household	48.89	14	36.12	0.001*
Occupation of Household Head	49.30	21	46.80	0.001*

*Significant; NS = Not Significant

The opposite tendency is also present; the higher the income, the greater the tendency to retain ownership of submerged land and thus to attempt to reduce loss if it reemerges in the future. Thus, in this context the better-off households can benefit from either strategy depending on their circumstances. As well, migration from villages vulnerable to erosion to safer places is related to the income levels (i.e., economic ability) of floodplain inhabitants.

Among the three landownership groups, the landless account for the lowest proportion in each of the adjustment strategy options. Since the landless do not legally own any parcel of cultivable land nor any valuable property, their concerns are more concentrated on survival strategies, namely, food and shelter. More than 60 percent of the landless view the salvaging of their houses as the only option available to reduce hazard loss. However, due to their lack of economic and social support compared to the middle and large farmers, the proportion of landless who opt to salvage their houses is lower relative to the other landownership groups.

The higher correlation between coping strategies and socioeconomic dimensions is also shown in the proportion of landowner groups. The large landowners have a number of options and are thus able to minimize their potential loss more effectively than are the small or marginal landowners. For instance, large landowners appear to opt more frequently for adjustment measures such as 'salvaging housing structure,' 'keeping ownership of land for reemergence,' and 'migration from village' than do smaller landowners and the landless. The range of options available to large landowners is increased by their ability and accessibility to information about safer locations for homestead land, their ability to pay land revenue taxes on submerged land, and the possibility to move to new land elsewhere. The landless and marginal landowners, in contrast, are compelled to accept the maximum loss potential because of their limited or non-existing options to move or sell their property in advance; they are instead likely to stay until the natural disaster occurs.

Nonetheless, the criterion of landownership alone cannot completely account for an individual's position in the social structure. A person can enjoy higher entitlements and accessibilities to socioeconomic opportunities by being engaged in white-collar or other non-agricultural occupations. This additional dimension can be analyzed by examining the occupational characteristics of floodplain inhabitants.

As was expected, laborers appear to be least concerned about land-related adjustment strategies. Only 25 percent of laborers view abandonment of land as a possible strategy—the lowest level of response among all the occupational categories. Likewise, only 7.4 percent of laborers see the retention of landownership as a means to reduce hazard loss; the respective figure for 'farmers and persons engaged in agricultural production' is 32.5. Laborers are the most 'disadvantaged' sector relative to other floodplain inhabitants; they consistently make up the highest proportion having to 'move their family to *bandh*.'

Experiential Dimensions

Both displacement status and displacement frequency appear to be significant explanatory variables; the null hypotheses are rejected at the 0.001 and 0.01 levels respectively (Table 8.12).

Displacees appear to be more pessimistic about active measures for coping with bank erosion than non-displacees. About 46 percent of displacees indicate that they will abandon their land, and thus accept maximum loss potential for future erosion; the respective percentage for non-displacees is 38.2. Meanwhile, more than 30 percent of non-displacees intend to maintain ownership of land by continuing to pay revenue taxes regularly for submerged lands in the hope that it will reemerge. However, only 17 percent of displacees perceive this as a potential adjustment strategy. Also, the more recurrent the displacement, the more is the tendency to become pessimistic and take only passive adjustment strategies. For instance, respondents who have experienced more than 12 displacements have a higher proportion of passive responses (such as abandoning land, praying to God) than have respondents with less than three displacements.

TABLE 8.12. Relationships between potential coping strategies and experiential dimensions, 1985

Experiential Dimensions (Explanatory Variables)	Types of Potential Coping Strategies (Dependent Variable)			
	Calculated Chi-Square	df	Critical Chi-Square	Significance Level
Displacement Status	26.53	7	24.32	0.001*
Displacement Frequency	48.63	28	48.28	0.01*

Locational Dimensions

The relationships between potential coping strategies and spatial dimensions can be discerned from Table 8.13. The percentage distributions of these variables are shown in Table 8.14. Evidently, the people in the CZ experience considerably more recurrent displacement and opt more for passive strategies, especially in relation to respondents in the KZ. A realistic view of adjustment strategies among respondents in more hazardous zones (i.e., the CZ and the BZ) is also apparent. As noted earlier, the geomorphological adoption of *preventive* measures to modify hazard loss is not feasible in the study area. However, 22 percent of KZ respondents view such preventive measures as potential adjustment measures; the respective percentages are 1.3 and 2.0 for the BZ and the CZ.

TABLE 8.13. Relationships between potential coping strategies and locational dimensions, 1985

Spatial Dimensions (Explanatory Variables)	Types of Potential Coping Strategies (Dependent Variable)			
	Calculated Chi-Square	df	Critical Chi-Square	Significance Level
Potential Hazardousness of an Area (i.e. Zone)	114.06	14	36.12	0.001*
Distance between Homestead and Bankline	195.52	35	66.62	0.001*

*Significant

Some interesting patterns are apparent in the distribution of potential coping measures in terms of the distance of a homestead from the bankline. The proportion of respondents who consider protecting their land by structural measures is higher among those located some distance from the bankline. Similar response patterns are shown for the strategy of retaining ownership of land that may reemerge. Also, a higher proportion of respondents living at some distance from the bankline (i.e., more than half a mile) consider migration to a safer place compared to those living in the vicinity of the river; in turn, this latter group commonly evacuates their families to the embankment (see Table 8.14).

The above response patterns may be attributed to a number of factors, but two are of principal importance. First, floodplain users have historically maintained an attachment to the aesthetic values of the riverine environment and to a sense of shared economic needs (see Chapter 5). Second, the continual hope for the reemergence of land is common among displacees, which reflects the crisis of losing land along with its inherent socioeconomic value.

TABLE 8.14. Potential coping strategies by zone, 1985 (%)

Type of Potential Coping Strategies (Multiple Response Possible)	Zones KZ n=73	BZ n=227	CZ n=247
i. BEARING THE LOSS			
Remain in the Homestead Land Until Affected by Erosion	5.5	-	3.6
Abandon Land	37.0	40.1	47.0
Pray to *Allah* (God)	37.0	5.3	23.5
ii. REDUCTION OF LOSS			
Protect Land	21.9	1.3	2.0
Dismantle Housing Structure	71.2	69.2	68.8
Keep Ownership of Submerged Land	32.9	23.3	19.8
iii. CHANGE USE/LOCATION			
Leave Village	54.8	24.7	37.2
Move Family to Embankment	6.8	18.1	6.9

8.3.3. INFERENCES FROM TEST RESULTS: A SUMMARY

Socioeconomic-Structural Dimensions

Socioeconomic-structural dimensions have generally been neglected in studies of responses to natural hazards. Only one socioeconomic variable, namely education, has received some attention in previous studies (Kates, 1962; Roder, 1961; M. Islam, 1974; Baker and Patton, 1974; Leigh and Sim, 1983). We may presume that the relative paucity in the use of socioeconomic variables in natural hazards studies has been influenced by the findings of some early studies which suggest that education does not play a significant role in responses to hazards. The present study reopens the debate by reintroducing socioeconomic-structural dimensions through the NAPP framework to explain variations in the perception of and coping responses to hazards; though education is again examined, the variables also include aggregate household income, landownership status, and occupational status.

Out of a total of sixteen chi-square tests, in five cases the null hypotheses could not be rejected at the 0.05 or lower level (Table 8.10). Of the four selected socioeconomic variables, educational attainment of respondents proved to have no relationship to either the perception of hazards or to the potential adjustment measures. This observation substantiates Kates' (1971) and Burton and Kates' (1964) findings in North America, and M. Islam's conclusions in his study of cyclone hazards in Bangladesh.

Collectively, the analysis of the socioeconomic-structural dimensions shows that the perception of hazards and the potential coping strategies undertaken in response to hazard-threat depend upon the ownership of the means of production and resource control. Ownership of the means of production is shown to be the most powerful explanatory variable among the socioeconomic-structural dimensions. As such, all components of the perception of hazards and adjustment measures are found to be significantly related to landownership status. Landless respondents are well aware of erosion hazard, but their lack of concern about the threat of erosion to land reflects their inaccessibility to ownership. Likewise, since the landless do not legally own any land nor possess any valuable property, their adjustment strategies are consistently more focused on survival strategies, namely seeking food and shelter. The large landowners, in contrast, have a wide range of options and are thus able to minimize their potential losses by taking corrective measures. These findings contradict M. Islam's (1974: 21-22) suggestions that insignificant relationships exist between 'socioeconomic class' and the perception of floods, and between 'socioeconomic class' and the attitudes towards floods.

The tests of household income and occupational status also show the significance of social and economic relations of production as explanatory variables of responses to hazards. The higher the income the greater is the tendency to take active measures with higher risk, and thus to attempt to reduce hazard loss. With regard to the landless, the analysis of potential adjustment strategies by occupational categories reveals that laborers are less concerned about land-related adjustment measures. Again, such findings refute M. Islam's (1974: 21-22) conclusions. The propositions established by this study affirm that socioeconomic-structural components significantly influence the range of options available to resource users and in turn affect their attitudes towards hazards and types of coping measures.

Experiential Dimensions

Our examination of the influence of experiential dimensions on responses to bank erosion relies on two variables, namely the displacement status (displacee or non-displacee) and displacement frequency (number of lifetime displacements due to bank erosion). In total, eight chi-square tests were undertaken to reveal the relationships between responses to hazard and experiential dimensions (see Table 8.10). From these, displacement status shows strong explanatory powers for both components of human responses—the perception of hazards and the potential coping strategies.

A relatively high level of hazard awareness among displacees is evidenced by their feeling threatened by ultimate effects, whereas non-displacees are more concerned with the immediate and visible threats. For instance, displacees emphasized the threats of erosion because it damages crops and creates food shortages, while non-displacees placed greater emphasis on the threat to land. In response to potential coping strategies, displacees are more pessimistic and undertake more passive measures when compared to non-displacees. All four chi-square tests of the components of human responses to hazards and displacement status are found to be significant at the 0.001 level. Furthermore, all but responses to 'perceived causes of erosion' according to displacement frequency are found to be significant at the 0.05 or lower level. From the survey data it appears that persons with higher displacement recurrence are more aware of the physical causes of erosion but attempt 'dissonance reduction' to eliminate the hazard threat (see Burton, 1972: 187). Analogous to the findings for displacement status, it is evident that as the number of displacements grows, the greater an individual's tendency to become pessimistic, and to refuse any active measure to help cope with the hazard.

Collectively, these findings suggest that attitudes toward nature, the perception of hazard threats, the causes of natural events and actual coping measures undertaken to cope with hazard effects, all significantly rely upon previous experiences of the event. The test results reinforce previous studies that have tested similar hypotheses using the behavioral-ecological paradigm (Kates, 1962; Saarinen, 1966; Kirkby, 1974; Oliver, 1975; Irish and Falconer, 1979; Brooks, 1972, 1973).

Locational Dimensions

In order to explain responses of floodplain dwellers to bank erosion, two spatial dimension variables are considered: (i) potential hazardousness of an area in terms of the risk zones of the study area; and (ii) the distance between current homestead and the bankline. All eight chi-square tests of responses according to these spatial dimensions support the hypotheses that these variables are not independent of each other; the null hypotheses are rejected at the 0.001 level (see Table 8.10). These findings imply that the erosion potential of the zones, as well as the level of respondents' familiarity with the source of the natural event relative to their geographical distance from the bankline are powerful explanations for differences in responses to erosion problems.

The above inferences take into account the fact that people residing in the high-risk areas (i.e., the *char* areas) expressed awareness of riverine hazards relative to those in the interior lands (i.e., the *kaem* zone). People in the *char* areas are more capable of perceiving the long-term effects of hazards, while people in the low hazard potential zone tend to perceive the more immediate effects. Evidently, people in the *char* zones experience considerably higher displacement frequency and opt for more passive strategies. Nonetheless, since preventive measures to modify hazard events are not technically or socially feasible in the study area, a passive strategy obviously reflects a realistic view among floodplain users. Overall, inhabitants residing in close proximity to the bankline demonstrate more realistic coping measures than those in distant locations.

In sum, these findings suggest that locational dimensions are indeed significant variables for explaining variations in responses to hazards. People in high-risk zones are more aware of the hazards than those in lower-risk zones; this finding is in agreement with the speculative hypothesis. Saarinen (1966) finds a similarly strong relationship between the relative aridity of Great Plains counties in the USA and farmers' perceptions of drought risks. Likewise, Brooks (1973) also supports this assertion with evidence of drought hazards in northeastern Brazil. These findings thus strengthen earlier postulations that peoples' attachment to a place in conjunction with the hazardousness of an area influence human perceptions and coping behaviors in hazardous environments.

CHAPTER 9

THE DISPLACED POOR IN URBAN ENVIRONMENTS
The Case of Squatters in Serajganj

9.1. Disasters-Induced Displacement and Rural-Urban Migration

9.1.1. THE RURAL TO URBAN MIGRATION OPTION: WHO MOVES TO TOWN?

Environmental disasters are usually characterized by the suddenness of their occurrence. Riverbank erosion hazard is no exception. As discussed in Chapter 4, the dynamic behavior of a braided river such as the Brahmaputra-Jamuna, especially in a deltaic environment, is difficult to theorize. Under such uncertainty, floodplain inhabitants seldom have the ability to anticipate the severity of an extreme environmental event, nor are they well prepared to react. Regardless, individual resource-users do not tend to undertake coping measures at times of stress due to the haphazard and abrupt onset of disastrous events. Contrastingly, in responding to a crisis that has occurred it has been observed that most disaster victims assess a wide spectrum of options (Hossain, 1988: 49; Haque, 1991; 1992). Obviously, coping options are usually not equally available to all diaster victims. Instead, the nature and number of options may primarily depend on the individual victim's abilities, entitlements and social network, the availability of transportation and communication, and emergency, relief and rehabilitation services.

In the case of riverbank erosion hazard, displacees who have lost their homestead presumably first consider the available options for shelter. As noted earlier, after their displacement the majority of displacees prefer to stay in the close vicinity of their place of origin. The rationality of such a decision can be partly attributed to the familiarity of disaster victims with the surrounding physical and social environment, a comfort which in turn eliminates or at least minimizes the stress that could emerge from the challenge of adjusting to a new location (Oliver-Smith, 1982; Coburn *et al.*, 1984). The tendency to stay close to a previous residence is also partly influenced by the hope of repossessing the lost land if it ever emerges from the riverbed. Among other options is to move to the rural interior located a safe distance from the large rivers. But such a response requires previous savings to buy homestead land or the assistance of a kin who would be willing either to donate a parcel of land for housing construction or provide a temporary shelter.

Urban centers also attract many displacees. The privileged social groups with higher income can often generate enough economic surplus to buy additional homestead land in towns and cities. In the event of losing a rural homestead they can readily move to houses in urban centers. A move of this kind is usually temporary as the rural elites tend to appropriate income from their rural assets and resources. In some cases, if the offspring are independent only the senior members of the family may decide to return to the countryside. Urban life has some appeal: sending children to urban schools and thereby entering them into urban, non-agricultural economic sectors elevates the social status and mobility of the family (Choudhury, 1978). Still, in terms of the total disaster-induced rural to urban migrants, the proportion of migrants belonging to a privileged social class is usually nominal.

Another group, the most impoverized and relatively 'helpless' displacees may not 'manage' to stay in rural areas. The lack of institutional supports and the required resources and network to move to other rural interiors leave them with no option but to move to urban areas as destitute migrants. Generally speaking, the disadvantaged displacees initially attempt to take shelter on public land, particularly on the flood control embankments. Either because of continuous harassment by public officials to induce voluntary movement or due to the lack of employment opportunities, many of the displacees decide to try their luck in towns and cities. Alamgir (1978) views the problem of rural-urban migration of the poor in the broader perspective of social processes. He notes that some social factors, such as exploitative production relations, lead to the social class configurations that sustain and aggravate the process of 'underdevelopment' in society and throw a growing number of people below the poverty line. It is such impoverished groups who become the first victims of the squeeze that comes in the wake of any socioeconomic, political, or environmental disasters (also see Eckstein, 1977). Many of these victims eventually find their way to urban centers.

Reasons for rural out-migration are also related to regional variations, especially in terms of the frequency and magnitude of different natural disasters. Qadir (1975) and Zaman (1984) observe that people move from riverine districts (e.g., Faridpur and Barisal) because of the threat of riverbank erosion and flooding, and from heavily populated rural districts (e.g. Dhaka, Comilla and Noakhali) mainly because they find difficulties in obtaining gainful work. Rahman (1985) notes that a mix of factors pushed hundreds of thousands of people out of rural areas to urban locations. These factors include the destructive impacts of the War of Independence, the famine of 1974, and adverse effects of frequent natural calamities such as cyclones, floods, riverbank erosion, and salinity due to salt water intrusion. Studying a slum in Dhaka city, Akhter (1984) cites that about a quarter of slum dwellers migrated from rural areas to Dhaka because they were uprooted by natural disasters. In his survey of three villages in Kazipur, Hossain (1984) records that about one-tenth of the riverbank erosion-induced displacees moved to urban areas in search of shelter, jobs, and other opportunities, joining the myriads of other voluntary, often economically motivated, rural-urban migrants.

9.1.2. URBANIZATION AND SQUATTER SETTLEMENTS

The flow of rural-urban migration has been so massive in the developing world in recent decades that urban areas have not been able to absorb all incoming migrants. As a result, urban areas of the developing world are faced with innumerable socioeconomic problems. Some obvious issues associated with rapid urbanization include the lack of housing and services, unemployment, illegal squatting, crime, overcrowding and the emergence of a polarized class structure. Due to the absence of a harmony between demand for an urban work force and the rate of urban influx (Lloyd, 1979), poverty has become endemic. Despite widespread and continually growing urban poverty, towns and cities continue to grow, both demographically and spatially, from both natural increase and in-migration.

The process of urban agglomerations in the above manner brings rise to slums and squatter settlements. Slums, in a general sense, are overcrowded squalid districts of an urban area inhabited by poor people. Squatter settlements represent similar housing conditions but the land that the housing sits upon is not legally owned by the inhabitants. Squatting is commonly recognized as an illegal act, but city officials often fail to address this problem due to the lack of alternative housing. Ulack (1978) suggests that slums and squatter settlements can be viewed in two ways. First, such settlements are a negative feature in urban life: they are a mere extension of rural areas and are more of a hindrance than a benefit (Lewis, 1966). From the squatter settlements pauperization spreads to the rest of the town; these areas are therefore described as the 'slums of despair.' A second view suggests that squatter settlements play an important role in urban development since their residents eventually become integrated into the rest of the urban area. Squatter settlements provide cheap accommodation for newly arriving migrants and thus allow for greater personal savings. They also create a cheap labor pool for the urban manufacturing and service sectors, thereby promoting macro-economic growth and development. The settlements become 'slums of hope' (Mangin, 1967), according to this view.

Bangladesh is still a country with a relatively low level of urbanization compared to South and Southeast Asia (Kosiniski and Elahi, 1985). According to the 1991 census enumeration, the percentage of urban population was 20.15 of the total population of the country (BBS, 1996). However, the rate of urban growth in recent decades has been very rapid and overwhelming. Elahi (1972) and Islam (1976) point out that the urbanization process which gained a momentum after the independence of Bangladesh in 1971 is likely to continue in the foreseeable future. The intercensal growth of urban population was recorded to be 137.6, 110.9 and 65.6 percent during 1961-74, 1974-81, and 1981-91 respectively. Khan (1982) asserts that the current growth rate is due more to rural-urban migration than a natural increase in urban population. This same phenomenon has caused mushrooming growth in squatter settlements in many urban centers. In general, more than one-fifth of the migrants fail to find standard housing facilities and are thus compelled to live in squatter colonies (see Jackson, 1979). In

1974 the squatter population constituted 21.6 and 16.9 percent of the total in-migrants to Dhaka and Khulna respectively (Islam, 1976). The Serajganj Municipality Authority, locally known as *Pourashava*, estimated that in 1985 about 18 percent of the town's population were squatters. Of these, about six percent were riverbank erosion-affected displacees (Hossain and Greenberg, 1985; Figure 9.1).

A number of questions can be raised here. Having experienced a sudden and traumatic uprooting and relocation, the riverbank erosion-induced displacees are required to cope with the consequent stresses of displacement. How do they cope with such stresses? The magnitude of economic, social and cultural constraints in the new urban environment are enormous for the in-migrants. How are these constraints addressed by the displacees in squatter settlements? A rural occupational background and skills are likely to be of less relevance to urban life. Even if the newcomers are absorbed into the informal urban sector, can they move to other occupations? What is the nature of occupational and social mobility? Are the displacee poor accepted as part of existing urban culture and social milieu by the permanent residents? What is the extent of economic, social and cultural integration of the displacee squatter dwellers within the prevailing urban society? The following sections will examine these issues, particularly focusing the squatters' reasons for moving to towns, and on their adjustment and assimilation processes.

9.1.3. STUDY SITE AND THE SAMPLE

The town of Serajganj was chosen as the site for studying riverbank erosion-induced displacees in the urban context. The severity of erosion in the Serajganj area and its resultant effects upon the locality provided adequate rationale for the selection. The history of Serajganj urban settlement dates back to 1809 when the *maroary*[100] traders began to develop it as a trading center. The new settlement evolved into a major riverport by 1930 as the old Jenai turned into a remarkably larger river as the Brahmaputra-Jamuna captured it. The Serajganj *Pourashava* was officially founded in 1869.

Since its inception as a small village market, Serajganj has been under the constant threat of riverbank erosion. Saha (1924) reported that the town had to shift its location on three occasions in the preceeding century. In 1848 the entire settlement had to be relocated to a new site to the west as a result of a significant westward shift of the Brahmaputra-Jamuna River. In 1876 the severe erosive nature of the Brahmaputra-Jamuna at Serajganj reappeared. Then between the 1860s and 1920s, the Serajganj *Pourashava* area was reduced from 18 square kilometers to 13.8 square kilometers due to encroachment of the river into the township. The neighboring areas of the town have also been affected in the same manner over the last century. More than one hundred villages in Serajganj district have been washed away during the last few decades (Hossain, 1984). Hossain (1988) reports that "a considerable number of local and re-

Figure 9.1. The 1995 flood victims who took shelter in the Serajganj District Headquarters building. Often such disasters-induced migrants cannot return to their rural homes, instead beginning life in urban squatter settlements.

gional erosion-affected displacees have moved to the town and live in its squatter settlements."

The Serajganj urban squatter survey was conducted as a pilot study to examine the nature and characteristics of adjustments of displacees to urban conditions, their social and occupational absorption and mobility, and possibility of their future relocation. Greenberg (1986) and Hossain (1988) conducted the survey in 1985 under the auspices of the REIS project. A reconnaissance survey of all squatters revealed that in 1985 there were 975 households who had lost their rural homesteads in the past and migrated to the town of Serajganj. These displacees resided in three major squatter concentrations within the municipality: (i) a flood protection embankment (520 households), (ii) a dried-up old riverbed (250 households), and (iii) an abandoned railway line (205 households). As a sample, a total of 207 displacee households were selected at random from these three squatter settlements. The distribution of the selected households was as follows: 133 from the flood protection embankment; 23 from the old riverbed; and 51 households from the abandoned railway line.

The sample household heads were interviewed using a questionnaire survey. In all cases except one, the household head was a male. The average family size was 5.7, which is slightly higher than the national average of that time (i.e., 1985). Most of the respondents were married (88 percent) and a vast majority were illiterate (73 percent). Statistics revealed that urban squatters in Serajganj consisted mostly of young adult household heads; about 47 percent of the respondents were below 39 years of age, including 5 percent who fell below 20 years of age. About three-quarters of the respondents experienced multiple displacements; the *mean* number of displacements was four with the average residency of squatters in Serajganj being 12 years.

9.1.4. MIGRATION CHARACTERISTICS

The survey confirmed that the squatters were displaced by encroachment of the Brahmaputra-Jamuna River onto settlements between 1925 and 1985. The highest percentage of residents (34 percent) migrated to Serajganj between 1965 and 1975, while about 15 percent migrated to the town in 1970 alone. In general, three broad phases in the migration flows can be identified: (i) a slow increase from 1925 to 1965; (ii) a rapid increase between 1965 and 1975; and (iii) a sharp decline between 1975-1985 (Hossain, 1988). These phases in migration suggest that most squatter settlers moved to the town before the current decade.

The slow influx of displacees into Serajganj town during 1925-1965 can be explained by the socioeconomic dynamics of that period. Inhabitants in the riparian environment have great affinity for their original place of residence and agricultural lands. Due to this attachment to ancestral land and traditional livelihood, rural settlers do not usually opt for a change in resource use and keep residing in erosion-prone riverbank areas. As noted earlier, the attachment to land is largely related to economic reasons. A year-round return from agricultural lands and associated activities forms the basis of rural household economy. During 1925-1965, such assets and income, along with material supports from kins and relatives possibly enabled most displacees to meet at least their basic household needs, thus discouraging them from moving to Serajganj. However, recurrent displacements and presumable failures in the local support system eventually prompted many displacees to move to the town.

The increasing number of displacees arriving in Serajganj town during the period 1965-75 was influenced by several factors. A scheme for protecting the town from erosion was initiated by the national government in 1962 (IECO, 1964; EPWAPDA, 1968; Engineering Consultants Inc., 1970). This structural-engineering scheme created a sense of security among displacees, while also opening up avenues for employment, although temporary in nature. These opportunities may have encouraged many rural displacees to migrate to Serajganj. In addition, the rural parts of Serajganj district were particularly hard-hit by erosion in 1970 and again in 1974, causing large-scale displacements. The 1974 famine also resulted in a mass exodus of rural people to urban areas (Bertocci, 1982; Maniruzzaman, 1975). These factors together contributed to a steady increase in rural-urban migration from riverbank erosion victims.

Only about 11 percent of the sample arrived in Serajganj during 1975-85. Hossain (1988) argues that a decline in rural-urban migration during the late 1970s and early 1980s relative to the 1965-75 period was largely associated with socioeconomic factors rather than a reduction in riverbank erosion. He notes that "the town is overburdened with unemployed or underemployed squatters; administrative negligence in providing displacees with relief or assistance; and displacees are not socially accepted by the urban community" (Hossain, 1988: 48). In sum, the factors have discouraged rural displacees to move to Serajganj town in recent years.

As suggested earlier, although the suddenness of the erosion activity causes immense stress for resource-users, disaster victims attempt to optimize their migration decision by taking the most 'suitable' option under their disposal. However, the primary choice is often not attained due to lack of resources, networks, information and social supports at their destinations. Table 9.1 presents the data concerning the primary choices displacees had during their last displacement. Notably, less than one-third of respondents thought of migrating to urban centers, which implies that about 71 percent of the sample either could not attain their first choice or remained passive. About one-quarter of the respondents considered evacuating their homestead and many of them may have eventually moved to Serajganj. Thus, there could be some overlap in options; for this reason we should be careful about explaining the decision-making process during and after the displacement. Some 17 percent did not consider any options at all but were forced to move to the town when displaced. Overall, the survey data concerning coping and migration options considered by the displacees suggest that the vast majority of displacees were not at first willing to move to the urban centers voluntarily. However, due to limitations and hindrances in making a livelihood in rural areas they eventually migrated to towns and cities to live in slums and squatters.

Both the 1985 and 1995 rural survey data of Kazipur reveal that the majority of displacees remained with 3.24 kilometers of their previous place of residence following a displacement caused by erosion. As elaborated in Chapter 7, the primary reasons for such short distance moves are identified as displacees' inability to afford the due costs, and the hope that their eroded land will reemerge. This finding is supported by Oliver-Smith (1982) in the case of Peruvian earthquake victims, and by Hossain (1984). Both Oliver-Smith (1982) and Mangalam (1968) explain short distance migratory behavior in terms of the migrants' familiarity with the social organization and willingness to continue with their cultural identity. Table 9.2 summarizes the distance moved by the displacees during their last displacement. The maximum frequency (*mode*) appears in the distance category of 0.80-3.24 kilometers; 54 percent having moved that distance. This pattern of distance based on the urban sample of displacees conforms with the rural surveys of Kazipur *thana*. Differences arise between the two in that the case of Kazipur is a rural to rural migration, while Serajganj represents a rural to urban migration. In the former case, the issue concerns how displacees succeed in making new economic and social spaces in a familiar environment. In the latter case, the key questions relate to how displacees adapt themselves to a new sociocultural context and infrastructural setting and how well the host societies attempt to integrate the newcomers into the town's social spheres.

TABLE 9.1. Primary options considered by urban displacees at the time of their last displacement

Primary Option Concerning Short-Distance Move and/or Migration	Frequency	Percentage
Move to urban area	60	29.0
Evacuate homestead	50	24.2
Move to the embankment or other area of free occupancy	34	16.4
Build new homestead on own land elsewhere	12	5.8
Move to mainland rural areas	12	5.8
Move to *char* areas	3	1.5
Considered no option	36	17.4
Total	207	100.0

Source: Hossain (1988: 49)

TABLE 9.2. Distance moved by Serajganj squatters during their last displacement

Distance in Km	Frequency	Percentage
Less than 0.80	12	5.7
0.80 - 3.24	112	54.0
3.25 - 6.48	41	19.7
6.49 - 12.96	30	14.4
More than 12.96	12	6.2
Total	207	100.0

Source: Hossain (1988: 51)

The place of origin of the displacee-squatter settlers does not vary significantly. In Serajganj town more than 84 percent come from within Serajganj *thana*, while only about 15 percent originate from beyond the district boundary. The displacees who crossed the district boundary all originated from adjoining districts, namely Pabna, Tangail, and Mymensingh. Notably, the principal cities of these districts are much further away than is Serajganj. While erosion-induced displacees searched for alternative locations to settle than only nearby rural areas, they opted for taking shelter as urban squatters in Serajganj, the town nearest to them. The factors influencing these displacees to select Serajganj as their destination need to be explored.

It is established that rural to urban migration, especially in developing nations, is predominantly a product of perceived economic opportunities at the destination (Connell *et al.*, 1976; Todaro, 1969). Migration flows are readily explained by the simplistic *push-pull* model, where negative factors act at the origin and positive factors at the destination. In the case of natural hazards-induced migrations, migrants are forced to move from their place of origin by *push* factors. We may thus presume that, although some attraction factors such as employment opportunities, eventually may *pull* displacees to Serajganj, the principal impetus is the disasters-induced *push* from the rural areas. The majority of the displacees (i.e., 61.8 percent) in squatter settlements answered that they chose to move to Serajganj rather than other locations because of availability of employment and/or the existence of *khas* land (Table 9.3). Other factors listed were influence of advice from neighbors (12 percent), availability of a food rationing facility (9 percent), and assurance of obtaining material supports from relatives and friends (9 percent).

TABLE 9.3. Primary reasons for choosing Serajganj as a destination

Reasons	Frequency	Percentage
Better job opportunity	99	47.8
Existence of *khas* land	29	14.0
Followed neighbors	24	11.6
Availability of food rationing	19	9.2
Expected material supports from friends/relatives	18	8.7
Education facilities for children	11	5.3
Availability of land to be leased	4	1.9
Availability of medical facilities	3	1.5
Total	207	100.0

Source: Hossain (1988: 59)

From the reasons offered for choosing Serajganj as a destination, it appears that on the whole economic factors were dominant for displacees, while considerations of some social factors and service facilities were also important. Such responses were not unexpected since the main concern of poor displacee households is to survive, particularly after losing land and other principal means of livelihood. What is noticeable in the urban squatter sample data is that unlike Todaro's (1969) proposition that a predominance of economic factors motivate voluntary migration, the availability of social and service facilities also appear to be influential in determining the destination of the disasters-induced migrants in Serajganj. In Bangladesh, riverbank erosion disaster initiates involuntary rural to urban migration which puts a strain on already impoverished towns and municipalities. Being landless, displacees are pushed into a cycle of impoverishment since they have few if any other options in agricultural sectors and limited opportunities in the towns where they take up residence. Psychological distress also often mounts due to prevailing conditions within squatter settlements and from the the pressure of the community's negative perceptions of them.

9.2. Marginalization and Poverty in Squatter Settlements

Marginalization is defined as a form of 'social entropy,' where the organization of society into complex social structures is attained at considerable cost to certain strata of society (Adams, 1974). As Lomnitz (1977) explains, these strata of society become marginalized in the sense that they exist increasingly at the periphery of the dominant economic and social system. Being compelled to live on the periphery of urban life, the impoverished displacees in Serajganj may gradually become 'marginals.' It is also important to determine the factors and processes that assist and prevent the displacees from becoming assimilated into the mainstream of urban socioeconomic and cultural life. The concept of marginality as used in this chapter refers to the squatters' peripheral status which has resulted from the deterioration of their socioeconomic situation since becoming displacees. Presumably, with a subsistence income, poor prospects for employment, dilapidated dwellings, and inaccessibility to other urban services, displacees turn into the most disadvantaged of urban inhabitants, entrapped in a vicious cycle of impoverishment. To elucidate these aspects, we now look to land occupancy patterns, occupational changes and mobility, income status, and other infrastructural and service facilities available to displacees as representative analytical variables.

9.2.1. LAND OCCUPANCY PATTERNS

Generally, in-migrants (particularly disasters-induced displacees) are unable to purchase land, build a house, or pay rent for accommodation in urban areas and thus tend to occupy vacant land as squatters. Stokes (1970) notes that the development processes of these settlements depart from the normal growth of a city. Peil (1976) adds that the growth of urban squatter settlements in West Africa was in response to oppressive public policy and practice. Bureaucratic decisions often ignore the needs and wants of low-income people who therefore have few options other than to locate in squatter settlements.

The urban squatter survey data indicate that the majority of respondents had not visited Serajganj prior to their arrival as displacees. They, however, did have some degree of prior information about the availability of shelter in the town. This was because their place of origin was seldom far from Serajganj, and relatives and friends passed along information about Serajganj as a migration destination. Such contacts further encouraged the displacees to relocate to Serajganj, and helped them erect their shacks at the present squatter locations. Several respondents reported that local administrators and townfolk did not prevent them from moving to Serajganj, perhaps mostly on humanitarian grounds, and actually permitted them to take shelter wherever they could.

As stated above, displacees often squat on unused public land, locally termed as *khas*land. The figures in Table 9.4 indicate that a majority of respondents (77 percent) reside on *khas*land. In the absence of any specific institutional effort for providing shelter or aid to displacees, finding a shelter on vacant government land is a common practice. Alternately, about 16 percent of the displacees owned, while four percent rented homestead land. These individuals resided in the same neighborhood, had a similar income profile, lived in poor housing, and had similar social experiences as the squatters who lived on *khas* land. In consideration of these characteristics, displacees who owned and rented homestead land in the town were not separated from the squatters in the analysis. Land is usually a premium in a populous, agrarian economy like Bangladesh, and in Serajganj there is threat of losing control over *khas*land. For this reason the townfolk are reluctant to make any permanent arrangements for the legal residence of displacees in their town. This state of affairs, along with the continuous influx of displacees in the squatter settlements, further accentuate the demand for already scarce land resources.

TABLE 9.4. Land occupancy patterns of urban squatters, Serajganj, 1985

Type of Occupancy	Frequency	Percentage
Khas land	160	77.3
Rented	8	3.9
Owned	34	16.4
Others (Floaters)	5	2.4
Total	207	100.0

Source: Hossain and Roopnarine (1992: 54)

In order to explore the existing resources of displacee households, their current agricultural land holdings are recorded and assessed. The lion-share of the displacees are landless; more than 82 percent of the respondents do not possess any agricultural or homestead land. Only eight percent have land between 0.1 - 1.01 hectares, and are grouped as "poor" peasants, while 10 percent own 1.02 - 3.04 hectares of land and are classified as 'middle' farmers. An examination of the amount of land lost by the displacees in the squatter settlements reveals that only six percent of the sample were landless prior to their last displacement (Hossain, 1988). The survey data also indicates that most displacee families lost all land they possessed, while the majority of whom were 'poor' and 'middle' peasants (82 percent) before their last displacement.

Who are the presently landless households in the squatter settlements? What was their landownership status before being dislocated by erosion disaster? A comparison of currently landless respondents with their pre-displacement landownership status indicates that half of currently landless households were classified as 'poor' peasants, followed by 'middle' (37 percent) and 'rich' (with more than 3.04 hectares of land; six percent) categories. These findings clearly indicate that the marginal and moderate farmers mainly move to the urban squatter settlements as destitutes.

9.2.2. OCCUPATIONAL STATUS AND MOBILITY

The occupational profile of the Serajganj squatters pre- and post-disaster is presented in Table 9.5. The majority of the respondents were previously employed in agriculture, whereas in Serajganj 62.3 percent were day laborers and service sector workers. These daily laborers are generally engaged in portering (*coolie*), earth cutting/digging, and helping in masonry works, activities which are physically arduous and low paying (Takas 30-40 per day, US $1.0), and for which employment is irregular. Service sector workers are employed in restaurants and tea stalls, factories and mills, and other low paying government jobs. An additional 18 percent of the respondents were involved in housekeeping, working as ferrymen, and selling goods such as ice cream, fish, and fodder; 10 percent of the respondents were rickshaw pullers; and only 6 percent were employed in technical work such as carpentry, masonry, tailoring, printing press composing, fish-net weaving and as cycle/tube-well mechanics. Meanwhile, about five percent of the respondents failed to secure any form of non-agricultural occupation in the town and worked as farm laborers in nearby villages; others worked as domestic servants, tutors, preachers, and restaurant helpers.

All the above occupations are generated within an informal sector of an urban economy (Nelson, 1969; Lomnitz, 1977; Amin, 1982) and as such do not provide the job security and the regular flow of income squatters once realized in their rural-agrarian setting. Todaro (1985) suggests that the primary motivation of workers in informal sector activities is simply to earn enough to survive. Informal sector jobs then serve as a useful entry point into the labor market (Prakash, 1983), but the instability of this work only adds to the magnitude of hardship and sacrifice on the part of the in-migrants. Wages are depressed due to the abundant cheap labor pool; it is not uncommon for laborers to work for 12 hour days but earn only Taka 40 (less than one US $), a sum hardly sufficient to sustain life on a daily basis (Hossain and Roopnarine, 1992). Moreover, with very few skills and their lack of awareness of the urban job market, squatter settlers are predisposed to economic exploitation. Not only is unemployment among squatters high (56 percent), but employers dictate the wage-bargain platforms. Squatters thus are forced to accept cheap wages which often contribute to their downward spiral (Greenberg, 1986).

The occupational composition of Serajganj squatters varies considerably from that in large cities. Amin (1982), for instance, observes in Dhaka that the majority (54 percent) of informal sector laborers are 'production workers' employed in the production, construction, and transport sectors of the city. In contrast, the survey of Serajganj squatters reveals that the higher percentage (41 percent) of squatters are engaged in day laboring activities, including construction occupations. Such regional variations are most likely due to differences in the scale of economy and city size. A much larger city than Serajganj, Dhaka can offer more diverse options in the job market, especially in the manufacturing and production sectors which play a relatively important role in the urban economy.

The above findings can be correlated with the observations of Lomnitz (1977). In the Mexican context she finds that marginals are engaged in unaffiliated manual labor, unpaid family labor, and small-scale family enterprises commonly termed as informal activities. Studying squatters in Tehran, Kazemi (1980) also cites similar trends in occupational status and composition. However, the findings of the squatter survey in Serajganj dispute Perlman's (1976) assertions that rural migrants to cities are universally above national averages in education, skills, and acquaintance with urban ways. Serajganj squatters are largely illiterate and they possess very limited urban job skills so necessary for the formal sectors.

An examination of employment demonstrates the appalling conditions that exist for squatters. When respondents were asked if employment conditions have changed since arriving in Serajganj, 55 percent said that they have deteriorated and 71 percent responded that their income has decreased despite a higher cost of living in the town compared to their rural home. The respondents who claimed that their employment status has deteriorated since arriving in Serajganj suggested that their prospects were particularly limited. Notably, squatters in other countries have not experienced the same decline in employment. For example, a 1977 study of urban migrants in Jakarta, Indonesia records that only nine percent of the sample experienced a decline in their occupational status (Krausse, 1977). Similarly, a study of urban migrants in the city of Ludhiana, Punjab, India, shows that "only" 25 percent of the sample experienced a decrease in income upon arriving in the city (Oberai and Singh, 1984: 509). In the Serajganj sample, 71 percent claimed that their income declined.

Both the Jakarta and Ludhiana studies find that occupational mobility potentially improves over time, a conclusion which is in keeping with other research on this issue (Lipset, 1959; Glass, 1954). Yet McGee (1971), who tested the assertion of occupational mobility among migrants in Kuala Lumpur, Malaysia, refuted the hypothesis concerning the positive effect of length of residence upon occupational mobility. In terms of in-migrants' adaptation and assimilation, occupational mobility is a very significant aspect. Rural to urban migrants arrive in cities with almost no possessions, inadequate education and few urban skills. The degree to which the city has a favorable economic climate for migrants to upgrade their economic status is important to their subsequent success.

One method of analyzing occupational mobility is to examine the occupational status of workers and establish whether they fall into the formal or informal sector

(Greenberg, 1986). In related studies, a common assumption is that the longer the length of residence, the greater the incidence of formal sector employment (Lipset, 1959; Glass, 1954; Oberai and Singh, 1984). Survey data presented in Table 9.5 show the effect length of residence has upon occupational status in the formal and informal sectors among heads of households in Serajganj squatter settlements. Theoretically, if the length of residence in a locality influences occupational mobility, the proportion of formal sector employment should indicate an increase over time, relative to the informal sector. The patterns in Serajganj do not appear to be this straightforward.

In the early years of migration to Serajganj, displacee-squatters obviously faced difficulty competing without the necessary skills and networks to enter into formal sector jobs. They overcame some of these constraints within their first decade of stay in the town—an attainment reflected in the fact that the majority of squatters with 6-10 years of residence in Serajganj entered formal sector employment (Table 9.5). What is conspicuous in the survey data is the declining numbers of respondents in the formal sector in the categories of 11-15 and 16 and more years. If one considers the long-term trend in occupational mobility of the displacee-squatters, the common assertion that length of residence has positive effects upon occupational mobility can be refuted. In general, riverbank erosion-induced displacees in Serajganj do not experience improvement in their occupational status with length of residence. The processes that create a favorable environment for economic growth, as enumerated by Lipset or as seen in Oberai and Singh's site in Ludhiana, are not operating in the same manner for the Serajganj squatters. It would be interesting to detect what factors and barriers exist that affect occupational mobility; these aspects warrant further research.

A typical displacee who was interviewed, namely Abdul Ali Mirza, aged 25, accounts for his situation in a way that typifies displacee-squatters:

> In 1973 when I was a boy of 13, my family lost our one acre of land (jute and rice) to riverbank erosion. We moved to Serajganj town and lived on the abandoned railway line. My father died so I pulled rickshaw and my brother became an urban daily laborer. Now, 12 years later, my household which consists of my wife, two sons and mother, live in the same hut on the same abandoned railway line. I am still pulling rickshaw and have made no economic progress (Greenberg, 1986: 58).

9.2.3. INCOME STATUS

While the majority of squatters (59 percent) are engaged in the informal sector in Serajganj, these irregular, low wage-earning activities tend to exacerbate their poverty. The *mean* monthly income of squatter households was Taka 945 (approximately US $24) at the time of the 1985 survey; the range of the monthly income of squatters is between Taka 360 (US $9) and Taka 3,000 (US $75). A positive correlation between meagre income levels and informal sector employment is unveiled when a comparison between the formal and informal sectors is attempted. Table 9.6 shows how poor wages of the informal sector—mean monthly income earned by a worker in the formal sector was Taka 1,112 compared to Taka 832 in the informal sector, a 25 percent difference. A statistical test of significance verifies whether the sample findings are valid for the displacee-squatter 'population' in general; chi-square calculations reveal that the income differences between formal and informal sectors for Serajganj squatters are significant at the 0.05 level [X^2(df=3; $p<0.05$) = 15.99]. This finding validates the hypothesis that displacees dependent on the urban informal sector are poorly paid relative to formal sector workers.

TABLE 9.5. Length of residence and occupational status of urban squatters, Serajganj, 1985

Length of Residence (Years)	Formal Sector		Informal Sector	
	Frequency	Percentage	Frequency	Percentage
0 - 5	8	25.8	23	74.2
5 - 10	23	51.1	22	48.9
11 - 15	40	44.4	50	55.5
16 and more	14	34.1	27	65.9
Total	85	41.1	122	58.9

Source: Greenberg (1986: 57)

TABLE 9.6. Monthly income as related to formal and informal occupations of the squatter settlers, Serajganj, 1985

Income Per Month (in Taka)	Formal Occupations		Informal Occupations	
	Frequency	Percentage	Frequency	Percentage
1,000 or less (US $25 or less)	55	64.7	101	82.8
1,001 - 1,500 (US $25.1 - 37.5)	17	20.0	19	15.6
1,501 and above (US $37.5 and above)	13	15.3	2	1.6
Total	85	100.0	122	100.0

Source: Calculated from Greenberg (1986: 62)

Overall, the findings reveal that in the absence of a dynamic economic environment which would include the expanding manufacturing and service sectors, the displacees cannot free themselves from dependence upon informal sector employment and its resultant poverty. The lack of employment opportunities, skills and training for jobs outside of the occupations traditionally occupied by displacees (e.g., daily laboring and rickshaw pulling) results in perpetual poverty and inhibits an adaptation to the urban economy and society. More importantly, the effects of a low income, including malnutrition, forces squatter households to face a 'long-term adaptation' to undernourishment (see Dasgupta, 1993: 441-472).

Gould and Lowentin (1979) observe that adaptation can occur at three hierarchical levels: (i) genetic; (ii) physiological; and (iii) behavioral. For instance, the low average height of pygmies is a genetic disposition, whereas stunted women and children in destitute households is an expression of physiological adaptation. Economizing on bodily movements in response to low energy intake is itself a behavioral adaptation. Most profound and visible among the Serajganj squatters is the physiological adaptation. These aspects need further research for a fuller understanding.

The effects of ill-health in association with malnourishment and undernourishment encompass a wide range of areas; the issues are more complex than they might at first appear. The general effects of persistent undernourishment and infections vary widely, but they all cause an impaired life. Several researchers (e.g., Adair, 1984; Mueller and Pollitt, 1982) confirm that in expectant mothers these factors affect the growth of the fetus and therefore the child's health status at birth. Nutritional anaemia, common among expectant mothers in poor countries, generates fatigue and lowers resistance to infections (Levin, 1986; Whitehead, 1989). Furthermore, chronic malnutrition in children causes muscle wastage and growth retardation, and thereby the future capability for productive work (Spurr, 1988; Martorell, 1990); it also eventually increases morbidity and childrens' proneness to infections (Scrimshaw *et al.*, 1968; Chen and Scrimshaw, 1983). Among adults, chronic malnutrition diminishes muscular strength, capacity to do physical work, and protection against a wide range of infectious diseases (Martorell, 1985), also bringing in its wake noticeable psychological changes manifested by mental apathy, depression, introversion, and lower mental concentration (Read, 1977; Dasgupta, 1993). The Serajganj squatters are subject to all of the above features and characteristics. The REIS urban survey did not specifically incorporate these aspects for in-depth investigation, but such symptoms were recorded in field notes and by the investigators' firsthand observations.

9.2.4. INFRASTRUCTURE AND SERVICE FACILITIES

Housing Conditions

The changes in housing conditions after displacement is one of the most significant indicators of poverty among displacees. Squatters reside in clusters of shacks constructed mainly of bamboo, straw, and grass (Figure 9.2). To build a shack, the squatters have no choice but to use indigenous methods and technologies as well as cheap, locally available materials. This results in very poor housing conditions. Although 60 percent of sample houses do possess a tin roof, the overall state of houses is very poor, with leaking roofs, withered walls, and muddy floors being the norm. Mathur (1984: 5) identifies similarly poor housing conditions among slum dwellers in India.

Furthermore, houses are small—five or six people share a single room, while 47 percent of the sample households have only a one-room house (Figure 9.2). Cooking, eating and sleeping are all performed in such a space. Since the majority of the squatters have been used to better housing and other infrastructural facilities during their pre-displacement period, adapting to life in a one-room house is psychologically and physiologically difficult. When asked about changes in their housing conditions compared to their pre-displacement life, over 81 percent of respondents reported that their housing conditions had substantially deteriorated. Only four percent of displacee-squatters succeeded in improving their housing structure and thereby their overall living condition. The inability to afford better houses combined with the unavailability of land for housing and the lack of service facilities together contribute to greater incidence of illness and overall poor health conditions.

Thirty-five percent of the squatters surveyed do not have access to a tube-well. Sanitation conditions are deplorable and no households have electricity. Lacking the resources to relocate permanently, about 71 percent of the squatters have been living in the same shack since arriving in Serajganj. Not surprisingly, these shacks are in dilapidated condition and need immediate repairs; yet squatters cannot afford to repair or rebuild. A survey of slum settlements in four large cities in Bangladesh reveals that one-fifth of the households had electricity and most had sanitation facilities (Islam, 1976; also see Qadir, 1975). Comparatively speaking, the squatters in Serajganj are among the most impoverished settlements in Bangladesh.

Health Conditions

Informal sector employment not only causes poverty for the displacees but also results in poor health. The inability to afford latrines, tube-wells or hospital care when necessary, together with lack of a balanced diet, contribute to appalling health conditions among the squatters. Social, economic and physiological deprivation undoubtedly results in a marked hindrance to adaptation. Aside from inadequate hygiene and the inability of squatters to afford basic health care, their adaptation to this new environment is further obstructed by emotional stress caused by relocation. Fried's (1963) "grieving for a lost home" syndrome applies as much to Serajganj squatters as it does to those who are displaced in connection with urban renewal in the United States. Thus, a combination of poverty and stress leads to a drastic deterioration in health conditions among displacees which in turn affects their ability to adapt to an urban social environment.

The fact that Serajganj has an adequate medical and healthcare infrastructure intensifies the dismal situation the squatters face. In spite of better access to medical and health services compared to their prior situation, health conditions of the squatters have declined drastically. For example, while Bangladesh's population to doctor ratio is

Figure 9.2. Squatter settlement in Serajganj Town. Squatters reside in clusters of shacks constructed mainly of bamboo, straw, and grass. In these settlements, drainage is poor and latrines are merely open pits (Photo: Charles Greenberg).

11,000:1 (de Vylder, 1982), in the town of Serajganj the ratio is 1,307:1 (BBS, 1983). Likewise, Bangladesh's (1974) ratio of population to hospital beds is 5,300:1, in Serajganj it is 778:1 (BBS, 1983). Evidently, these riverbank erosion-induced displacees reside in a place which possesses the urban 'luxury' of health care. Unfortunately, they are excluded from the benefits of these resources.

A significant reason for poor health conditions is the lack of adequate service facilities. A proper drainage system is absent from the squatter settlements. The squatters' shacks sit on irregularly sloped plots that gather stagnant water during monsoon season. Toilets are equally inadequate and unsanitary, and latrines are merely open pits; meanwhile, sewage connections and organized garbage removal facilities are non-existent. In fact, 18 percent of the households have no access to a latrine. In comparison, a 1979 study of slum populations in Dhaka, Rajshahi, Khulna and Chittagong found that only four percent of the households had no access to a latrine (Islam, 1979).

Access to controlled and clean water is also limited --there is no tap water in the Serajganj squatter settlements and 35 percent of households do not have access to a tube-well. Many squatters draw water for washing, cooking and drinking from ponds and the river. Those who have tube-wells compete with numerous other households for access to them. The ratio of population to tube-well is very high, with some servicing hundreds of households. The result is excessive pressure on the pumps and wells that frequently malfunction or dry up.

The stress associated with relocation also increases the intensity of illness. Relocation coupled with disaster deepens the stress even further for squatters. Barton (1970) argues that disasters represent stress situations which occur when members of a society fail to receive expected conditions of life. The fact that Serajganj squatters are not only migrants, but are also involuntarily dislocated by a disaster, increases their life-change and subsequent illness. The literature on the effect of disaster upon illness in the 'remedy[101]' period is vast (Spiegal, 1957; Chapman, 1962). However, only a few studies have investigated the nature and occurrence of secondary illnesses on post-disaster populations, years after the phenomena. Melick (1978) examines post-flood illnesses among white males in Wilkes-Barre, Pennsylvania, USA, finding that flood-respondents were significantly more likely to report that their health was worse than before the flood than were non-flood respondents. Scudder (1968), researching populations displaced by human-made lakes in Africa, reports that despite improved medical services, mortality and morbidity rates appeared to rise. He presumes that a combination of environmental, epidemiological, demographic (increased population densities), and psychological factors were involved. Also, the stress of disaster and dislocation combined with a temporary loss of identity in a new environment contributes to physiological breakdown.

Table 9.7 exhibits various crude death rates in Bangladesh. Considering the appalling conditions among the squatters of Serajganj, coupled with the intense stress they have experienced, the results, although disturbing, are not too surprising. Also, infant mortality rates are shockingly high in the squatter settlements, especially compared to national levels. An infant mortality rate of 216 per 1,000 represents one of the highest in the world.

TABLE 9.7. Mortality rates in Bangladesh by location

Location and/or Source	CDR (per 1,000 population)	IMR (per 1,000 population)
Rural areas, Bangladesh (BBS, 1975)	-	162
Slums in Dhaka City (Islam, 1979)	19.14	-
Bangladesh (BBS, 1980)	-	153
Bangladesh (PRB, 1985)	17.00	133
Squatters in Serajganj (Greenberg, 1986)	27.00	216

Another aspect worth investigating is the change in fertility among the displacees. In the agro-based rural areas a large family is beneficial for successful farming and economic stability. In urban areas, on the other hand, a large family is not essential and is often detrimental. In this respect it could be anticipated that length of residence and household size would be negatively correlated. However, in a coefficient of correlation test an 'unexpected' finding emerges ($r=0.11$). This observation demonstrates that the economic stimulus and motivation for smaller families is not present among the squatters. With high unemployment and a high cost of living, one would suppose that the average family size would decrease over the years in Serajganj. The fact that fertility remains high concurrent with significant infant mortality rates in the urban environment demonstrates the inability of the displacees to shed previous rural customs. Abu-Lughod (1961), in studying migrant adjustment in Egyptian cities, suggests that "With a lower capacity for assimilation, they tend to build for themselves within the city, a replica of the culture they left behind."

High fertility, stress, and improper sanitation and service facilities all characterize the prevailing health conditions among the displacees. Also, rigorous working conditions can affect health. Abu Bakar, aged 19, a squatter residing on the abandoned railway line, relates his tale of failing health:

> In 1970 my family lost our small plot of land and we moved to the Serajganj squatter settlements. In 1982 my father died and I was forced to pull rickshaw to support my mother, two brothers and sister. After only three years of pulling, I desperately need a new job. My health is continuously deteriorating because of the strain pulling puts on my body. I need a salary job with better working conditions. I like Serajganj and will probably live here permanently but I urgently need a healthier job.

9.3. Resource Conflict and the Exclusion of Squatters

9.3.1. SOCIAL MORASS

With a growing number of people displaced by homestead land-loss due to riverbank erosion each year, the town of Serajganj has witnessed a steady influx of the destitute. The municipal authority, *Serajganj Pourashava*, views the displacees as endangering the 'normal' and 'peaceful' life of the urban dwellers; squatter settlements are perceived as temporary and the people are treated as non-locals. In conveying such notions the *Pourashava* has deliberately created distinctions between the townfolk and displacees. Local administrators and influential residents are reluctant to accommodate in-migrants into the social domain of Serajganj. Public officials, planners and administrators are supposedly the agents with responsibility for alleviating poverty and integrating squatters. All evidence shows that the Serajganj squatters are excluded by local government agencies from any attempt at integration and assimilation; these officials appear not to comprehend the scale or nature of the displacees' plight. In fact, the administrators create barriers that make integration difficult.

Smith (1976), in his study of environmental quality in Windsor, Ontario, Canada, finds that urban residents and city planners have a different spatial awareness of various urban environmental hazards. He argues that policy-makers are not fully aware of some of the problems and risks faced by the public. Chapin (1965: 67), in connection with the same issue, states that "There is a strong possibility that some planning may reflect more the values of the city planners themselves than of the people of the community." Redclift (1973) claims that often in Latin American squatter settlements the administrators and policy makers, mostly consisting of politicians, purposefully keep squatters in poverty to maintain their political support. The powers of patronage are often bait for political parties to secure support among squatters in return for promises of financial or material assistance. Some political offices are even elected mainly by the votes of squatters. In most cases, however, elected politicians bring about few improvements, their elections founded on "empty promises" (Bamberger, 1968).

The degree of such political manipulation in Serajganj is not precisely known, but there is clear indication that administrators are not paying proper heed to the squatters' situation. Firstly, the *Pourashava* office lacks critical demographic and spatial statistics on the squatters: their ignorance and exclusion of the squatters is evident. When an investigator inquired about information and questions concerning the displacees, the responses of *Pourashava* officials were grossly inaccurate. For example, they claimed that there were approximately 500 homeless displacees (floating population) in Serajganj. The REIS investigators recorded only as few as 10 displacee 'floaters.' Also, the Serajganj *Pourashava* office suggested that a small prostitution village on the outskirts of town was heavily populated by displacee women. A survey of the village, however, revealed that not a single prostitute was a victim of riverbank erosion disaster. This type of prejudice is likely founded in officials' deliberate attempts to undermine the credibility of displacees and subjugate them socioeconomically.

Secondly, though the *Pourashava* office reported that there was no rehabilitation program for squatters, a rationing program does exist. Various food items are regularly purchased from government stores at subsidized rates and distributed among the squatters—a claim made by the Serajganj *thana* Relief Office. Despite these apparently good intentions, the survey data do not exhibit that this distribution is as extensive as the relief office suggests. Over 80 percent of squatters have never received assistance. What happens to all the relief that is allocated for squatters? While local administrators in Serajganj recognize the benefits assistance can render, they are clearly not conducting proper and fair distribution. It is no secret that corruption is common among many administrators and politicians at all levels of government; there is a good probability that much of the relief is ending up in the hands of some corrupt officials.

Thirdly, as stated earlier, there exists strong evidence that local authorities refuse to admit the permanent nature of squatters in Serajganj. The fact is, nonetheless, that displacees have resided in Serajganj for an average of 12 years, while a few have been there as long as 50 years. The survey further indicates that 89 percent of households consider themselves permanent residents of the town. Also, 61 percent responded that even if they had sufficient money to purchase land in rural areas they would not for fear of repeating their displacement experience. It is important to stress the permanence of the displacees in Serajganj. Their condition of destitution and lack of options are marks of their permanency. Yet, local authorities do not provide them with assistance, amenities, nor with opportunities to assimilate. For example, in 1982, when a Global Asian Development Bank (GADB) project made 71 slab latrines available to the Serajganj *Pourashava*, none of these were distributed among the squatters—the area of the greatest need in Serajganj. The *Pourashava* Public Health Engineer suggested that since displacees reside on *khas*land and because they will eventually return to their emerged land in the rural areas, the squatter settlements should not receive any of the slab latrines. The Public Health office is also not authorized to sink tube-wells or set up latrines on *khas*land, yet a 1969 municipal law states that all households must have a latrine. The contradictory nature of these two ordinances is evident. Again, it is the squatters who pay the price for this kind of social deprivation.

9.3.2. EXCLUSION AND CONFLICT

Today, social segregation and exclusion of displacees from mainstream activities in Serajganj are well established norms. In this respect, displacees appear doubly disadvantaged. On the one hand, they are the poorest of the poor in the town; as such, they cannot afford to participate in urban communal functions. On the other hand, the townsfolk have a tendency to treat all displacee-squatters as a subordinated and distinct group, alienating them in this way from the mainstream community. The townsfolk, specifically the people involved in urban functions and decisions, regard the displacee-squatters, the *nade-bhanger lok* (people of the broken riverbank), with low status. Such a pervasive attitude largely determines social interaction between townfolk and squatters (Hossain and Roopnarain, 1992). A similar social distinction is also recognizable in rural areas between *char* and mainland villages. Zaman (1988: 126) finds that "the social significance of these distinctions goes beyond individual *izzat* (honor) and position in a status system to the determination of social relationships between *char* and mainland villagers."

The above exclusion of disadvantaged urban in-migrants is not unique to Bangladeshi squatters. A study of *favellas* (slum) in Rio de Jeneiro, Brazil, reveals that 75 percent of males and 80 percent of females believe that the state governor does not give sincere help to the slum population (Bonilla, 1961). The existing conflict between administrators and squatters is a serious issue and may lead to violent events. Attempts to reclaim *khas*land by government officials often result in conflicts between the state and squatters. Even though squatters appear to have established a permanency over the land they occupy, they are frequently ordered by local government offices to evacuate. The REIS in-depth field investigation shows that many townsfolk, themselves potential investors and users of *khas*land, often instigate local government officials to evict squatters from the *khas*land. The alliance between the state, here represented by bureaucrats, and the local elite forms a common ground against the subordinated (in a socioeconomic sense) displacee households.

In villages, local leaders maintain links with the local administration to take possession and ownership of depositional land (Zaman, 1988). Zaman finds that "these links are critical both for manipulating the local administration and for continued dominance over the peasantry" (Zaman, 1988: 125). In 1982, the squatters living on the flood protection embankment mobilized themselves against one such eviction order that subsequently led the government to abandon its eviction efforts on 'humanitarian' grounds (Rahman, 1984; Zaman and Wiest, 1985). The conflicts over land by squatters are widespread and the potential for violent clashes clearly exist. A displacee-migrant, Abdul Latif, exemplifies the resource-conflict over the land occupied by squatters:

> I have been living in this squatter settlement for 12 years. But local administrators refuse to admit to the permanent nature of my residence as I live on *khas* land. Recently, a wealthy town resident who claims that he owns this land by lease (from the government) told me to evacuate it. I lack options to relocate yet, local administrators do not provide me with any assistance. How can they be so cruel to me!

In sum, riverbank erosion-induced displacees in Serajganj are not well assimilated and integrated into their urban environment. Their occupational status, housing and health conditions are testimony to this assertion. The prevailing institutions and structures, which have created several strong barriers, hinder displacees' adaptation and assimilation.

9.4. Conclusions

An examination of the status and conditions of the displacee-squatters in Serajganj reveals that they live on the fringes of everyday life in this town. Squatters have been living here for decades, yet the natives (i.e., permanent residents) and public officials are reluctant to recognize their permanent residency in Serajganj. Prejudices and discriminatory policies and practices, economic exploitation and deprivation, and reciprocity networks among squatters may serve to slacken the pace of social assimilation and integration into mainstream urban living. Although squatters remain on the lowest rung of the economic ladder, they undoubtedly participate in the urban economic system through their daily economic activities. Improvement in economic well-being may yet provide them the impetus toward upward social mobility in the urban environment. Despite social and economic hardships, displacee-squatters have been able to develop reciprocity networks within themselves and enter into low-earning urban economic functions. This is indeed indicative of the potential of the squatters, and supports Mangin's (1967) notion that the squatters are 'ambitious.' A human resource development approach which could help squatters emerge as an independent group is imperative if we are to realize the energy and human spirit that exists in these communities on the periphery.

Part Four

Emerging Policy Issues: Towards Sustainable Reduction of Disasters and Floodplain Development

CHAPTER 10

PUBLIC POLICY ISSUES
Water Management, Hazard Mitigation and Resettlement

Human actions in mitigating disasters can be taken both individually and collectively. In both cases such actions are profoundly influenced by the social distribution and delivery of responsibilities, the perceived notions of public and private liability, the magnitude and scale of events, and the resource and technological capacity of those involved. Yet the roles and responsibilities of individuals and collectivities vary from society to society. In some instances they are relatively well-defined while in other cases ambiguities are not uncommon. In this regard, Burton *et al.*, (1993) show that an individual householder may be reluctant to take precautions against a snow avalanche because he believes that precautions should be carried out by the community or because community views or policies do not favor such adjustment actions.

Collective authorities, which include all levels of government and NGOs (non-government organizations), deal with three main policy areas of disaster prevention and mitigation. First, they provide services which control or modify natural events or reduce or prevent damage potential or distribute the burden of damage. Responsibilities of this kind are left with collective authorities primarily because the required actions exceed the capacity of individuals. Second, institutions or governments make decisions concerning the development of 'common property resources,' such as water, air, and forests, which may have direct implications for hazard vulnerability. Third, macro-level entities which include local and national governments set laws, regulations, incentives and penalties to guide citizens' behavior in a 'private' capacity (Burton *et al.*, 1993: 126).

In a developing country like Bangladesh with the absence of a well-developed private sector, the public institutions or the state assume the major roles of controlling socioeconomic imperatives and regulating natural resources. Throughout the 1970s and 1980s, for instance, the public sector accounted for more than 80 percent of the total development expenditure in Bangladesh (Rahman, 1992). Since flooding and riverbank erosion are chronic in the region, their adverse impacts are a recurring phenomenon and cause heavy losses for individual farmers and the economy in general. Humanitarian relief can at best offer temporary measures towards averting the impact of such disasters; a long-term solution to this recurring problem requires that natural disasters mitigation strategies be incorporated into the overall rural development and water resource management plans and policies. The acceptance of such a strategy is already reflected in the national government's attempt to use 'relief-wheat' as a means of building flood control and drainage infrastructure through the Food for Works program aided by the World Food Program (Rahman, 1992). To assess fully the public policies of riverine disaster prevention and mitigation in Bangladesh we must evaluate the plans, policies and programs in the areas of (a) water management, agriculture and irrigation, (b) 'flood-control' and town-protection from riverbank erosion, and (c) relocation and resettlement of disasters-induced displacees.

10.1. Rural Development, Food, Riverbank Erosion Prevention and Control Policies

The pre- and post-Independence rural development and water policies of Bangladesh can be broadly categorized in four distinct phases. The first phase (1955-1971) is characterized by post-disaster recovery and *la Grande* schemes; the second phase, a shift to small scale projects; the third phase, the return of *la Grande* schemes; and the fourth phase, the search for new directions. These water policies each have addressed either of two areas. First, a set of policies have been formulated to deal with floods, droughts or low water supply during the dry season. Another group of water management strategies have joined with national agricultural policies to improve the *status quo* food situation.

10.1.1. FIRST PHASE (1955-1971): POST-DISASTER RECOVERY AND *LA GRANDE* SCHEMES

Historical Roots

Historical evidence shows that most countries in delta areas have attempted to construct some form of infrastructure for flood protection, irrigation and drainage. The territory which is now Bangladesh is no exception. Evidence suggests that in the last millennia ditches and ponds were excavated by the state and local rulers to provide materials to build dykes. Numerous canals were also constructed to drain off surplus rainwater and to prevent meandering streams in the Tippera Surface region (Rashid, 1977).

During the British colonial period, modern hydraulic engineering helped irrigate the floodplains of the complex river systems of Punjab. The success of this technology in preserving water for the dry season gradually permitted colonization of virgin land, augmented cash crop production, and encouraged other interventions in large fluvial systems. For example, the absence of regular inundation in the zones west of the Gorai-Madhumati river led to widespread incidence of communicable diseases such as malaria and the plague, water logging in the marshes, and a short supply of silt to replenish cultivable land (Bentley, 1945; cited Ahmad, 1968). In order to address these problems, plans were made to irrigate the Moribund delta area covering the Kushtia, Jessore and

Khulna zones and to resist saltwater encroachment into arable land. The feasibility studies of this multipurpose scheme, known as the Ganges-Kobadak (G-K) Project, were carried out in the earlier decades of the 20th century, but the actual diversion of water could not commence during colonial rule.

La Grande Master Plan

The efforts by the public sector to prevent, control and mitigate floods in Bangladesh date back to the 1950s. During the first phase (1955-1971) of development and water management policies based chiefly on only post-disaster recovery strategies, the focus was on building large-scale flood prevention and irrigation schemes. Some specific cases are worth elaboration.

Following the severe flood of 1955, the government of Pakistan sought advice from the United Nations Technical Mission on flood prevention strategies. On the basis of the Krug Mission Report of 1956, a project planning, implementing, and monitoring agency, entitled the East Pakistan Water and Power Development Authority (EPWAPDA), was created in 1959. Under the auspices of the EPWAPDA and the guidance of the Krug Mission, the government began the construction of several massive riverbank and coastal embankments.

Similarly, after the 1962 floods the reports of two United Nations consultants (the Hardin Report of 1963 and the Thijsee Report of 1964) examined options for flood control, recommending various types of large-scale engineering works for the confinement of flood waters within the main channels of large rivers. These measures, which included embankment construction, channel improvement by dredging, river training and cutoff, and the construction of bypasses or floodways, were incorporated in the Master Plan of the EPWAPDA in 1964. The Master Plan was prepared for EPWAPDA by the International Engineering Company (IECO), based in San Francisco, USA. "The major emphasis in the IECO Master Plan," Bari and Shahjahan note:

> was on reducing flood damages, which was a major articulated national objective at that time. The Master Plan has been described as basically a massive scheme for empoldering large areas of the country into some 50 major projects which would provide flood protection and drainage facilities to about 12.1 x 10^6 acres and irrigation facilities to about 7.9 x 10^6 acres by 1985 (Bari and Shahjahan, 1989: 258).

Following the 1962 floods, a review of development policies also led to some efforts to integrate infrastructure development with flood control efforts. A few such important flood control projects include the Kurigram Embankment, Old Brahmaputra Phase I and II, Belcuchi, Bogra, Chalan Beel, Dhaleshwari and Bangshi Channel Improvement, and Ganga-Kaputakkhya. Other projects included the Southern Rajshahi, Dhaka Southwest, Chandpur, Gorai, Arial Khan, Boral, Haor schemes, Kushiyara, Meghna-Dhanagoda, Titas, Gumti, Khowai, Manu, and Matamohari (Tarafdar, 1974).

As a result of this new integrative approach, the Rural Works Program (RWP), which was supported by wheat supply under United States Public Law 480, began to construct large-scale engineering works including flood protection embankments. One of the most impressive structural efforts to control floods was the completion in 1968 of the Brahmaputra Right Bank Flood Protection Embankment (BRBFPE). This project has significantly decreased overbank spillage (Tarafdar, 1974), creating potential flood protection for about 2.3 million hectares of which 157,000 hectares of agricultural land are within its immediate vicinity. Rahman (1984) records a production increase of 131,856 tons (or more than 68 percent) over the pre-project total of 192,650 tons.

The First Five Year Plan (FFYP) document from the Government of Bangladesh reports that as of the 1970s embankment construction provided flood protection for about 1.2 million hectares of land along the coastal areas and floodplains. These works included the construction of more than 3,200 kilometers of coastal embankment to repel saline water intrusion and about 320 kilometers of riverbank embankment to control overbank spills. Irrigation-related physical works included a total of over 1,600 kilometers of main, secondary, and tertiary irrigation canals, some 4,600 sluices and regulators, and three major and 85 minor pumping stations (Bangladesh Planning Commission, 1973; Haque, 1994; Haque and Rahman, 1994).

The rationale for undertaking large-scale development schemes was rooted in the concept of rapid economic growth to attain the so-called 'take-off' stage of development (see Rostow, 1960). As Rutton (1986) succinctly explains, there were several reasons for large-scale infrastructure investments during the 1950s and 1960s. A major economic rationale was that public benefits exceed private benefits by such a wide margin that only the public sector could afford to undertake them. Spillover or secondary development effects were believed to "substantially exceed the benefits that could be captured in the form of price or user charge" (Rutton, 1986: 39).

Implications for Local-Level Participation

Nandy (1993) evaluates the dynamics and effects of large-scale infrastructure and flood protection projects upon their stakeholders and beneficiaries. He finds that the entire process of a project, from its selection to implementation, deals mainly with 'technicalities' and macro-level socioeconomic returns while ignoring the environment and the people affected at the local level. He cites an example to illustrate how public works gradually diminish the role of local-level participation. Under the coastal embankment plan the 'islands' were extensively embanked and sluices were constructed to allow water intake and drainage. In following years, farmers began to make cuts in the embankments. This was often done within certain organizational arrangements. In many cases, the local *beel* committees took the lead to arrange an informal understanding with WAPDA that should there be any failures, individuals would be responsible for repairs. This arrangement was actually illegal from the point of view of WAPDA. Regardless, cuts were usually made on low spots, and wooden structures—locally called *kols*—were put in place to manipulate the water. In some cases, cuts were made without informing WAPDA officials and in many instances caused noticeable damages to the embankments. Nandy (1993) concludes that the above conflict arose because the affected people were never consulted nor supported the public sector interventions.

Addressing the Riverbank Erosion Problem

In Bangladesh, floods and riverbank erosion hazards had never been spelled out in national planning documents and programs. Until the formation of Bangladesh in 1971, EPWAPDA was responsible for power generation and distribution, and for programs in the field of flood control, irrigation and drainage. From the onset, EPWAPDA was technically oriented and its mandate encompassed mainly permanent-corrective measures (i.e. flood and bank erosion control) involving structural interventions and permanent preventive measures (i.e., floodplain development and regulation). During the post-Independence period, water and power sectors were separated with the creation of the Bangladesh Water Development Board (BWDB) and the Power Development Board (PDB). Currently, the BWDB functions under the Ministry of Irrigation, Flood Control and Water Development.

As stated earlier, the concerns of national water policies focused on protection of agricultural land and rural settlements only from abnormal floods; riverbank erosion-related threats to rural resources were either ignored or embodied within the scope of flood control-related measures. However, WAPDA, which later became BWDB, eventually undertook numerous 'town protection' schemes to address riverbank erosion and scouring hazards, along with the problems of drainage congestion and overbank spilling during high flood stage. According to a report prepared in 1974 by BWDB, a total of 73 schemes were initiated to protect 11 district headquarters, 16 subdivision headquarters, 37 commercial centers and nine communication links (BWDB, 1974). In general, such engineering schemes to alleviate riverbank erosion are of two types: (1) small-scale structural works which are essentially river training measures; and (2) large-scale structural engineering measures which are related to flood alleviation schemes that include large embankments, flood-relief channels, intercepting channels and flood storage reservoirs. In order to prevent local bank erosion and slumping, the former type of intervention is often preferred.

Partial cross dams are designed to deflect the main current of a river to prevent bank erosion within a given course in order to save important urban, historical or commercial areas. This measure allows controlled stream currents for certain areas and provides relative safety for those areas immediately downstream from the structure. Still, the success of such schemes largely depends on the geographical characteristics of the floodplain as well as the current adjustment capacity of the reach.

Parallel embankments alongside the channel flow are another kind of structural measure. This type of scheme is basically designed to control bank erosion by mitigating the main current of a channel from direct action on the concave bank slope so that bank erosion is halted at least temporarily. There are, however, disadvantages to this measure. In an area of Quaternary or Recent alluvial deposits like in the floodplains of Bangladesh, a subaqueous process of erosion may eventually lead to catastrophic bank failures, development of sinkholes and rapid river channel encroachment into the land.

The other engineering-technical works for alleviating riverbank erosion involve flood-mitigation measures. For example, flood-relief channels control flood flows by the use of a series of channels in addition to natural channels, by means of dividing, diverting, or delaying the flood waters and thereby preventing the flooding of surrounding areas.

Measures undertaken by the EPWAPDA and Bangladesh WDB to protect towns and other commercial centers have involved both 'permanent-corrective' responses (e.g., cross dams, parallel embankments, earthen and concrete groynes, spurs and revetments) and 'temporary-corrective' adjustments (e.g., placing of brick-filled wire-net cages, brick mattressing, and steel 'jack-jatty' fields). In general, construction of embankments, groynes and river training works along the banks of the small rivers have been effective (Bari and Shahjahan, 1989; BWDB, 1974). However, structural-engineering interventions along the major braided and meandering rivers of Bangladesh—rivers characterized by non-cohesive soil, unstable banks and beds, and a high degree of variability in discharge and sediment load—have had very limited success (Rahman, 1984; Haque, 1983b; Rasid, 1966). The case of the Serajganj town protection scheme is worth assessing here. In 1830, the right bank of the Brahmaputra-Jamuna River was 10 kilometers away from the Sailabari and five kilometers away from the Jailkhana area of Serajganj town. By 1963, the river encroached 7.5 kilometers towards Sailabari and one kilometer towards Jailkhana. Then that same year the riverbank erosion approached Serajganj town very rapidly resulting in four kilometers of river encroachment; the river thalweg moved towards the right side of the channel which created a scour and eroded nearly 30 meters of the bank. During the floods of 1969 and 1970 several private and public buildings were washed away, including the criminal and civil courts, two rest houses, the general post office, land registration office, municipal office, veterinary hospital, and the public library.

The technical studies carried out by Satter (1980) indicated that a permanent solution to westward movement of main channel was not feasible, both from the engineering and economic perspectives. However, under immense public demand for structural-engineering intervention, the EPWAPDA authority prepared a scheme in 1970 aiming at 'permanent' protection of Serajganj town. These measures included the placement of the steel 'jack-jatty' fields (Figure 10.1) and cement riprap blocks (Figure 10.2) to reduce velocity at an estimated cost of Taka 241 million; however, only a total of Taka 75 million was spent up until 1980 (Satter, 1980). My own field observations in 1983 and 1985 suggested that the 'permanent-corrective' measures in protecting Serajganj town were in vain. In recent years the westward movement of the Brahmaputra-Jamuna along Serajganj town continues despite increasing efforts to arrest it through structural means.

Recently, I examined another politically motivated engineering scheme to deal with severe riverbank erosion. During 1992-93, settlements along the left bank of the Gan-

Figure 10.1. Steel 'jack-jatty fields' were installed along the right bank of the Brahmaputra-Jamuna to protect Serajganj Town from the encroaching channel. Although the measure succeeded in buying some time, it eventually failed to arrest the advancing channel.

ges in Munshiganj district faced rapid encroachment by the river. Since they were not aware of the limitations of structural intervention, the people of the area demanded immediate response from the national government. The BWDB, in this case, formulated a project for constructing an underwater groyne along the banks in Louhajang *thana* at an estimated cost of Taka 90 million; it was suggested that this intervention would save public and private properties and resources amounting to Taka 2 billion. Following a reassessment of the project viability by the then Prime Minister, a total of Taka 10 million was allocated. Then, in 1995 the entire groyne was washed away, along with a good portion of the town of Louhajang (Haque, 1995b).

10.1.2. SECOND PHASE (1971-1988): A SHIFT TO SMALL-SCALE PROJECTS

Economic Rationalization of Water Sector Projects

By the late 1960s people began to realize that the proportion of costs to benefits in large-scale projects was too high. Maintenance emerged as a huge, uncontrollable task—rent collection from major irrigation projects was slow and in some cases unsuc-

Figure 10.2. Cement riprap blocks are being placed to reduce water velocity and to mitigate bankline shifting near Serajganj Town. Financial corruption among bureaucrats and contractors leads to a poor quality of blocks often making this a meaningless measure.

cessful. Because implementation of large-scale projects required a much longer time, the emphasis on the sponsorship from the public sector lost its momentum and significance over time. This is not an unfamiliar scenario. Since the independence of Bangladesh, concern over limitations of economically viable large-scale projects has influenced flood control and irrigation strategies. In the early 1970s, World Bank land and water sector studies recommended changes in strategy, which led the government to shift to small-scale, low-cost, and quick-return projects in flood control, drainage and irrigation development (World Bank, 1970; 1972). The advent of low-lift pumps, tubewells and diffusion of dry-season cropping helped to promote this new agenda. This trend continued until the major floods of 1987 and 1988 devastated Bangladesh's infrastructural facilities and crop production.

Since Independence, the focus of development strategy in Bangladesh has been concentrated on achieving self-sufficiency in food supply. Agricultural extension programs, such as irrigation development and input and credit supplies to farmers have received utmost priority in public policy measures. Because the country has had an annual deficit of about two million tons of food grain for the last two decades, the main thrust has always been on increasing food crop production (Hossain, 1980; Huq, 1980). As such, the problem of natural disasters has been considered a secondary issue, only discussed to the extent that it might affect the crop production levels.

By the early 1980s, the need for a long-term water resources development plan was realized and the Government of Bangladesh established the Master Planning Organization (MPO) under the Ministry of Irrigation, Water Development and Flood Control. The Organization prepared a National Water Plan (NWP)—Phase I, in 1986, focused on the assessment of water resources and future demand. The Plan developed a range of models and analytical tools, and prepared a draft water code. Despite successes in compiling a considerable volume of information and planning tools, the NWP failed to integrate the intra-sectoral and interministerial programs and to address environmental concerns (FPCO, 1995). Further, the requirements of fisheries, navigation, public health, industries and municipalities were treated by the Plan as constraints rather than as part of an overall water sector demand status.

Still the strategies adopted at this time for achieving a self-sufficiency in food production showed some commendable successes in increasing food grain production at a rate faster than population growth. For example, during 1975-1980 food grain production in the country grew at an annual rate of about 3.2 percent compared to a growth in population of about 2.6 percent; the overall agricultural production grew at an annual rate of 2.8 percent during this period. These remarkable increases in production were made possible by replacement of long-stem local varieties of rice with the adoption of High Yielding Varieties (HYV).

Among the major irrigation, drainage and flood control projects implemented by the BWDB, the Northern Tube-Well Project and the Chandpur Project had the potential for irrigating 28,740 hectares and 51,400 hectares of land, respectively. Besides developing other infrastructural facilities (such as rural roads and communications), about 3 million hectares of agricultural land were protected from floods and a provision was made to irrigate more than 2 million hectares. Overall, given such constraints as political instability, and the untimely allocation and procurement of funds, a commendable number of works were completed under different Public Works Projects (PWP) such as Rural Works Programs and Food for Works Programs (FFWP; Alamgir, 1983; BBS, 1985; Haque and Rahman, 1994; also see Table 10.1).

Infrastructure and Drainage Dilemma

Due to increased impediments to the natural drainage systems caused by the newly placed infrastructure (e.g., embankments and dams) and the lack of an ecologically sustainable water resource policy, Bangladesh became more vulnerable to unusual floods. When the abnormally high floods struck in 1987 and 1988, the country lost more than 2 billion US dollars. Ahmed's (1989) investigation reveals that the lack of coordination among the various government authorities and NGOs in implementing infrastructure programs was largely responsible for the difficulties with monsoon season drainage. He writes:

> The Upazila Parishad, L.G.E.B., CARE, etc. give attention only to constructing the village roads without keeping proper openings for free passage of flood water . . . Sometimes these agencies even go for reconstructing village roads by closing the existing flood path. Such indiscriminate works done under F.F.W. schemes create flood hazards. In doing such unplanned construction, the spill area of flood path has greatly been restricted. These man-made barricades inevitably cause enormous reduction in the flow of flood water into the sea, resulting in the highly aggravated situation of inundation within the country (Ahmed, 1989: 202-3).

TABLE 10.1. 'Food for Works' projects and achievements, 1974/75 - 1983/84

Year	No. of projects implemented	Roads (km)	Embankments (km)	River and canal excavation constructed (km)
1974/75	21,479	3,824	1,402	1,756
1975/76	1,554	1,448	1,246	1,611
1976/77	2,328	1,735	3,067	2,694
1977/78	2,087	2,366	488	3,639
1978/79	2,113	1,770	3,487	3,082
1979/80	2,124	3,301	3,262	4,212
1980/81	3,927	4,981	5,064	6,693
1981/82	3,431	6,910	6,437	1,292
1982/83	3,029	9,213	8,582	1,724
1983/84	4,292	22,083	8,405	3,333

Source: BBS (1985).

In sum, it appears that during the second phase an emphasis on small-scale flood control and irrigation schemes, rationalization and expansion of the private sector, and a liberalized credit system for cultivators all contributed significantly to the food supply increase. Such policies, however, simultaneously made the country more susceptible to natural disasters.

10.1.3. THIRD PHASE (1988-1993): THE RETURN OF *LA GRANDE* SCHEMES

The 1988 floods of Bangladesh received overwhelming attention and responses from the international community. Flood studies were commissioned by the World Bank, the United Nations Development Program (UNDP), the United States Agency for International Development (USAID), and the national government agencies of France and Japan, in collaboration with the Bangladesh Government. The resulting reports offered varied viewpoints. The most dramatic and costly proposals were made by the French engineering consortium. Promoting the construction of embankments to prevent, divert and regulate waters, it proposed an elaborate set of new constructions along both the Ganges and the Brahmaputra and their four major distributaries, as well as seven other smaller rivers—a total of 3,350 kilometers of high embankments to prevent abnormal flooding. An implementation of these projects, which could take twenty years to construct, would cost an estimated US $10 billion, not including the annual maintenance cost ranging between US $200 million and $600 million (Bingham, 1989). Flood study teams from the United States and Japan supported a less-grandiose policy strategy, proposing to allow the river to overflow into relatively unimportant or under-populated areas while protecting the important and densely populated areas.

The major aid donors to Bangladesh discussed the flood mitigation problem at the G-7 summit meeting in Paris in July 1989. They agreed that the World Bank should coordinate the efforts of the international community so that a sound basis for achieving a real improvement in alleviating the effects of flood can be established. In a follow-up meeting held in London, UK, in December 1989, they endorsed a proposal to formulate a Flood Action Plan (FAP), involving a set of 26 studies and pilot projects (Figure 10.3). This Plan commenced with 11 main components and 15 supporting studies and pilot projects.

The components of the FAP are shown in Figure 10.3 superimposed on a plan depicting the future embankments and other works proposed in the UNDP-Government of Bangladesh Flood Policy Study. The main features of the FAP, to be implemented over 15-20 years, include: (1) strengthening of the Brahmaputra-Jamuna right embankment and improving flood control and drainage in the northwest; (2) intervention to control flooding on the left bank of the Brahmaputra-Jamuna and the development of protected areas; (3) measures to control flooding on the right bank of the Ganges, the lower Meghna, and the left bank of the Meghna; (4) developing a regional water management program for the northeast; (5) rehabilitation of the coastal embankments to protect against cyclone-related tidal surges; (6) flood protection and drainage works for the Greater Dhaka area and bank protection for towns such as Chandpur, Serjganj and Bhairab Bazar; (7) strengthening of the flood forecasting and early warning system; and (8) the development of a flood preparedness program. In addition, several supporting activities and pilot projects are suggested. Among them: (1) surveys of the operation and maintenance, agricultural, social and environmental aspects of existing flood control and remotely sensed data; (2) pilot projects in compartmentalization, bank protection, river training, fisheries and flood proofing; (3) expansion of hydrological data collection; (4) adaptation of existing surface water simulation models; and (5) a study of the institutional arrangements.

The Action Plan for Flood Control, prepared by the World Bank in 1990, accommodates both the 'flood prevention' and 'flood control' approaches (IBRD/World Bank, 1990). Actually, this plan is directed to create preparatory arrangements for the eventual implementation of a Flood Master Plan proposed in the UNDP Policy Study. As such, it apparently represents a 'conservative mix' of various studies. Nonetheless, as Boyce (1990) succinctly states, "[d]espite grave and well founded reservations as to the wisdom of massive embankments, the long-term strategy envisioned in the [World] Bank's action plan incorporates central elements of the French scheme." For instance, although the study does not completely agree to embank four major distributaries of the Ganges and the Brahmaputra-Jamuna and seven smaller rivers, it proposes to build embankments all along the major river courses in Bangladesh. In addition to these structural measures, the plan intends to create 'compartments' inside the main embankments by constructing internal embankments to allow controlled flooding by means of regulators and sluices in both the main and internal embankments. Notably, total flood prevention does not allow the natural fertility benefits to the soil and the necessary moisture supply to continue. In view of the potential damage and destruction to the existing natural resource base, there has been growing opposition to the proposal of the French consortium to construct preventive embankments (Haque, 1988; Islam, 1989; Rasid, 1989; Rogers *et al.*, 1989a).

Ironically, the French proposal ignores the fact that local rainfall often leads to high floods—compartmentalization is rather likely to aggravate the problem of drainable in many cases. Moreover, it has already been observed that the confinement of runoff to river channels intensifies the bank's erosive capacity during the monsoon period in both the Brahmaputra-Jamuna and the Ganges rivers (Haque, 1988c; Rahman, 1984). The Brahmaputra Right Bank Protection Embankment, a 217- kilometer flood control measure, has already been cited as a major cause of rapid westward migration of the Brahmaputra-Jamuna channel. Even if the proposed embankment succeeds in controlling floods, it is probable that the accelerated bank erosion of these braided rivers might outweigh the benefits for this land-scarce country. Compared to the temporary effects of seasonal flooding, the permanent features of riverbank erosion.

10.1.4. FOURTH PHASE (1993-PRESENT): SEARCH FOR NEW DIRECTION

Concerns about the potential adverse impacts of the FAP are expressed by many scholars. In a comprehensive review of the issue, Haque and Zaman (1993) suggest that the application of major preventive structural adjustments would create a number of serious problems for Bangladesh. First, the construction of embankments along the major rivers requires a capital cost of as much as US $10 billion (Bingham, 1989), this in addition to annual operating costs in the order of more than US $500,000 (Rogers *et al.*, 1989a). Since the bulk of the national development budget continues to come from foreign sources, a costly preventive program would result in a steady rise in the nation's foreign debt. Second, preventive embankments tend to provide a false sense of security and ironically may actually lead to an increase in flood damage. Consider here the findings of a recent study in southern Bangladesh by Stewart (1988) that show that the average material damage was worse in areas inside the embankment than in areas outside of it. Third, embankments impede the movement of freshwater fish and cut off their spawning areas, which then threatens the livelihood of more than a million fishermen and jeopardizes the already marginal supply of animal protein in the diet of the majority of the population. The recently completed FAP 17 study of several flood control and drainage projects reveals an inconclusive impact on fisheries (FPCO, 1995: 7). Economic studies of FAP 12 which deal with the existing flood control and drainage projects indicate that the project benefits are sensitive to fisheries losses induced by project intervention. Fourth, the coastal embankments and polder projects have seriously affected shrimp cultivation in the area. The shrimp cultivators, in need of brackish water, are breaking the embankment to allow water to penetrate to their land (Zaman, 1991b). Finally, embankments deprive the farmers of the beneficial effects of normal seasonal floods.

Normal annual flooding in Bangladesh is part of peasant life; it is a beneficial phenomenon producing the leaching and eluviation processes that have rejuvenated the soil-water properties and soil fertility in the region for centuries (Paul, 1984). New inorganic materials from the Himalayan region drain downstream and alongside floodplains to accumulate in this deltaic plain; this silt deposited by floods, however, remains infertile for several years before it begins to improve soil fertility. Brammer (1990b) notes that flood-related soil fertility is generated "from nitrogen-fixing, blue-green algae, from decomposing, submerged, plant remains living in the water, and from the increased availability of phosphorus and other nutrients in the chemically reduced, submerged top-soils." Alam and Samsuddin (1988) assess the differences in soil properties in flood-protected and unprotected areas of the Meghna-Dhonagoda flood control, drainage and irrigation project. Specifically, the particle sizes in protected zones are found to be clay dominated, with an average pH value and a higher concentration of carbon, organic matter and nitrogen, relative to unprotected zones.

FAP Supporting Activities

As stated above, the FAP consists of both short- and long-term components and 'supporting activities.' Some of these components are based on ongoing projects while others focus on research studies and pilot projects (Figure 10.3). Three 'supporting activities' of the FAP are of direct relevance to the main issues of this book: (a) the compartmentalization pilot project (FAP 20), (b) bank protection (FAP 21), and (c) river training and active floodplain management (FAP 22).

Compartmentalization. Compartmentalization is a technique designed to regulate flood levels in different parts or compartments of the floodplain by constructing secondary embankments (IBRD/World Bank, 1990; UNDP/GOB, 1989). This is to a degree analogous to the principle of poldering, in which the flood level is regulated inside an enclosed area by using drainage regulators, such as sluice gates and pumps. Nonetheless, the primary goals of these two techniques are different. The key objective of compartmentalization is to generate controlled flooding on the floodplain; in this way it functions as a component of a comprehensive floodplain management strategy, whereas polders are isolated water-management structures. The FAP calls for an elaborate construction of compartments involving: (a) secondary embankments, both

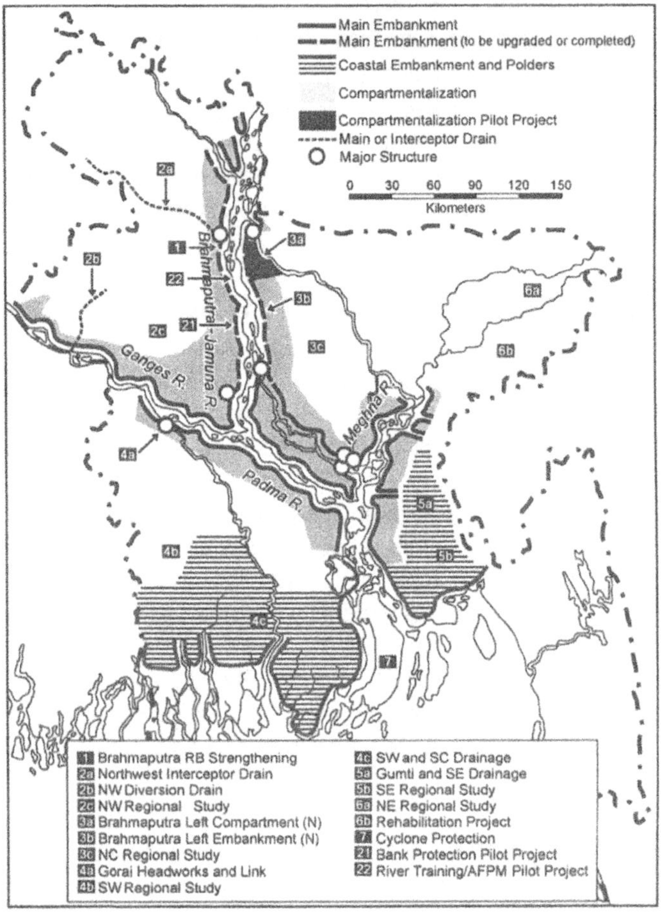

Figure 10.3. Flood policy study proposals (1989) and the Flood Action Plan (FAP) activities (1990-1994).

perpendicular and parallel to the main embankments, (b) drainage channels along the main embankments on the landward side, and (c) inlet/outlet regulators along the main embankments and drainage regulators along the secondary embankments on the landward side. Laboratory models show that with this system the peak water levels within a number of selected compartments could have been reduced from seven meters to four meters during the 1988 abnormal floods. Thus, a controlled flooding could have been kept below the danger level (UNDP/GOB, 1989: 5-39).

Rasid and Mallik (1993; 1996) further examine the advantages and limitations of compartmentalization. Based on the lessons from the Dhaka-Narayangaj-Demra (DND) case study, they identify one concern to be the potential structural instability of the embankments. BWDB has so far built about 7,000 kilometers of flood control embankments, along with several thousand kilometers of polders, especially in floodplain areas that were formerly deeply flooded during the monsoon season (Brammer, 1992). These river embankments have frequently been breached—naturally by intense flood currents and bank erosion (Figure 10.4), or more often by human-made cuts. A second problem that may arise from compartmentalization is the stagnation of water resulting from enclosed embankments. Due to blockage of the drainage channels, waterlogging was experienced in several sections of the DND projects. Rasid and Mallik (1993) argue that even deep flooding may occur in the compartments if a given flood exceeds the design flood levels in rivers, necessitating the opening of regulators and flood 'fuses' in embankments. Brammer (1990b) adds that heavy rainwater with high flood levels in adjoining rivers may also result in deep flooding within compartments; inequitable distribution of water between and within compartments could also create a serious operational constraint.

This issue of control of water distribution extends into larger conflict between wealthy and powerful landowners and marginal farmers (see Adnan, 1991). The World Bank (1990), in a review of inequity in water level regulation, reports that the regulation of freshwater polders was manipulated by the most influential farmers for the requirements of their crops. Besides these specific concerns, a more general environmental issue is raised with respect to potential ecological consequences of the conversion of open floodplains into a series of enclosed aquatic environments (see Rasid and Mallik, 1993).

Bank Protection and River Training. Two major components of the FAP are bank protection and river training (FAP 21/22), which together utilize one third of the total funding allocated to the plan. The final planning study report completed in January 1993 (FPCO, 1993) identifies two goals of the project: first, to investigate ways of refining the design criteria and improving the construction and maintenance of bank protection works, and second, to study the methods of stabilizing the courses of the river channels. In this respect, the bank protection component (FAP 21) deals with protective works where local measures are feasible and in which the river training component (FAP 22) comprises an 'active' approach, an attempt to divert the river courses from threatened reaches, and in the long-run to change the characteristics of the river. During the first phase (1991-1993; i.e., the planning study period), a review of existing 'experiences' determined that although bank protection measures in many alluvial rivers in the Indian subcontinent have been successful due to the unfavorable subsoil conditions, these same measures are not applicable to the conditions of a large braided river like the Brahmaputra-Jamuna. Also, this review found that since no deterministic principles are available to stabilize the courses of large braided sandbed rivers, a 'trial and error' approach may be necessary. Training a river like the Brahmaputra-Jamuna requires a long-term commitment since the time scale for the work exceeds half a century. Evidence further suggests that the river contains no natural 'nodal points' representing

Figure 10.4. A section of the flood protection embankment that breached during the 1984 flood. Earthen embankments often fail to withstand intense flood currents, resulting in high maintenance costs.

hard points; the only apparent hard points are seemingly the outcome of previous human intervention through engineering works. Based upon the above evidence, two test sites were selected in the Brahmaputra-Jamuna reaches: Kamarjani, to construct three groynes, and Bahadurabad, to construct eight slope revetments by 1995. These trial schemes, under FAP 21, were expected to provide insights on the real costs and effectiveness of alternative bank protection techniques. Preliminary results from the pilot projects under FAP 21/22 clearly revealed the need for a shift in overall floodplain management and river training approach, particularly in the context of the Brahmaputra-Jamuna River. FPCO states:

> The classical approach to prevent erosion and to narrow the river is by training it with 'hard' measures (e.g., groynes and revetments), which contradicts strongly the river's nature. FAP 22 considered . . . the modification of the river's planform by applying 'softer' recurrent measures, for example by closing or deferring particularly aggressive outflanking channels. This would not result in completely changing the river planform from a braided into a meandering river, but rather in reducing the braided index and finally result in a still braided river but with a reduced width. The FAP 22 study indicated that a river training

> scenario where recurrent measures and permanent structures are complementary to each other is attractive from a socioeconomic and environmental point of view (FPCO, 1995: 2).

Among the recurrent measures under FAP 22, surface screens (also known as surface vanes) and artificial channel cutoffs were identified as the most promising controlling measures. An inclination towards the application of recurrent measures in pursuing an Active Flood Plain Management (AFPM) policy was apparent among the consulting team of the study since they "make use of the forces of the river which will maintain its dynamic character to some extent but in a more controlled way" (FPCO, 1995: 2). However, there was a lack of consensus to what extent the Brahmaputra-Jamuna River can be trained and how much contribution can be achieved from the recurrent measures to river training strategies.

Another dimension of human interventions for controlling large rivers is the concerns about environmental impacts. One common technique to address these concerns is to attempt environmental impact assessments of proposed projects. In the following section these issues are analyzed in the context of the FAP.

Environmental Assessment

Although the original goals of the FAP were geared towards structural solutions, non-structural, flood proofing responses could not be ignored because of growing academic and policy debates on the long-term economic and environmental implications of the plan. The debate eventually turned into a dichotomized institutional position between the World Bank and the Government of Bangladesh—the government in favor of the construction of massive embankments along the major river systems, and the World Bank and other donor agencies "growing increasingly resistant" to such insistence (Pearce and Teickell, 1993).

After the disaster along the Mississippi River in the summer of 1993 and receiving feedback from the feasibility studies of the FPCO, momentum has already developed toward limited intervention in the Brahmaputra-Ganges floodplain ecosystems. The preliminary results from the three years of feasibility studies into the physical dimensions of the FAP have revealed "the practical impossibility of sealing off the rivers" (Pearce and Tickell, 1993: 4). Before implementing the final plan, these assessment efforts cost only US $150 million for the pilot and experimental projects, and undeniably worked as a safety valve for the massive US $10 billion to US $15 billion capital investment. The overall exercise has provided a very valuable lesson in the environmental assessment (EA) process.

The accumulated results so far for the five regional and other sectoral and supporting projects all include an EA component; this mandatory provision for all FAP projects was put in place before the commencement of the pilot studies. The FAP 16, or environmental study, includes EA guidelines for ongoing and future flood control and water management projects. The relevant variables are itemized in the first through fifth level impacts (for details see Islam, 1993). The preliminary outcomes of the pilot studies, which largely follow the EA guidelines, offer some significant lessons for policy makers. The most significant ideas stemming from these exercises include:

- All hazard mitigation and development plans should be flexible and adaptive (that is, the original plans should set up the broader 'mission objective' and establish a provision for a series of pilot experiments to identify the best alternative means of achieving the stated goals).
- The specific goals should be formulated through a dynamic process involving feedback from the pilot projects and the changing social, political and economic objectives.
- The incompatibility between the time frame of the economic plans and the ecological process, which makes the EA process quite difficult, should be recognized during the earlier planning phases.
- Public participation and direct input into the decision-making process should be regarded as a vital component of the EA process.
- It should be realized that the outcomes of short-term pilot projects (for example, the three-year experimental schemes) can only demonstrate the *tendencies* of possible impacts of the proposed plans; formulation of the EA with a high degree of confidence requires considerably longer-term ecological research. Generation of longitudinal databases of all major physical, biophysical, and socioeconomic parameters would be a significant step in this direction.

10.2. Public Perception and Preference of the Future Government Role

10.2.1. THE NEED FOR PUBLIC INVOLVEMENT

The debate concerning Bangladesh's riverine hazards is far from over, and will likely continue in the foreseeable future. The engineering profession and the BWDB favor massive structural intervention in the form of embankments, polders, and compartmentalization projects (Boyce, 1990; Brammer, 1990b; Rasid, 1993). The environmentalists and social scientists, however, oppose large-scale intervention because of their concerns about the potentially adverse ecological and social impacts of such projects (Khalil, 1990; Adnan, 1991; Rasid and Mallik, 1993; Haque and Zaman, 1993). One of the ironies of this controversy is that the supposed beneficiaries of the proposed schemes, that is the floodplain inhabitants, have been left out of the debate (Paul, 1997; Rasid, 1993). Rasid (1993) points out that neither the BWDB nor the international consultants heard any major input from the floodplain residents regarding their choice of flood control techniques. Notably, in many developed nations, the requirement for public involvement (PI) in the national environmental decision-making process appeared as early as in the 1960s. For example, by the late 1960s and into the 1970s the US Congress had attached PI requirements to every important piece of environmental legislation (Brudge and Robertson, 1990; Haque, 1996b). Following this trend, in 1980 the World Bank formulated a new policy on population relocation and resettlement associated with development projects in which the Bank calls for a greater role of PI in the decision-making process (Cernea, 1989).

10.2.2. PUBLIC PERCEPTION AND ATTITUDE: SURVEY RESULTS

In recognizing the unacknowledged public awareness and attitudes towards various aspects of the FAP, several scholars have attempted to determine floodplain inhabitants' perceptions of and preferences for governmental responses (Paul, 1984; 1995; 1997; Alam, 1990; Rasid, 1993; Rasid and Mallik, 1995; 1996). Paul (1995) surveyed 170 male household heads of farms in two *thanas* in Manikganj and Tangail districts to study their opinion on the possible impact of the proposed embankments as outlined in the FAP. The farmers supported structural solutions to the flood problem; however, the study revealed that they were not familiar with the multitude of ecological and environmental impacts which might be generated due to the embankment projects. Respondents were also not aware of the cost of the construction of embankments, nor of the fact that the large-scale project would cause a steady rise in the nation's foreign debt. Insufficient knowledge about non-structural measures of flood control, such as zoning, regulation, flood insurance, flood forecasting and early flood warning systems, was also found to be common. Paul (1997) carried out a follow-up study, consisting of 293 female respondents to determine the presence of any significant gender variation in attitudes towards the FAP. The results revealed that female responses were not significantly different than their male counterparts.

In a study of 195 households in Tista and the Ganges-Brahmaputra floodplains, Rasid (1993) also found that all structural measures were favored heavily. However, the majority of the Tista respondents, already benefitting from embankments and from partially regulated flood levels by the Tista barrage, could not envision benefits from additional embankments. Notably, Rasid had little success in obtaining spontaneous responses, and therefore used a pre-determined checklist to record respondents' preferences. This method of prompting may largely misrepresent the actual attitudes of respondents, especially in regions with low literacy and a low or limited access to objective information (Haque *et al.*, 1985a). The study further revealed that nearly all respondents, irrespective of their geographical location, preferred regulated flood levels (i.e., from 0.3 to 2.0 meters) on the floodplain, instead of confining floodwaters within the main river channels. Such a preference was influenced by the belief that regulated flood levels would secure the monsoon crops, particularly transplanted *aman* rice.

A relatively larger survey, consisting of 486 respondents from 23 sample villages representing five major floodplains, reaffirmed the similar types of attitudes among farmers across the country (Rasid and Mallik, 1995). In this study Rasid and Mallik (1995) were interested in finding out whether the practising farmers' perception of the impact of compartmentalization schemes conform with the 'official' assumption of simultaneous gains from flood control and increased rice production. The findings revealed that the majority of the respondents were against the idea of preventing flooding entirely. However, the overwhelming majority of them preferred some forms of engineering interventions, particularly for maintaining the regulated flood levels during the monsoon. This study concluded that the objective of the compartmentalization scheme, as well as farmers' preferences, may have been focused on flood regulation, but lacked adequate analysis of the constraints that accompany such regulation.

TABLE 10.2. Preferred government role in alleviating future high flood (*bonna*) problems, Kazipur, 1995 (multiple response possible)

Preferred future government action	Number (n=210)	Percentage
(a) STRUCTURAL MEASURES		
Construct embankments along the rivers	120	57
Construct polders/compartments	51	24
Dredging of river channels	68	32
Dams and barrages across smaller channels	15	7
(b) NON-STRUCTURAL MEASURES		
Increase relief efforts	78	37
Building flood shelters	21	10
Increase agricultural rehabilitation work	44	21
Make agreement with India	5	2

I also surveyed some selected villages from Kazipur *thana* in the Brahmaputra-Jamuna floodplain to determine inhabitants' perceptions of and preferences for the government role in responding to flood and riverbank erosion hazards. A departure of my survey from the previous studies was that the 210 sample respondents were asked questions without prompting so that their spontaneous responses could be recorded. From these, a total of four structural and four non-structural measures were identified as preferred future government action to address high floods (*bonna*). A majority favored construction of flood protection embankments along the rivers (Table 10.2). None of the other measures, however, received overwhelming support. Nearly one-third of the respondents desired increased dredging of river channels, as well as greater flood-relief efforts by the government.

Structural measures were also favored by the Kazipur respondents. While several investigations have confirmed that embankments along the river channel confine water within the channel and aggravate bank erosion (FPCO, 1995; Rahman, 1985; Haque, 1988c), the majority of the Kazipur sample still preferred construction of more embankments (68 percent) and cross-dams (48 percent; Table 10.3). Altogether, four types of structural-engineering interventions were suggested by respondents to mitigate bank erosion, these included strengthening existing embankments and stabilizing banks using soil-cement blocks. Some respondents suggested that the government get more involved with relocation of displaced families (39 percent) and with the construction of temporary shelters (16 percent). In general, the floodplain inhabitants showed a lack of awareness of the role of non-structural measures in disaster mitigation.

TABLE 10.3. Preferred government role in alleviating future riverbank erosion disaster, Kazipur, 1995 (multiple response possible)

Preferred future government action	Number (n=210)	Percentage
(a) STRUCTURAL MEASURES		
Construct more embankments along the rivers	143	68
Strengthen existing embankments	59	28
Construct cross dams	100	48
Stabilize banks using soil-cement blocks	77	37
(b) NON-STRUCTURAL MEASURES		
Relocate displaced families	81	39
Construct shelters (temporary)	34	16

As noted above, floodplain inhabitants' access to information, particularly on development and environmental projects is limited. Only one-quarter of the sample (n=55) were found to be knowledgeable about the proposed left bank flood-protection embankment. Since Kazipur residents have had flood protection embankments in their area since the 1960s, it was reasonable to assess their perception of the potential impact of the left bank embankment. The findings revealed a mix of beneficial and adverse effects. For instance, more than 60 percent answered that embankments would control undesired high floods, but would also enhance bank erosion along the river (Table 10.4). Similarly, more than one-third of the sample respondents knowledgeable of embankments perceived that such interventions both increase agricultural production and aggravate the risk of breaching and material loss (see Figure 10.4).

TABLE 10.4. Perceived impact of the proposed Brahmaputra-Jamuna left bank embankment, Kazipur, 1995 (multiple response possible)

Perceived impact of the Brahmaputra-Jamuna Left Bank Embankment	Number (n=55)	Percentage
Beneficial due to flood control	38	69
Increase agricultural production	21	38
Facilitate settlements	8	15
Increase bank erosion	34	62
Increase risk of breaching	19	35
Chars will be threatened	16	29
Land fertility will be adversely affected	27	49

Overall, survey-based studies exhibit that the cognitive and perceptual abilities of people are not well developed without adequate access to education, information and resources. Inconsistencies and ambiguities thus often arise, especially when experts' opinions and those of project-affected people are compared. Many of the project-affected peoples' have perceptions and preferences rooted in their experience. In order to minimize the gap between the objective and subjective conditions, particularly concerning any proposed project, a process of informed consent should be formulated.

10.3. Policies Concerning Riverine Disasters-Induced Displacees and Resettlement

The national government policies of Bangladesh on riverine-disasters induced displacees, especially regarding their temporary shelter and resettlement, are limited to three specific programs: (a) establishment of relief 'camps,' (b) an inter-district relocation or population redistribution, and (c) resettlement on locally available *chars* and/or *khas*lands.

10.3.1. RELIEF 'CAMPS' AND TEMPORARY SHELTERS

The efforts of the national and local governments to set up temporary camps are often not adequate to accommodate the disasters-induced displacees (Maloney, 1990-91). During the floods of 1988, 476 relief camps were established in the city of Dhaka which provided shelter to 1,068,959 persons. The displacees were living in all types of public and private establishments—schools, business premises, and unfinished buildings. A study conducted by the Bangladesh Institute of Development Studies (BIDS) found that because these shelters were insufficient, displacees set up lean-tos nearby, along the roads and even on barges in the river, keeping within range of the gruel kitchens and distribution centers (Bhattacharya, 1989). The average camp held 2,800 people.

"There was hardly any planning," Maloney (1990-91: 9) describes, and in most cases nobody set up these camps, "but the refugees just burst into whatever places they could find, and the government had to accept that it was a camp and distribute food and supplies" (Maloney, 1990-91: 9). The study led by Bhattacharya (1989) of the 54 relief camps during the floods of 1988 records that over four million taka (US $100,000) of relief goods had been supplied, including raw rice and cooked foods, provided by government, professional agencies and relief organizations. This research notes the common misappropriation of a substantial part of the donated food and materials by the authorities in charge of the camps.Flood-induced displacees could have endured any temporary conditions, but for many in the camps the gnawing problem was that they had nowhere to return. This problem grows in cases where flood-victims have lost their homesteads due to intensified riverbank erosion during the monsoon period. Staying in the camps or moving to embankments or elevated roads are the immediate options available to them. Resettlement of these disasters-induced displacees is an immense problem in Bangladesh, a country with close to 900 people per square kilometer. Should the displacees living on the embankments be allowed to stay or be relocated? Who should decide the relocation destination? These questions lead to crucial policy issues.

Riverine hazards-induced displacees who live in squatter settlements on the Brahmaputra Right Bank Flood Protection Embankment remain in constant fear of eviction by the authority, that is, the Bangladesh Water Development Board (BWDB). These squatter-settlers are homeless due to the destruction by the mighty river, and have no place to go. In 1982, the BWDB attempted to evict the illegal occupants under the then Martial Law Regulation No. 15, but later abandoned the plan on 'humanitarian' grounds (Rahman, 1984; Zaman, 1989; 1995). The zonal water development board officials concede that proper maintenance of the embankment requires eviction; they argue that construction of houses on the embankment weakens the structure and creates problems of seepage and formation of numerous rat holes leading to breaches of the embankment. The embankment, they argue, hence should not be allowed for permanent use by the displacees.

10.3.2. INTER-DISTRICT RELOCATION

Inter-district relocation seems to ease the land-population ratio by reducing pressure in the sending area. But in practice it turns out to be extremely difficult, and often leads to serious conflicts and even violence between the native-locals and new settlers. Inter-district relocation in Bangladesh generally follows two different strategies: (1) relocation of landless and displaced population from Noakhali, Lakshimipur, Feni, Dhaka, Munshiganj, Manikganj, Comilla, Feni, and Chandpur districts to selected zones in Khgrachhari and Rangamati districts in the Hill Tracts region as part of a government plan to diffuse the tribal political conflict (Mey, 1984; Zaman, 1982b; Maloney, 1990-92); and (2) the relocation of squatters (*bastee*) from Dhaka city to new *chars* in Faridpur, Noakhali and Barisal districts for resettlement. The former policy was vigorously pursued by the former military governments of Bangladesh, especially during 1975-1990. Under this policy of redistributing Bengalis from the thickly populated areas to Chittagong Hills, each settler family was promised two hectares of land, a cash subsidy, and government military protection (Maloney, 1990-91: 10). As noted by Maloney (1990-91), the ulterior motive was to transform the Chakmas and other 'tribes' there into minorities in their own region. The policy gave impetus to the 22-year war in the Hill Tracts districts between the Chakmas and the Bangladesh armed forces, leading to the flight of 70,000 tribal refugees to India, and to continuing violence. In the last two decades, several thousands of tribes people and new Bengali settlers died in violent fights in the region (Paul, 1982). The present government of Bangladesh is now negotiating with the tribes' representatives to resolve the civil war and conflict in the Hill Tracts districts, and has discontinued the policy of Bengali resettlement.

10.3.3. RESETTLEMENT ON *CHARS* AND *KHAS* LANDS

The other possibilities for resettlement in the country are the unassigned government lands (*khas*) and the sandbank islands (*chars*) emerging in the rivers. Ironically, these *chars* are at risk of being eroded again in the near future. During the late 1970s (i.e., the regime of Ziaur Rahman), many groups of displaced families from Demra *bastees* in Dhaka city were take to Faridpur district for resettlement on *chars*. This was heavily publicized by the government as an example of rehabilitation for the landless and displacees. Ali (1980) argues that inter-district relocation and land-resettlement on a large scale may eventually prove unsuccessful because of strong 'local feelings' that characterize Bangladeshi rural society. Citing cases of violence between 'local-natives' and 'non-locals' in Dinajpur and Dhaka districts, Ali (1980: 177) writes that the "rehabilitation of the landless population of one district in another district is sure to unleash bitter feelings based on narrow regionalism which no bureaucratic machinery will be able to control." For example, the settlers from Dhaka *bastees* were unable to stay on the *chars* for a sustained period because of atrocities committed by the *jotedars* and their paid *lathiyals*. Many resettled families eventually returned to Dhaka (Zaman, 1991).

Implementing a policy of resettlement for displacees on previously unassigned public lands in Bangladesh is a complex and difficult task. There was an order from the head of state in 1972 that all newly emergent lands should be given not to original owners, but to the government, to be recorded as *khas*lands. The aim was said to be the recovery of these lands from local control for the resettlement of displacees. In reality, nevertheless, local 'power bosses' took authority into their own hands. Thus, "violence, dispossession, murder, rape and confiscation of crops and animals have become established practices of *char* life in Bangladesh" (Zaman, 1989: 202). Under these present conditions socioeconomic decline is likely irreversible for poor peasant displacees.

Zaman (1988; 1989) assessed 29 hectares of *char*land, of which 18 hectares had been allocated for resettlement to 32 families who received it under the above-mentioned Presidential Order. He found that although the landless and smallholding peasants were the major recipients of *khas*land, two families who owned over 3.3 hectares (and therefore not entitled to any *khas*land) also received land. The records provided by the local revenue office (known as the *Tahasil* office) were ambiguous and generally unreliable. For example, 10 of the 32 names listed as recipients did not even exist in the study area. Furthermore, the eight landless families who apparently received land according to the land records office stated that they had neither applied for nor received any *khas*land. The records presumably were fictitious, arranged by collusion between locally powerful people and officials of the local revenue department. "While working in the village," Zaman (1989: 203) writes, "I was repeatedly told that *khas*lands are for those who can handsomely bribe officials."

Three problem areas can thus be identified with regard to allocation and resettlement of displacees on *char*lands: (1) delays in the bureaucratic process and inaccurate records and surveys, (2) corrupt practices of the revenue department officials, and (3) violence on the part of powerful local landowners who attempt to grab the land, illegally or through manipulation of public laws and officials.

In the 1985 REIS survey of Kazipur sample villages, a total of 394 heads of displaced households were interviewed on their preference for relocation if land and relocation assistance were made available to them. In the 1995 survey, similar questions were asked of displacee household heads from a sample of 154 units. From this survey there is clearly a high preference for relocation on *khas*, or unassigned government land within the *thana* area (more than 80 percent), and on nearby *char*land (more than 60 percent; Table 10.5). Preference for relocation to *khas*land in other districts and distant *char* appeared to be modest, with an increasing tendency between 1985 and 1995. Displacees showed least interest in resettling to coastal *chars* in the southern delta or to the Hill Tracts districts. Again, these percentages increased between 1985 and 1995. Overall, it is evident that the displacees have been facing more acute problems of shelter in recent years than ever before. The preferences given indicate that the displacees want to remain in their own locality in particular and in rural areas generally. In addition, the desire for relocation to nearby *chars* or within the *thana* suggests that the displacees intend, as far as possible, to remain close to their *kin* and local *samaj* groups. These patterns appear to be consistent when compared between the 1985 and 1995 surveys (Table 10.5).

TABLE 10.5. Preference for relocation destination, Kazipur, 1985 and 1995 (multiple response possible)

	1985		1995	
Preferred Destination	Number (n=394)	Percentage	Number (n=154)	Percentage
Khas land within the *thana*	341	87	128	83
Nearby *char*	250	63	126	82
Distant *char*	100	25	80	52
Khas land in another district	105	27	71	46
Coastal *char* (reclamation area)	38	10	58	38
Hill Tracts region	38	10	63	41

Considering displacees' options local resettlement using available *khas*land and newly emergent *char*land becomes a viable policy option. Yet, this relatively difficult and complex task must still take into account many sociopolitical and administrative factors which play a crucial role in the resettlement decision and process (see Zaman, 1991c). As elaborated above, government apparatuses, under the present bureaucratic and sociopolitical structure cannot ensure implementation of the proclaimed national policies.

Thus, there is an urgent need to reformulate the existing land management and resettlement policies to ensure that the displacees and the needy get access to emergent lands. To accomplish this goal, a bottom-up sociopolitical structure should be established which would empower the disadvantaged social groups. Village committees representing various interests, but consisting mostly of the most needy, should be encouraged to form to deal with these problems. The difficulty that may arise here is that displacees are often cut off from their larger community and are treated as "outsiders" even by the local poor (Maloney, 1990-91). Even more so, local committees may not likely approve the intrusion of displacees from other regions of the country competing for *char*lands in their locality. The specific policy recommendations concerning resettlement of riverine disasters-induced displacees will be presented in the next chapter.

CHAPTER 11

TOWARDS A SUSTAINABLE FLOODPLAIN DEVELOPMENT STRATEGY

This book argues that the environment can be made safer through interventions in the geophysical and social determinants of nature-triggered disasters—by means of improving technical, social, political, economic and cultural conditions. The Nature and People Process (NAPP) framework shows that human coping ability can be enhanced considerably; it also reveals the limits of modelling the complex interrelationships of natural environments and social factors, such as socioeconomic inequities, cultural prejudices, political injustice and legal manipulations. In this concluding chapter, I attempt to shed light, first, on the conceptual and policy implications of this study on the specific problem of riverine hazards in Bangladesh; second, on the issue of sustainable livelihood development and disaster reduction in floodplain ecosystems; and finally, on the broader application of the conceptual framework proposed in this study.

11.1. Dealing with Riverine Hazards: Research and Policy Implications

The study of riverine hazards in Bangladesh advances our existing knowledge concerning hydrological and meteorological parameters, the disaster-impact process and human coping responses. This section considers these contributions and suggests several areas where this new research will be valuable. Three specific areas will be addressed: (1) the nature and extent of temporal and spatial variability in flood and riverbank erosion events, and the problem of geophysical modelling of these phenomena; (2) factors influencing human coping responses to natural hazards and disasters; and (3) the relative vulnerability of social groups to natural disasters.

Much of the existing work on the nature of river channel behavior focuses on small and medium-sized rivers in the Western world, while research on large braided rivers is limited. Not surprisingly, the regime type width-depth relationship developed for a single channel river is not applicable to a braided river as a whole, although separate relationships for the branch channels can be developed. Several explanatory variables influence the braiding of a river, including the plentiful supply of sediment and the abundance of water. In turn, the existence of non-cohesive bank material along a river also has considerable influence on its braiding pattern.

Several constraints prevent the complete formulation of a predictive geophysical model of braiding patterns. These factors include the unstable character of bedforms, mid-channel bars and islands, and the erratic nature of bankline recession due either to the liquification and flowage of material or the shearing away of bank material. Furthermore, little progress has been made in predicting the sporadic changes in water and sediment distribution, particularly in terms of splitting, joining or other substantial changes in the distribution of sediment over branch channels. Yet a fair degree of success has been achieved in predicting the joining events of adjacent channels; for example, the recently developed theoretical model by Khan (1995) is capable of predicting outer bank movements for the following year with a standard error of approximately 100 meters. Overall, although efforts on modelling large braided rivers are still minimal, primarily constricted by their massive scale and numerous control variables, recent successes are indicative of the making successful short-term predictions of river channel changes and bank erosion in a large braided river system.

Throughout this study I have explored the possibility of accommodating and integrating approaches in the social and physical sciences for a better consideration of the relationships between nature and society. The findings reveal that none of the established frameworks alone can adequately explain this relationship. Alongside the experiential and spatial dimensions, the social-structural dimensions surrounding the control of and entitlements to resources also help in determining human coping responses to hazards. In examining the significance of the existing paradigms in nature-society relationships, this research agrees with Johnston's call to emerging approaches to "marry the ability to develop realist theories of infrastructure with methods of studying human action within the superstructure" (Johnston, 1983a: 198).

The existing literature on human coping responses to natural disasters speaks mostly from a behavioral stance, focusing on experiential, spatial and personality traits of individuals. By examining coping responses according to the former two dimensions, the present study reveals that the differentials in perception of and coping strategies for environmental hazards occur partially as a result of *previous experience* and the *hazardousness of an area*. Although the *personality traits* of resource users have been considered elsewhere as significant, this variable was not included in the present study. Effective application of the existing methods of personality measurement is doubtful in context of the predominantly agrarian, developing societies; moreover, contemporary social-psychological literature poses serious questions about the independent entity of personality (Brody, 1970; Gergen, 1982: Gergen and Davis, 1985). Future research might be directed at assessing the influence of personality in coping responses to natural hazards in a cross-cultural context. Such an attempt, nevertheless, will require substantial development in behavioral-psychological research methods for regions where a considerable proportion of population does not have access to education and basic survival means.

The results of analyzing the perception of hazards and coping strategies must be interpreted with caution. We must consider that several assumptions are embodied in the testing of a hypothesis. In the present study inferences on relationships between dependent variables and explanatory variables were based on 'zero order' differences computed without controlling variables outside the bivariate cross-tabulations. Also, this study dealt with nominal and ordinal data because of the perceptual content of the research problem and the wider scope of the study. Future research might take the form of a multivariate analysis in which human coping response is the dependent variable and experiential, spatial, and social structural variables are the explanatory variables.

The behavioral-ecological viewpoint, one of the established approaches to hazards research, restricts analysis primarily to the micro level. This research focuses on the underlying mechanisms which produce the forms and patterns of human behavior. By placing emphasis on the assessment of both the geophysical and social dimensions of natural disasters within the NAPP framework, this study has reintroduced the need for understanding the relationship between micro- and macro-level variables to explain natural disasters and their associated human behavior (see Palm, 1990).

The specific points of departure of this study from the existing body of literature are then as follows. First, while the predictability of geophysical events are often thought to be dependent entirely on the type of concerned phenomenon, this study asserts that these events are also related to the way society treats nature and its processes. Second, socioeconomic status has been regarded as an *a priori* independent variable in behavioral-ecological hazards studies and has been used as an explanatory factor for human responses to hazards. This present research, in contrast, suggests that an individual's characteristics are linked to a larger social environment; thus, in part, they are the products of the historical development of social, legal, and economic relations. For this reason, micro-level forms are here analyzed in terms of the development and functions of the state, and its legal, political, and economic institutions in which individuals act and react to prevailing conditions.

In the present study, inferences about the impacts of natural disasters are based chiefly upon two cross-sectional surveys. Some data were compromised here due to the large coverage of the questionnaire, sensitivity of information and ethical restrictions. A longitudinal study would be effective in filling the gaps in understanding the process of disaster impacts.

This research reaffirms that inhabitants residing in the floodplains of large rivers are prone to hazards. In quantitative terms, the privileged resource owners, such as large landowners, are more susceptible to natural hazards due to their greater share in resource control and legal ownership than the impoverished sections. But the relative loss of privileges and vulnerability to hazards are minimal for large landowners. First, they possess a wide range of options to choose from in adopting effective coping strategies; second, they are capable of manipulating the disrupted situation created by a sudden scarcity of resources; and third, their accessibility to the social, legal, and administrative authority is institutionally ensured. Our understanding of the vulnerability to hazards of the so-called disadvantaged population is reinforced by the study through an examination of the marginalization and pauperization of the small and marginal landowners, and the landless occupants of the floodplains. Vulnerability is thus empirically exhibited in terms of an individual's position and relationships within a broader social spectrum. Future research on relativity in hazard impact might be useful with the consideration of other natural events and social formations. Such efforts would provide the scope for better comparison and generalization in disasters research regarding the concept of vulnerability to nature-triggered events.

Implications for Public Policy Formulation

The findings of this study have a number of implications for public policy formulation in Bangladesh. These eleven points are outlined below:

a. Given the finding that a significant proportion of floodplain users are subject to displacement by riverine hazards and disasters, and that displacement due to floods and bank erosion cause serious deterioration in the socioeconomic conditions of disaster victims, the formulation of a comprehensive public policy on disaster prevention, preparedness and mitigation is needed. At the present time, the planning and management of riverine hazards are administered by the Water Development Board, which is largely inclined towards engineering-structural responses. Although the Disaster Management Bureau, under the Ministry of Relief, has begun (since 1993) to administer the emergency and post-disaster recovery programs, its activities are mainly confined to cyclones and tidal surges. It is recommended that in order to incorporate non-structural measures in flood and riverbank erosion disasters preparedness and mitigation, the activities and management mandates of the Bureau should be broadened.

b. Cross-sectional hydrological data should be collected more frequently as it is done for water level and discharge. In order to facilitate studies on bank movements using sediment budgeting, micro-level investigations, frequency and the number of stations to collect sediment transport data should be increased considerably. Upon availability of sufficient spatial and temporal data of water and sediment discharge and of the cross-sections, a total hydrograph of water discharge, instead of a single value, should be incorporated into the model. This will improve the performance of the model in terms of bankline movement prediction.

c. An overriding issue in the whole study is the complexity of the physical processes contributing to the erosion problem. The sheer magnitude of the problem and the complex interrelationships between all the processes prevent the development of methodologies to model the individual processes. Instead, a 'black box' procedure, whereby outputs are directly related to inputs, is the best that can be obtained. Inclusion, in some form, of physical parameters which are significant in the erosion processes will provide some recognition of the range of physical processes at work.

d. Given the dearth of data on migration in general and involuntary migration in particular, the development of a data bank of these aspects is a necessary step towards meeting planning needs. Also, the generation of a data base on multiple natural hazards is required to facilitate regional and national planning to minimize hazard loss.

e. While engineering-structural measures are the principal means of mitigating flood and riverbank erosion, they require enormous capital commitments. Low levels of technology, lack of adequate capital, and dearth of large-scale private inputs make it unlikely that high floods and erosion can ever be prevented. More emphasis is therefore needed to better prepare riparian areas for minimizing the loss to disasters. Alternatively, greater international cooperation with the upper riparian countries, namely, China, India, and Nepal is required if more effective river monitoring is to lead to clearer early-warning systems.

f. For generations floodplain users have made various types of indigenous coping responses to natural hazards. Since preventive measures have generally been found to be ineffective in reducing loss or to even pose threats to long-term sustainability of human ecological systems, should not scarce internal resources be directed more towards reinforcement of the non-structural, social coping measures than the current practice of engineering interventions?

g. The introduction of disaster mitigation policies must take into account the sociopolitical structures that are in place; there exists a small, elite section that controls large tracts of land and holds the less fortunate and newly dispossessed in bondage. The relatively skewed distribution of land, along with the manipulation of landownership during erosion and reemergence, underscores the call for formulating serious land reform policies. The confinement of such land reform policy to the legal sphere alone would immensely jeopardize its effectiveness. Participation of the rural disadvantaged social sections in decision-making and administering the execution of policy is necessary.

h. People, including the disadvantaged, must be mobilized to form social organizations and to exercise their rights to participate in and demand accountability of public practices. As such, village-based self-help programs are most effectively implemented through locally constituted organizations. Although the issue of raising the consciousness of the rural population is not new to the discourse of rural development in Bangladesh, the rural population in general, and the displacees specifically, need encouragement and institutional assistance in creating local-level organizations that can effectively articulate their concerns and needs to macro-level structures. Through such organizations the extent of passive loss-acceptance can be reduced and replaced by more effective actions which promote loss-reduction coping measures. Emphasis should be placed upon confidence-generating reeducation, encouragement, and protection of self-help endeavors, and the strengthening of cooperation among kinsmen and community members.

i. The general problem of impoverishment is not a direct product of riverine hazards in the floodplains, yet a significant proportion of floodplain users become marginalized and impoverished due to land-loss. The formulation of policies towards rural income diversification is thus needed to address such problems. No land regulation alone can change, or for that matter, turn the already fragmented and ever decreasing sizes of holdings into viable farms for all needy peasants. A pragmatic reform will relieve agriculture from bearing the undue burden of supporting the majority of the rural population. The agricultural labor force would be served by creating an effective demand for agricultural products.

j. Restructuring the form and nature of local government would likely help to minimize the manipulation of submerged, *khas*, and government-acquired land. The formation of a local-level government primarily by representatives of rural interests-groups and collective organizations instead of distant bureaucrats is a possible direction. The 'on-the-spot' survey and settlement operations by agencies involving local public representatives would help reduce the violence, corruption and injustice associated with access to reemerged land.

k. A greater level of political mobilization of landless displacees is desirable if they are to retain their land resources. It is only through positive political action that their right to land can be assured. Along with this, a unified mobilization of social resources (family ties, neighborhood relations, *samaj*, etc.) would also help to safeguard the common pursuits of equitable access to land. It is in this area that voluntary organizations have a vital role to play.

11.2. Approaching a Sustainable Floodplain Development

This book set out to explore critically the nature-society relationship, focusing on the empirical problem of riverine hazards in Bangladesh. The first chapter suggested that the country as a whole can be epitomized by a composite of precarious ecological and socioeconomic conditions. Indeed, no other country has had as frequent and colossal human loss and experience of natural disasters as has Bangladesh. Similarly, with the exception of China, no other country has had to cope with so much devastation of people's livelihoods from disasters. The threats and risks that Bengalis are faced with today from environmental events arise from a complex human-ecological relationship that has evolved in the Ganges-Brahmaputra deltaic region over its entire history. For this reason, the nature of people's opportunities and hazards here, like other social formations, cannot be explained solely by either the geophysical factors or the societal variables. The thesis of this book is that disasters are intersects of geophysical, biological, social, political, economic and cultural spheres; a holistic approach is therefore necessary for studying natural hazards and disasters.

An analysis of the human ecological aspects of the region reveals that the Quaternary geological formation and associated processes of the Bengal delta themselves show a large magnitude and frequent variability in geophysical parameters. In addition, many other geographical factors, such as site and locational characteristics, have compounded the problem of extreme environmental events involving soil and land, water and wind. The abundance of alluvial soil and adequate water supply during the paddy and other major crop cultivation periods, have historically attracted people to colonize and form a sedentary population on the riverine plains. Through centuries of accumulated experience and adaptive measures, people have successfully lived, struggled against, and coped with the environmental threats that have prevailed in Bengal's plains.

Yet the scale and nature of current environmental hazards in Bangladesh are quite different than those of the past. The demographic history of the 20th century has shown phenomenal population growth without accompanying progress in the economic and spatial spheres. The absence of an established entrepreneurial class and the lack of rapid industrialization have thus generated enormous pressure on arable land and the relatively few cities and towns. Consequently, the expanding population, through fierce competition, has attempted to take advantage of any resources and associated opportunities that have appeared from ecological changes and economic growth. Such efforts have been made even when there is an immense risk involved.

Because humans strive to produce more resources and to safeguard existing ones, human intervention and modification of the physical environment (e.g., the alluvial plains) are growing at a dramatic rate. This trend is expected to continue into the foreseeable future, and with it will come conflicts—river-training, bank protection, flood prevention, and other interventions into the natural environment. Among many others, Scudder (1991) identifies some of the adverse impacts of African dams on floodplain users and local ecosystems. Similarly, Rahman's (1984) determination of the many adverse effects of flood protection embankments in Bangladesh's floodplains includes aggravated scouring and bank erosion, frequent breaching, and increased risk and disaster-loss. In addition, Minkin (1989; cited in Boyce, 1990: 423) reports almost a total loss of open water fish in flood control projects in the Ganges-Brahmaputra-Meghna floodplains. Without significant attention toward more sustainable forms of floodplain development, it is quite probable that such severe damage to ecological and cultural habitats will accelerate.

The notion of sustainable development provides hope for the floodplains of Bangladesh. This concept was reinforced by work done by Raymond Dassman while he was with the International Union for the Conservation of Nature and Natural Resources (IUCN) in the 1970s (Dassman *et al.*, 1973). It was later promoted by the World

Commission on Environment and Development (WCED, 1987) in the 1980s. The Commission defined the concept as "development that meets the needs of the present without compromising the ability of future generations to meet their own needs" (WCED, 1987); this broad theme generated a semantic debate that continued in several works (e.g., Manning, 1990; Howarth and Norgaard, 1990; Lele, 1991).

Overall, sustainable development is an anthropocentric approach to the globe—a human perspective relating to human use of the biosphere. This notion is a step forward towards a balanced treatment of the nature-society relationship as it explicitly recognizes interdependencies that exist among environmental, economic, social and cultural phenomena. However, in order to emphasize the underlying unity of humans with other beings and processes, an emerging biocentric view has been proposed by Naess (1983), Tobias (1984) and Sylvan (1985). The adherents of Deep Ecology suggest that the search for unity of phenomena should be performed at the metaphysical level. They argue that humans are not the sole holders of value in the world, "and not all things of value are valuable because they answer back in some way to human concerns" (Sylvan, 1985: 5). One major contribution of this emerging idea is its reminder to decision-makers and practitioners of the impossibility of incorporating environmental values fully within the economic paradigm (Redclift, 1987).

The formation of the Bengal delta through the deposition of alluvial deposits from the mighty rivers and their associated geophysical dynamism have always brought about opportunities and risks for the ecosystem and its residents. Historically, village after village has been swallowed by channel changes of the river. But when the river gives the land back in the downstream reaches or in the estuaries in the form of emerging islands, people try to take full advantage of these new opportunities. However, along with the threats from the dynamic deltaic geophysical systems, the manipulation of the natural hazards by the powerful landholding class compel many of the disadvantaged to move to unsafe areas. Despite these social and political difficulties, people continue to live in tandem with the environment. For instance, the annual inundation during the monsoon has always been beneficial to riverine populations and ecosystems. Indeed, annual flood regimes have historically flourished and sustained the economies of the floodplain users without any significant environmental degradation. Thus, river channel changes, riverbank erosion, and flooding themselves are not an evil but a source of life and livelihood for the vast majority of people in this deltaic landscape. Two broad areas of public policy regarding riverine hazards are relevant here: (1) a holistic approach—unlike the technocratic-engineering strategy—to human and ecological aspects of water resource management, and (2) a sustainable floodplain development plan that will ensure a livelihood to its inhabitants and reduce future disasters.

Because the delta-building process and flooding are so critical to the economy, society and ecology of Bangladesh, a technocratic policy stressing the prevention of geophysical events, such as river channel migration and flooding, is definitely not an appropriate solution. Instead, an engineering-structural solution represents a partial view and, perhaps at its best, a somewhat piecemeal solution with all kinds of possible adverse ecological and environmental impacts. In my view, the current emphasis of the Bangladesh water management policy should be changed from event prevention to event adaptation and management. Such an approach may still require some limited protection of important cities and commercial centers by building new embankments. The primary policy objectives, however, should be to restructure and reform the existing institutions and to encourage and assist the actual resource users in the villages to identify the best ways to protect themselves, their crops, and their livestock by developing improved measures for flood and bank erosion management and preparedness.

I would like to assert that the riverine hazards problem in Bangladesh is not a problem of mere hydraulic dynamics. It is rather an issue which reaches into macro-level societal concerns involving the economy, ecology, society, demography, settlement, politics and culture. As such, riverbank erosion and flooding issues must be addressed both in terms of sustainability and long-term development. Since high disaster-loss relates to a number of developmental factors (e.g., increased population pressure, perpetual economic marginalization of rural cultivators, and ill-conceived infrastructure), addressing disaster impact requires going beyond the concept of sustainable floodplain development. For instance, there is a need to develop strategies which retain the unquestioned benefits of annual flooding. Moreover, the general problem of impoverishment is not a direct product of riverine hazards in the country; instead, a significant proportion of floodplain users become marginalized and impoverished due to loss of income, assets, and an increased debt burden. Policies which lead to rural income diversification thus need to be considered. The generation of capital stock from economic surpluses in agriculture and the redistribution of existing local resources, including land, are fundamental to such reforms.

11.3. Broader Applications of the NAPP Framework

The questions surrounding environmental hazards are deeper and broader than the scope of this book. The traditional scientific-technocratic approaches have regarded hazards as 'special,' 'abnormal,' and 'extreme' phenomena, and classified them according to the appropriate type of geophysical events. A challenge to this notion was vigorously pursued under the leadership of Gilbert White. He revealed that despite huge capital investment and significant improvement in technological devices to monitor and modify natural events, actual loss and damage was increasing from nature-triggered disasters. As a reaction to this observation, the focus of hazards studies broadened "beyond the set of natural physical hazards," and shifted to the actual experience of 'extreme events' and to the adjustment responses of individuals and national governments (Burton *et al.*, 1993: 241).

Critiques of this ecological-behavioral approach in natural hazards research have projected two general sets of limitations (see Waddell, 1977; Torry, 1979b; Hewitt, 1983; 1997; Watts, 1983; Palm, 1990). First, it has been argued that the use of extreme geophysical events as an *a priori* phenomenon for analysis is questionable, because it detracts from comprehending the human ecology of everyday existence. The everyday and extraordinary are intertwined in reality; the vulnerability of people is imbedded in the struggle of both everyday existence and in the rare and extreme event. Second, it has also been asserted that the early, established approaches ignore the social, economic and political constraints that serve to limit the choice of people and their institutions. The common use of choice and decision models and the concentration on such psychological constructs as individual perception imply that "humans are masters of their fate to a much greater extent than is valid" (Burton *et al.*, 1993: 242). This stress on individual autonomy and separation from macro-structures in choice-adjustment ignores the reality of the constraints and structures that are part of social existence. The call of this alternative approach has been to address the prevailing 'constraints' directly and vigorously, in this way releasing the 'pressure' that leads to vulnerability and disasters (Blaikie *et al.*, 1994).

I have argued that none of the above existing approaches to environmental hazards can sufficiently explain the complex structures and relationships concerning nature and humans. Specifically, these approaches fail to recognize that humans are an integral part of nature and meanwhile a differentiated, autonomous subset of it. Disasters, in my view, are the outcome of a complex process involving both the natural and social realms. The NAPP framework presented in this book provides a general outline for explaining these complex processes and spelling out the subsets and relationships between various component parts. Furthermore, this framework can be applied to all types of environmental hazards. In studying the state of coping characteristics to earthquakes, cyclones, or to technological hazards, such as fall out from nuclear power plants or chemical spills, both the environmental and societal spheres must be considered simultaneously to explain the problem and to direct management responses. Equally important, an approach to hazards must include both the roles of macro-level structures, such as social norms, institutions, and legal systems of property ownership, and micro-level characteristics, such as personal preferences, intentions, and tendencies.

The unifying framework put forward in this book seeks to explain not only patterns and forms (i.e., appearances—what is happening?), but also to explore the underlying processes concerning the nature-society relationship. It calls upon researchers investigating hazards and disasters to encompass both the geophysical and human spheres. Thus, we should consider the following questions when conducting hazards research: (a) What are the mechanisms (natural laws) of physical processes? (b) What are the parameters of environmental variability and change? (c) What are the existing patterns of people's livelihood and how are they being affected by economic, social, political and cultural factors? (d) What are the existing coping options to modify threatening physical events? (e) Considering the social formation, what type of event-modifying responses would be appropriate to mitigate environmental hazards in the short-term as well as lead to convergence with natural systems in the long-run? (f) What type of international, national and regional policy measures would reduce vulnerability of people to natural hazards and disasters? (g) How can development and disaster issues be integrated and addressed? (h) What should be the direction of institutional and social-structural intervention to transform disastrous environments into sustainable resources and to harmonize human action with nature's tendencies? These questions and issues are not exhaustive. Additional concerns should look to an epistemological paradigm that would allow us to continue advancing our knowledge, while also searching for an understanding of who we are and how we relate to nature.

[illegible]

NOTES

1. In this book, the meaning of *nature* refers to an ahistorical usage, implying the objective university and its laws independent of human existence. This view thus differs from Von Weizsacher's (1949) notion that nature is "historic" as it is "being within time," which has a phenomenological underpinning.

2. The terms *natural environment* and *nature* are used in the text to mean the physical biological universe. The term environment implies *phenomenal environment* (see Kirk, 1963; Billinge, 1981), meaning a broadened version of the usual concept of environment to include not only natural phenomena but the manifestations of human activities, involving the physical and biotic environment together with those phenomena altered or created by human endeavors.

3. *Variability* here implies differences that take place within the "expected" range. For example, in climatology, *climatic variability* refers to the variation of the climatic parameters *within* the typical averaging period. In contrast, climate change refers to the condition when the differences between successive averaging periods exceed what *noise* (i.e., the variance of climate attributable to short-term weather changes) can account for; in case of a *climate change*, a distinct signal exists that is apparent above the noise (Hare, 1985; also see Hare, 1995).

4. For example, by studying only the United States, White and Hass (1975) estimated 5,192 person-years of research opportunities on natural hazards over a ten-year period, which is indicative of a high degree of emphasis in this particular area of research.

5. The Brahmaputra River is known by the same name in the upper catchment areas of southwestern China and northeastern India. It takes the name Jamuna after entering northern Bangladesh. Analysts, however, use both names to refer to the Bangladesh portion of the river. For instance, Morgan and McIntire (1959), Coleman (1969), Chowdhury (1973) call it Brahmaputra from a generic viewpoint, while Rashid (1966), Galay (1980), Paul (1984) call it Jamuna following the local preference. A dual name—the Brahmaputra-Jamuna—is used in this book to refer to the Chilmari-Aricha reach of the river to integrate both the international and local names.

6. In this connection, Johnstone (1983c: 130) points out that "in physical geography, process is generally associated with answers to the question 'how'?: research is concerned with mechanisms . . . existence of processes (physical laws, etc.) underlying the mechansims is taken for granted. In human geography,

on the other hand, there is concern not only with mechansims—answers to 'how'?—but also with processes—answer to 'why'?"

7. The *mode of production* or economic base is the key determinant in the complex network on interconnections involving the state of institutions, patterns of behavior, beliefs and other make-up of society. The *mode of production* consists of *productive forces* or capacity to produce involving labor, resources and instruments of labor and the *relations of production*. The latter involve individual positions in social structure and production process (see Desai, 1974).

8. The process of biological adaptation has been relatively very slow and cannot play any remarkable role in the short-run response to natural hazards (Burton *et al.*, 1993).

9. It has been generalized that pre-industrual adjustments are often "mystical, arational," or imbedded in the broader cultural context of life and livelihood, are flexible and easily abandoned, and include a high rate of adoptions among individuals and communities (Kates, 1971). In the modern industrial state, acceptance shifts from bearing loss personally to sharing with the wider society through institutional (e.g., relief, insurance) means; this includes, control of nature through a limited range of technological actions and uniform, inflexible response in application. Post-industrial adjustments adopt a greater flexibility, variety of capital, and institutionalize a broadened choice (Burton *et al.*, 1993; Kates, 1971).

10. Section 2.2.3 includes an elaborate discussion of this problem.

11. Okrent (1980; cited in Smith, 1992) clarifies this distinction with an example. We, say, consider two persons crossing a lake, one in a liner and the other in a rowing boat. The death of drowning, that is the *hazard*, is equal in both cases but the probability of drowning, that is the *risk*, is very different.

12. This implies that environmental events are the outcomes of underlying natural processes, regulated by natural laws which operate, according to Sayer (1979) and Johnston (1983b), at the *superstructural* level.

13. The term refers to extreme environmental events which are triggered by natural (i.e. geophysical, climatic and biological) forces such as earthquakes, cyclones, floods and landslides. The controversies and problems of defining natural hazards and disasters are discussed in Chapters 1 and 3.

14. The most recent and comprehensive work on this issue is the monograph prepared by Myers and Kent (1995) for the Climate Institute, based in Washington, D.C., USA.

15. "Carrying capacity" is the maximum number of population that can be supported and sustained for an unlimited time period by a given environment, society or human ecology.

16. The figures are adjusted for inflation.

17. The term "vulnerability" is used specifically to refer to the characteristics of a person or group in terms of their capacity to predict, cope with, resist, protect and recover from the impact of a natural hazard. It involves a combination of factors that influence the degree to which someone's life and livelihood is put at risk by a discrete and identifiable event in the environment or in society (Blaikie *et al.*, 1994: 9).

18. This alternative approach, called *the vulnerability analysis*, will be elaborated further in the latter part of this section.

19. This approach has been developed by the Natural Hazard Research Group of the University of Chicago. Based on Barrow's concept of human ecology, it is often referred as the "Chicago Approach."

20. Underdevelopment of a country refers here to the geographic territory or nation characterized by (a) sectoral inequalities in productivity, (b) disarticulation of the economic system, and (c) external domination. An underdeveloped country therefore must be viewed as the result of a particular kind of historical evolution of the concerned country, but not as deterioration in the aggregate pattern of a series of quantitative indicators. For further details, see Stavenhagen (1964) and Frank (1966).

21. The term "entitlement" refers to the ability of different classes or of an individual in a society to establish command over resources (see Sen, 1981a; 1981b).

22. Several general overviews of the research exist including those Burton *et al.*, (1993), Cutter (1993; 1994), Alexander (1993), Smith (1992), Blaikie *et al.*, (1994), Gold (1980), Walmsley and Lewis (1984), and Hewitt (1983). Hewitt (1983) and Blaikie *et al.*, (1994) provide the most extensive review of the dominant and alternative work in the field.

23. The semantic differential technique makes use of linguistic encoding as an index of meaning by using a combination of association and scaling procedures in accounting for the psychological meaning of *concepts*. The concepts or phenomena are evaluated in terms of a *bipolar adjectival scale* (see Osgood *et al.*, 1957). The responses are then analyzed by adopting the *factorial analytical technique* to reveal the underlying dimensions of factors which comprise the meaning and image of phenomena.

24. The objective of the methodology is to abstract themes of stories told by respondents on a set of standardized pictures.

25. Also see Schiff (1977), and Gold (1980: 206) for a discussion of the limitations of personality factors to explain variations in human perceptions and responses.

26. Wiest (1979: 180) points out that the functionalist view concentrates on the interrelationship of variables at one point in time. "Thus the approach is subject to shortsightedness, and the analysts are usually content with making judgements on the basis of short-term, immediate and obvious, yet superficial, effects."

27. Maddock (1969) points out that with discharge and sediment load as independent variables, resistance and debris size as semi-independent, and water-surface width, flow depth, velocity, slope, and the pattern of flow as dependent variables, the required number of equations would never be available.

28. See Coleman (1969: 131) for further elaboration on these features.

29. The lower part of the Ganges takes its name *Padma* after the bifurcation with the Bhagirathi near Farakka, and the Meghna near Chandpur.

30. Notably, the Bangladesh Water Development Board and the Inland Water Transport Authority of the Government of Bangladesh also use this dual name for the Jamuna portion of the river.

31. A detailed description of the diversion of the Brahmaputra River in the 18th century is given in Section 4.3.3.

32. See Sen (1968), Viz and Shenoy (1968), Desai (1968) and Goswami (1985) for studies on the Brahmaputra in Assam and Meghalaya in India. See Ascoli (1910), Auden (1949), Chatterjee (1949), Latif (1969), Majumdar (1942), Rennell (1781, 1793), Strickland (1940), Rasid (1966), Chowdhury (1973), Rahman (1984), Bristow (1987) and Khan (1995) for studies on the lower Brahmaputra or the Brahmaputra-Jamuna channel.

33. For the typical gradation curves of suspended sediment and bed material based on samples collected at Serajganj station, see Latif (1969) and Rahman (1984).

34. Knighton (1984) classifies the timescales of landform change as:
    Instantaneous timescale = $<10^{-1}$ years
    Short timescale = 10^1 - 10^2 years
    Medium timescale = 10^3 - 10^4 years
    Long timescale = $>10^5$ years

35. See Leopold and Wolman (1957), Knighton (1984: 124), and Richards (1982: 182) for a discussion of the problems of such a classification.

36. See Knighton (1984: 125-9) for a detailed discussion of the role and relationships of pertinent variables in identifying channel patterns. He notes that slope-discharge relationships have been most widely used to distinguish between types of channel patterns (also see Ackers and Charlton, 1970; Schumm and Khan, 1972).

37. The term "present" is meant here "as the time period over which input and output conditions have remained reasonably constant on average" (Knighton, 1984: 162).

38. Also see Mosley (1975) and Lewin and Brindle (1977) for an application of historical maps and survey to the study of planform changes in the rivers of Britain.

39. An ancient series of Hindu scriptures.

40. See Rudra (1981) for a critical analysis of the methodology adopted by Ptolemy for calculation of the latitude and longitude of the mouths of the Ganges River.

41. For additional information, see Gait (1902), Walters (1905), and Bhattacharyya (1977).

42. For example, Burkham's (1972) study of historical changes along the Gila River in Arizona, USA, confirmed that, between 1905-1917, the river widened from about 19 meters to 610 meters as a response to large winter floods.

43. Rennell called it Tista Creek.

44. Flood Action Plan (FAP), a mix of projects to study technical, socioeconomic and environmental aspects of flood prevention and mitigation in Bangladesh, was

sponsored by the World Bank in 1989. A total of 26 components were identified and the plan was implemented during the 1990-1996 period. Further discussion on FAP is available in Chapter 10.

45. The term *occupance* refers to the entirety of human use of the surface of the earth, particularly as it changes through time (see Burton, 1962: 1).

46. See Chapter 1 for a detailed discussion of this issue.

47. The region between the Ganges and the Brahmaputra, which stretches from the Himalayan piedmont or Tarai region in the north and to the Ganges in the south, is called the "Ganges-Brahmaputra Doab" (see Bagchi, 1944: 18; Bhattacharyya, 1977: 12-3). The doab is bounded by the Kosi and Mahananda rivers to the west and the Old Brahmaputra to the east. Physiographically, it comprises both the old and new alluvial lands of Bengal.

48. In contrast, the early European cities were characterized by the existence of "natural capital" developed independently of landed property (Merrington, 1975; also see Haque, 1987b).

49. Gait (1902: 2) in this respect argues that: "reasons are there . . . for thinking that at a [particular] period the Brahmaputra had flowed, as it now does, west of the Madhupur jungle, and that its diversion to the east was due to elevation of that tract and the simultaneous subsistence of the land now covered by the Sylhet *jhils* [depressions]."

50. See Desai (1959), and Hindess and Hirst (1977) for a useful description of the interrelationships between the feudalistic mode and technology progression.

51. For details on the "human use system," which basically refers to location, livelihood, and social organization characteristics, and the "natural event system," which implies the array of wind, water, and earth processes, see Burton *et al.*, (1993: 31-34).

52. Buchanan (1966: 13) points out that for centuries, "simple agricultural equipment and hand manipulated tools for manufacture were all that were known. Even wind-mills and water-wheels were seldom employed. The sickle and plough, the saw and chisel, the spinning wheels and pit-loom, were of thrifting amount of material in a very short time, but sometimes gave service for generations."

53. The attempts to increase revenue and tax were reflected in many state policies. Some such instances were the conversion since Akbar's regime of grain demand

as land-tax to cash demand as well as legalization of excess rent (*abwabs*) by the then governor (*subadar*) Murshid Kuli Khan.

54. Following Davis and Blake (1956), given such high mortality conditions in the region, high fertility was a necessary adjustment for maintenance of the size of population. Early marriage, usually before menses, was one of the principal institutional adjustments to maximize womens' reproductive span.

55. Members of one of the primitive communities of northern India.

56. Currey's (1979) findings will be presented in detail in Chapter 7, Section 7.1.1.

57. These *zamindars*, however, were not all heirs to the Mughol land tenure system. Some were descendants of the old Hindu and Muslim rulers of the country; others represented the large landholding families who had become *de facto* rulers in their own territories during Mughol regimes "on account of its sufferance or favour" and had paid a fixed tribute only. The rest of the *zamindars* consisted of those whose ancestors had held the office of revenue collector for generations (see Panandikar, 1926: 80-1).

58. An independent *talukdar* was a person to whom a *zamindar* sold his rights of rent-collection for some part of his estate; the independent *talukdars* henceforth paid land revenue directly to the state (Abdullah, 1976: 68).

59. The purposes for implementing the scheme were manifold: namely, (1) privatization of land, (2) the collection of land revenue from few landlords rather than from a legion of small peasant properties or whole communities, and (3) the creation of supportive local agents for political and administrative goals (see Desai, 1959: 35-6; Mamun, 1975).

60. There was a redistribution, but it was not from rich land owners to poor *raiyats*. The main effect of the Act was a redistribution of income from a class of rural intermediaries to the (urban middle-class) provincial governments (Abdullah, 1976: 85).

61. For a complete description of the context and implications of the political changes in Bangladesh since independence, see Khan, 1983.

62. This trend of landholding consolidation can further be substantiated by the prevailing patterns in land transactions by various landholding groups. Khan (1977) shows that the marginal peasants, owning less than one acre of land, sold more than 50 percent of their remaining land each year. On the other hand,

Alamgir (1978) records that the buyers of land as a proportion of total land transactions increased for higher landholding groups.

63. Also see Bertocci (1972), Alamgir (1975), and Abdullah *et al.*, (1976) for comparable statistics and analysis.

64. This embankment is locally known as WAPDA *bandh*.

65. Saha (1923: 20) also classifies the physiography of the Pabna district into these three divisions. See Saha (1923) for further details.

66. In the succeeding text, the term Kazipur will indicate Kazipur *thana* unless otherwise stated.

67. See Haque *et al.*, (1985b) for further details on the sampling design of this study.

68. Chapman (1963: 158-9) put forward two major advantages of such random sampling: first, it provides scope "to test the sample to see how closely it represents the universe, and second, it eliminates the possibility of inaccurate results from investigation of the universe with resource constraints."

69. Moser and Kalton (1977: 111), in the context of advantages of using PPS, note that this procedure "leads to greater precision than would a simple random sample of the PSUs and a constant sampling fraction" since it considers the relative changes of the PSUs being selected.

70. The concept of *confidence level* relates to the theory of sampling distribution. The sampling distribution in general terms can be defined as the distribution of all possible sample means. In relation to this, for example, a 95 percent *confidence level* can be described as follows: however large the variation in the possible samples, one would always expect 95 percent of the sample means to be closer to the true population mean with a distance of 1.96 times the standard error. In other words, 95 percent *confidence level* means a probability of 5 percent of making a statement which might be wrong about the relationship between the estimate and the parameter.

71. The distance from the sample mean (in either direction) within which one would expect to find the population value is the *confidence limit* (Dixon and Leach, 1978: 6).

72. Kish (1965: 522) calls it "interviewer bias" on the responses (also see Warwick and Lininger, 1975: 199-201).

73. The KZ is excluded from the analysis concerning displacement because it contained only one observation in 1985 and none in 1995.

74. *Samaj* is a relatively large intravillage group "based on common social and religious interests in the village" (Zaman, 1982a: 39). The functional structure of a *samaj* is usually informal, but its influence on decisions over social, ritual, ceremonial and religious matters is significant.

75. The *gosthi* is a group form based on kin relationships that determine political, ritual, and ceremonial ties in the village. As Zaman (1982: 41) puts it, "a gosthi is a patrilineal kin group reckoned through a remembered 'common' male ancestor in a group of people" (also see Ellickson, 1974.

76. The analysis involved a questionnaire survey of village elders, retired government officials, foreign aid distributors, and technical experts. The composite map was drawn combining the mental maps of these groups and the available spatial data. Thus, the study basically relied upon non-physical sources.

77. See Haque *et al*,. (1985a), and Haque and Zaman (1991) for details of the methodology adopted.

78. By comparing the changes in vital rates in local spatial units with the regional or national rates, the volume of population migration can be measured. This indirect method can be recommended only for cases where direct data on migration are not available. For details, see Shryock *et al.*, 1979: 377-396.

79. *Talukdar* and *jotedar* are generic names used to refer to a class of rural landlords associated with the presently defunct *zamindari* (landlord) system of Bengal. Despite the abolition of the system by legislation in 1950, the land-based power and the authority of landlords remains very much alive today.

80. It is expected that the socioeconomic attributes of the floodplain users will show significant variations between displacees (i.e., involuntary migrants) and non-displacees (i.e., natives). The survey data on demographic characteristics (e.g., age, sex, and marital status) indicate that attributes are generally homogeneous among the floodplain inhabitants (see Halli, 1988) and are primarily determined by the prevailing social and cultural norms, beliefs and practices.

81. Amin (1982: 130) cites an example that "access to education is determined, primarily, by immediate family access to land."

82. Landlessness is defined as "a rural household that claims ownership of no land, either homestead or other land" (BBS, 1982: 787).

83. The "index of concentration" (I) is a measure used to describe the extent of relative concentration of a given distribution from a hypothetically even distribution. This is particularly useful as a method of comparing a number of distributions. In a Lorenz curve, the *index of concentration* is the ratio of the area of the right-angled distribution to the area contained by individual curves. The index (I) can be found by:

$$I = \frac{A - V}{M - V}$$

where, A = the area cumulative percentage total; V = the variable cumulative percentage total; and M = the maximum cumulative percentage total assuming 100 percent of frequencies in rank 1. Applying a Lorenz curve, we get:

$$I = \frac{A - 550}{450}$$

The index will range from a maximum value of 1 (maximum concentration, when the Lorenz Curve deviates as much as possible from the diagonal) to a minimum value of 0 (minimum concentration, when the Lorenz Curve corresponds exactly to the diagonal). Thus, the index (I) shows the relative magnitude of concentration, depending on its position in a contiuum from 1 to 0. See Hammond and McCullagh (1991) for further details.

84. $[X^2(p<0.05, DF=4) = 7.26]$.

85. These are religious schools that originated during the Mughol rule (pre-1757) over Bengal.

86. The term "other displacee" is used interchangeably with "displacee sample of the main survey."

87. For comparability with national census and labor for survey data, the Census Commission's classification is followed in the present study. The "manpower" of an area is the sum of persons who could produce and supply services if they desired to participate in such activity. A distinction from "labor force" is notable

here. The "labor force" is that part of the manpower which actually engages, or attempts to engage, in the production of economic goods and services (Haque, 1983a: 176).

88. This classification, nevertheless, may not reveal the occupational pecularities and diversities of the rural economy.

89. Farm wage laborers, *rakhal* (common laborer), boatman, rickshaw peddlers and other unclassified workers are included in this category.

90. This is because the n-th dummy variable is entirely determined by the first n-1 dummies entered into the equation (Halli, 1988: 10).

91. The term *concept* is used here in a psychological sense. Golant and Burton (1976: 120-121) adopt this term to refer to an individual's mental comprehension of the common nouns that relate to natural hazards phenomena such as "floods," "earthquakes," "droughts," etc.

92. There are more than two thousand varieties of paddy cultivated in the floodplains of West Bengal, India and Bangladesh. *Aus, aman* and *boro* are the three most predominant traditional varieties. Some High Yielding Varieties (HYV) of *boro* paddy are also currently being sown in Bangladesh.

93. For details on agricultural adjustments of varying magnitudes of floods, see Paul (1984).

94. See Burton *et al.*, (1978) for a discussion on the distinction between "cultural adaptation" and short-term "purposeful adjustments" to hazards.

95. $X^2 = 21.11$; $p<0.001$; df=2.

96. Note that the yield of *boro* and *aman* paddy per unit of land is much higher than *aus*, while the prices per unit of *boro* and *aman* rice are higher than *aus*.

97. With a low level of technology in Bangladesh, agricultural occupations usually involve material labor, whereas non-agricultural occupations are characterized by mental labor because of the predominance of the commercial and service sectors (see Haque, 1983c).

98. The *cognitive dissonance theory* asserts that individuals reject information and suppress facts that might be damaging to their prevailing value system (see Festinger, 1957). Thus, the theory can explain why hazards are often ignored by

individuals living in a threatened area but exaggerated by people in neighboring but yet unaffected areas. Both groups use information about hazards to justify their own choice.

99. A Type II error means a failure to reject a null hypothesis when it is in fact false. See Taylor (1977: 107) for some examples of Type II errors.

100. Non-local businessmen who migrated from West Bengal, India. However, these merchants normally speak Urdu or Hindi rather than Bangla.

101. Powell and Raynor (1952) have designated "remedy" as the period immediately following the disaster.

REFERENCES

Abbas, B.M. 1963. Utilization and development of the deltaic area of East Pakistan. In *Proceedings of the regional symposium on flood control, reclamation, utilization and development of deltaic area*, United Nations Water Resources Series No. 25, 133-151. New York: United Nations.

Abdullah, A. 1976. Land reform and agrarian change in Bangladesh. *The Bangladesh Development Studies* 4 (1): 67-114.

Abdullah, A. 1980. The peasant economy in transition: the rise of rich peasants in permanently settled Bengal. *The Bangladesh Development Studies* 8 (4): 1-20.

Abdullah, A. *et al.* 1976. Agrarian structure and the IRDP - preliminary considerations. *The Bangladesh Development Studies* 4: 209-166.

Abu-Lughod, J. 1961. Migrant adjustment to city life: the Egyptian case. *American Journal of Sociology* 67: 22-32.

Ackerman, E.A. 1945. Geographic training, wartime research, and immediate professional objectives. *Annals of the Association of American Geographers* 35: 121-143.

Ackerman, E.A. 1963. Where is a research frontier? *Annals of the Association of American Geographers* 53: 429-440.

Achers, P. and F.G. Charlton 1970. The slope and resistance of small meandering channels. *Proceedings of the Institute of Civil Engineers* 47 (supplementary paper 7362-S): 349-370.

Adair, L.S. 1984. Marginal intake and maternal adaptation: the case of rural Taiwan. In *Energy intake and activity*, E. Pollit and P. Amante (eds). New York: Alan R. Liss.

Adams, R. 1974. Harnessing technological development. In *Rethinking modernization: anthropological perspectives*, J.J. Poggie *et al.* (eds). Cunnecticut: Greenwood Press.

Adelman, H. (ed.) 1980. *The Indochinese refugee movement: the Canadian experience.* Toronto: Operation Lifeline.

Adnan. S. 1991. *Floods, people and the environment.* Dhaka: Research and Advisory Services.

Adnan, S. and A.H. Monsoor 1976. *Land, power and violence in some Barisal villages.* Working Paper No. 6. Dhaka: Village Study Group, University of Dhaka.

Ahmad, N. 1968. *An economic geography of East Pakistan.* London: Oxford University Press.

Ahmed, M. 1989. Food for work program and its interference to the drainage system in Bangladesh. In *Floods in Bangladesh*, M. Ahmad (ed.), 199-207. Dhaka: Community Development Library.

Ahmed, M. and A. Jenkins 1991. Traditional land grabbing and settlement pattern in the Southeastern delta. In *Riverbank erosion, flood and population displacement in Bangladesh*, K.M. Elahi *et al.* (eds), 290-306. Dhaka: Riverbank Erosion Impact Study, Jahangirnagar University.

Akhter, R. 1984. Causes and consequences of migration to city: a case study of slum dwellers in Dhaka, Bangladesh. Unpublished manuscript. Dhaka.

Alam, M.S. 1992. Rural development in Bangladesh: quest for an alternative strategy. In *Bangladesh, geography, environment and development*, K.M. Elahi *et al.* (eds), 80-93. Dhaka: Bangladesh National Geographical Association.

Alam, M.S. and D. Samsuddin 1988. Effect of flood control embankments on properties of soil. *Jahangirnagar Review* 12 (Part A): 29-34.

Alam, S.M.N. 1990. Perceptions of flood among Bangladeshi villagers. *Disasters* 14 (4): 354-357.

Alamgir, M. 1975. Some aspects of Bangladesh agriculture: review of performance and evaluation of policies. *The Bangladesh Development Studies* 3(3): 261-300.

Alamgir, M. 1977. *Famine 1974: political economy of mass starvation in Bangladesh.* Statistical Annex, Part I. Dhaka: Bangladesh Institute of Development Studies.

Alamgir, M. 1978. *Bangladesh: a case of below poverty level equilibrium trap*. Dhaka: Bangladesh Institute of Development Studies.

Alamgir, M. 1980. *Famine in South Asia - political economy of mass starvation.* Cambridge, Mass.: Oelgeschlager, Gunn and Hain.

Alamgir, M. 1983. A review of the public rural works programmes in Bangladesh 1960-78. *The Bangladesh Development Studies* 11 (1 & 2).

Alexander, D. 1991. Natural disasters: a framework for research and training. *Disasters* 15 (3): 209-226.

Ali, S.M. 1980. Administration of charland in Bangladesh. *Asian Affairs* 2 (2): 259-303.

Amin, N. 1982. An analysis of labour force and industrial organization of the informal sector in Dacca. Doctoral dissertation. Winnipeg, Manitoba: University of Manitoba.

Amin, S. 1974. *Accumulation on a world scale - a critique of the theory of underdevelopment.* New York: Monthly Review Press.

Amin. S. 1977. *Unequal development: an essay on the social formations of peripheral capitalism.* Sussex: Harvester Press.

Anderson, M. and P.J. Woodrow 1989. *Rising from the ashes: development strategies in times of disaster*. Boulder, Colorado: Westview Press.

Ang, A. H-S. 1978. Comment in the US-Southeast Asia symposium on Engineering for Natural Hazards Protection, Representative of the National Science Foundation. Urbana: University of Illinois.

Arthur, W.B. and G. McNicoll 1978. An analytical survey of population and development in Bangladesh. *Population and Development Review* 4 (1): 23-80.

Ascoli, F.D. 1910. The river of the delta. *Journal and Proceedings of the Asiatic Society of Bengal* 6 (New Series): 543-546.

Auden, J.B. 1949. Geological discussion on the Satpura hypothesis and Garo-Rajmahal gap. *Proceedings of the National Science Institute* 15: 315-340.

Awanohara, S. 1982. In the shadow of death. *The Far Eastern Economic Review* 118 (42): 42-43.

Aysan, Y. 1987. Homeless in 42 square metre. *Open House International* 12: 21-26.

Aysan, Y. and P. Oliver 1987. *Housing and culture after earthquakes*. Oxford: Oxford Polytechnic.

Baden-Powell, B.H. 1882. *The land system of British India*. London: Clarendon Press.

Bagchi, K. 1944. *The Gages delta*. Calcutta: Calcutta University Press.

Baker, E.J. and D.J. Patton 1974. Attitude toward hurricane hazards on the Gulf Coast. In *Natural hazards - local, national, global*, G.F. White (ed.), 30-36. New York: Oxford University Press.

Bamberger, M. 1968. A problem of political integration in Latin America: the Barrias of Venezuela. *International Affairs* 44 (4).

Bangladesh Planning Commission 1973. *The first five year plan 1973-78*. Dacca: Ministry of Planning.

Bari, M.F. and M. Sahjahan 1989. Water resources development activities in Bangladesh: survey and appraisal. In *Flood in Bangladesh*, M. Ahmad (ed.), 254-270. Dhaka: Community Development Library.

Barrows, H.H. 1923. Geography as human ecology. *Annals of the Association of American Geographers* 13: 1-14.

Barth, F. 1958. Ecological relationships of ethnic groups, North Pakistan. *American Anthropologist* 58.

Bartocci, P.J. 1979. Structural fragmentation and peasant classes in Bangladesh. *Journal of Social Studies* 5: 43-70.

Bartocci, P.J. 1982. Bangladesh in the early 1980s: practorian politics in an intermediate regime. *Asian Survey* 22 (10): 988-1008.

Barton, A.H. 1970. *Communities in disaster*. New York: Anchor.

BBS (Bangladesh Bureau of Statistics) 1975. *Statistical pocketbook of Bangladesh 1975*. Dacca: Statistics Division, Ministry of Planning.

BBS (Bangladesh Bureau of Statistics) 1978. *Master survey of agriculture* (seventh round, second phase). Dacca: Statistics Division, Ministry of Planning.

BBS (Bangladesh Bureau of Statistics) 1980. *Statistical yearbook of Bangladesh 1979*. Dhaka: Statistics Division, Ministry of Planning.

BBS (Bangladesh Bureau of Statistics) 1982. *Agricultural yearbook of Bangladesh 1982*. Dhaka: Statistics Division, Ministry of Planning.

BBS (Bangladesh Bureau of Statistics) 1983. *Pabna district statistics*. Dhaka: Statistics Division, Ministry of Planning.

BBS (Bangladesh Bureau of Statistics) 1984a. *Bangladesh population census 1981: analytical findings and national tables*. Dhaka: Statistics Division, Ministry of Planning.

BBS (Bangladesh Bureau of Statistics) 1984b. *Statistical yearbook of Bangladesh 1983-84*. Dhaka: Statistics Division, Ministry of Planning.

BBS (Bangladesh Bureau of Statistics) 1985. *Statistical pocketbook of Bangladesh 1985*. Dhaka: Statistics Division, Ministry of Planning.

BBS (Bangladesh Bureau of Statistics) 1995. *Statistical yearbook of Bangladesh 1993-94*. Dhaka: Statistics Division, Ministry of Planning.

BBS (Bangladesh Bureau of Statistics) 1996. *Statistical yearbook of Bangladesh 1994-95*. Dhaka: Statistics Division, Ministry of Planning.

Beaujeur-Garnier, J. 1976. What is geography? In *Methods and perspectives in geography*, J. Beaujeur-Garnier (ed.), 1-22, J. Bray (tr.). New York: Longman.

Bell, D. 1975. Technology, nature and society.In *The Frontiers of knowledge*, The Frank Nelson Doubleday lectures, 27-78. New York: Doubleday and Company.

Belshaw, C. 1951. Social consequences of Mount Lamington eruption. *Oceania* 21 (4).

Bennett, J.W. 1969. *Northern plainsmen: adaptive strategy and agrarian life*. Chicago: Aldine.

Berry, J.W. 1989-90. Refugee acculturation and adaptation: an introduction. *High Plains Applied Anthropologist* 9 & 10: 211-221.

Beverly, H. 1872. *Census of Bengal 1872*. Calcutta: Bengal Secretariat Press.

Beyer, G.A. 1989. *Improving institutional response to humanitarian situations*. Washington, D.C.: Refugee Policy Group.

Bhaskar, R. 1975. *A realist theory of science*. Leeds: Leeds Book Ltd.

Bhaskar, R. 1978. Social scientific knowledge. In *Issues in Marxist philosophy*, Vol. 3, J. Mepham and D.H. Ruben (eds), 107-139. Brighton: Harvester Press.

Bhaskar, R. 1979. *The philosophy of naturalism*. Brighton: Harvester Press.

Bhattacharya, D. 1989. *Social responses to natural disaster: insights from 1988 flood relief camps in Dhaka City*. Dhaka: Bangladesh Institute of Development Studies.

Bhattacharya, N.D. 1978. Changing courses of the Padma and human settlements. *The National Geographical Journal of India* 24: 62-76.

Bhattacharyya, A. 1977. *Historical geography of ancient and early medieval Bengal*. Calcutta: Sanskrit Pustak Bhandar.

Billinge, M.D. 1981. Phenomenal environment. In *The dictionary of human geography*, R.J. Johnston *et al.* (eds), 251-252. New York: The Free Press.

Billinge, M.D. 1986. Environmental perception. In *The dictionary of human geography*, R.J. Johnston *et al.* (eds), 106-107. Oxford: Basil Blackwell.

Bingham, A. 1989. Floods of aid for Bangladesh. *New Scientist* 1693: 42-46.

Black, R. 1994. Forced migration and environmental change: the impact of refugees on host environments. *Journal of Environmental Management* 42: 261-277.

Blaikie, P. *et al.* 1994. *At risk: natural hazards, people's vulnerability, and disasters*. London: Routledge.

Blair, H.W. 1978. Rural development, class structure and the bureaucracy in Bangladesh. *World Development* 6: 65-82.

Bloesch, U. 1995. Impact of humanitarian crisis on ecosystem. Paper presented at the ECHO expert consultation on Policy Issues on Environmental Impact of Displacement during the Emergency Phase, Brussels, September 23-24.

Bonilla, F. 1961. Rio's favelas: the rural slum within the city. In *Peasants in Cities*, W. Mangin (ed.), Boston: Houghton Mifflin Company.

Bourdillon, J.A. 1883. *Report on the census of Bengal 1881*, Vol. I. Calcutta: Bengal Secretariat Press.

Box, G.E.P. and G.M. Jenkins, G.M. 1976. *Time series analysis: forecasting and control*. San Francisco: Holden Day.

Boyce, J.K. 1990. Birth of a megaproject: political economy of flood control in Bangladesh. *Environmental Management* 14 (4): 419-428.

Brammer, H. 1990a. Floods in Bangladesh I. geographical background to the 1987 and 1988 floods. *The Geographical Journal* 156 (Part I): 12-22.

Brammer, H. 1990b. Floods in Bangladesh II. flood mitigation and environmental aspects. *The Geographical Journal* 156 (Part II): 158-165.

Bras, R.L. and R.G. Prinn (eds) 1993. *World at risk: natural hazards and climate change*. Oxford University Press.

Bristow, C.S. 1987. Brahmaputra River: channel migration and deposition. In *Fluvial sedimentology*, F.G. Ethridge *et al.* (eds), 63-74. Tulsa, Oklahama: Society of Economic Paleontologists and Mineralogists.

Brody, E.B. 1970. Migration and adaptation: the nature of the problem. In *Behavior in new environments: adaptation of migrant populations*, E.B. Brody (ed.), 13-22. Beverly-Hill, California: Sage Publications.

Brookfield, H.C. 1964. Questions on the human frontiers of geography. *Economic Geography* 40: 283-303.

Brookfield, H.C. 1973. Explaining or understanding? the study of adaptation and change. In *The Pacific in transition*, H.C. Brookfield (ed.), 3-22. New York: St. Martin's Press.

Brooks, R.H. 1972. Flight from disaster: drought perception as a force in migration from Ceara, Brazil. Doctoral dissertation. Boulder, Colorado: University of Colorado.

Brooks, R.H. 1973. Differential perception of drought in northeastern Brazil. *Proceedings of the Association of American Geographers* 5: 31-35.

Brudge, R.J. and R.A. Robertson 1990. Social impact assessment and the public involvement process. *Environmental Impact Assessment Review* 10 (1 & 2): 81-90.

Bryant, E.A. 1991. *Natural hazards*. Cambridge: Cambridge University Press.

Buchanan, D.H. 1966. *The development of capitalist enterprise in India*. London: Frank Cass & Company.

Bunting, T.E. and L. Guelke 1979. Behavioral and perception geography: a critical appraisal. *Annals of the Association of American Geographers* 69: 448-462.

Burger, J.W., G. Klassen and A. Prins 1988. Bank erosion and channel processes in the Jamuna River. Paper presented at the international symposium on the Impact of Riverbank erosion, Flood Hazard, and the Problem of Population Displacement. Dhaka, April 11-14.

Burgess, R. 1978. The concept of nature in geography and Marxism. *Antipode* 10(2): 1-11.

Burkham, D.E. 1972. Channel changes of the Gila River in Safford Valley, Arizona, 1846-1970. *United States Geological Survey Professional Paper* 655-G.

Burton, I. 1962. *Types of agricultural occupance of flood plains in the United States*. Department of Geography Research Paper No. 75. Chicago: University of Chicago.

Burton, I. 1972. Cultural and personality variables in the perception of natural hazards. In *Environment and social sciences: perspectives and applications*. J.F. Wohlwill and D.H. Carson (eds), 184-195. Washington, D.C.: American Psychological Association.

Burton, I. *et al.* 1968. *The human ecology of extreme geophysical events*. Department of Geography Research Working Paper 1. Toronto: University of Toronto Press.

Burton, I., R. Kates and G. F. White 1978. *The environment as hazard*. New York: Oxford University Press.

Burton, I., R. Kates and G.F. White 1993. *The environment as hazard*. Second edition. New York: The Guilford Press.

BWDB (Bangladesh Water Development Board) 1974. *Report on town protection schemes in Bangladesh*. Dacca: Office of the Chief Engineer, Planning.

BWDB (Bangladesh Water Development Board) 1978. *Morphological features of major rivers of Bangladesh*. Dacca: Directorate of River Morphology.

Cain, M.T. 1977. The economic activities of children in a village in Bangladesh. *Population and Development Review* 3: 201-227.

Cain, M.T. 1981. Risk and fertility in India and Bangladesh. *Population and Development Review* 7 (3): 435-474.

Campbell, S.G. 1878. The famines which affected Bengal in the last century. In *Sessional papers*, Great Britain, House of Commons report of the select committee on East India on the expediency of constructing public works with sums raised on loan, as regards to financial results and prevention of famine with proceedings, minutes of evidence, appendix and index, 12: 455-466.

Cannon, T.G. 1977. 'Natural' disasters and the Third World. In *Geography, social welfare and underdevelopment*, N. Smith *et al.* (eds), 73-88. St Andrews: University of St. Andrews.

Cannon, T.G. 1994. Vulnerability analysis and the explanation of 'natural' disaster. In *Disasters, development and environment*, A. Varley (ed.), 13-30. London: John Wiley & Sons.

Cardy, W.F.G. 1994. *Environment and forced migration: a review*. Nairobi: United Nations Environment Programme.

Carello, C. 1993. Realism and ecological units of analysis. In *Human ecology: fragments of anti-fragmentary views of the world*, M. Nauser (ed.). London: Routledge.

Carper, T.R. 1990. The national flood insurance program - beating a retreat. *Natural Hazards Observer* 14 (4): 1-2.

Carstairs, R. 1895. *Human nature in rural India*. London: William Blackwood & Sons.

CBJET (China-Bangladesh Joint Expert Team) 1991. *Study report on flood control and river training project on the Brahmaputra River in Bangladesh*, Vol. 1. Analysis and hydrology and river morphology of the Brahmaputra River. Dhaka.

Cernea, M.M. 1988. *Involuntary resettlement in development projects: policy guidelines in World Bank-financed projects*. Washington, D.C. The World Bank.

Cernea, M.M. 1989. Anthropology, policy and involuntary resettlement. Paper presented at the meeting of the Resettlement Task Force, American Anthropological Association, Washington, D.C.

Cernea, M.M. 1991. Involuntary resettlement: social research, policy and planning. In *Putting people first*, M.M. Cernea (ed.), 188-216. New York: Oxford University Press.

Chang, H.H. 1979. Minimum stream power and river channel patterns. *Journal of Hydrology* 41: 303-327.

Chapin, F.S. 1965. *Urban land use planning*. Urbana: University of Illinois Press.

Chapman, D.W. 1962. A brief introduction to contemporary disaster research. In *Man and society in disaster*, G.W. Baker and D.W. Chapman (eds), 3-22. New York: Basic Books.

Chapman, M. 1963. Sampling methods for geographers. *Pacific Viewpoint* 4 (2): 155-163.

Chatterjee, S.P. 1947. River problems in Bengal. *Punjab Geographical Review* 2 (1): 12-20.

Chatterjee, S.P. 1949. *Bengal in maps: a geographical analysis of resource distribution in West Bengal and East Pakistan*. Bombay: Orient Longmans.

Chen, L.C. and N.C. Scrimshaw (eds) 1983. *Diarrhea and malnutrition: interactions, mechanisms and inventions*. New York: Plenum Press.

Chorley, R.J. 1973. Geography as human ecology. In *Directions in geography*, R.J. Chorley (ed.), 155-169. London: Methuen.

Choudhuri, R. 1943. Physical and historical geography. In *The history of Bengal*, Vol. 1: Hindu Period. R.C. Majumdar (ed.), 1-34. Patna: N.V. Publications.

Choudhury, R.H. 1978. Determinants and consequences of rural-urban migration: evidence from some villages in Bangladesh. Paper presented at the IUSSP conference on Economic and Demographic Change: Issues for the 1980s, Helsinki.

Chowdhury, A. 1987. Bangladesh: a semi-feudal, semi-colonial society. In *Bangladesh: politics, economy and society*, C.E. Haque (ed.), 33-48. Winnipeg, Manitoba: Bangladesh Studies Assemblage, University of Manitoba.

Chowdhury, K.R. and A.H. Bhuiya 1990. Environmental processes: flooding, river erosion, siltation, and accretion - physical impacts. In *Surface water systems of Bangladesh*, A. Rahman *et al.* (eds.), 93-103. Dhaka: University press Ltd.

Chowdhury, M.I. 1973. Morphological process in the lower Brahmaputra valley. *Journal of the Bangladesh National Geographical Association* 1 (1): 19-29.

Chowdhury, M.I. 1977. The maps of Bangladesh during the Muslim period. *Institute of Bangladesh Studies Journal* 1: 145-154.

Chowdhury, M.I. 1982. Topography of Bangladesh. *Bangladesh Quarterly* 2 (3): 7-15.

Church, M.A. 1972. Baffin Island sanders: a study of Arctic fluvial processes. *Geological Survey of Canada GSC Bulletin* 216.

Cicourel, A.V. 1981. Notes on the integration of micro- and macro-levels of analysis. In *Advances in social theory and methodology: towards an integration of micro- and macro-sociologies*, K. Knorr-Cetina and A.V. Cicourel (eds), 51-80. Boston:Routledge and Kegan Paul.

Coburn, A.W., J.D.L. Leslie and A. Tabban 1984. Reconstruction and resettlement 11 years later: a case study of Bingol Province, Eastern Turkey. In *Earthquake relief in less industrialized areas*, S. Schupisser and J. Studer (eds.), 49-58. Rotterdam: A.A. Balkema.

Colebrooke, R.H. 1803. On the course of the Ganges through Bengal. *Asiatik Researchers* 7: 1-32.

Coleman, J.M. Brahmaputra River: channel processes and sedimentation. *Sedimentary Geology* 3 (2/3): 129-239.

Colson, E. 1971. *The social consequences of resettlement: the impact of the Kariba resettlement upon the Gwembe Tonga*. Manchester: Manchester University Press.

Connell, J. *et al.* 1976. *Migration from rural areas: the evidence from village studies*. Delhi: Oxford University Press.

Consulting Consortium 21/22. 1993. *Bank protection and river training (AFPM) pilot project FAP 21/22*. Final report, planning study, Annex 2: River Morphology. Dhaka: Flood Plan Coordination Organization.

Copans, J. 1983. The Sehelian drought: social sciences and the political economy of underdevelopment. In *Interpretations of calamity*, K. Hewitt (ed.), 83-97. Boston: Allen & Unwin, Inc.

Covello, V.T. and J. Mumpower 1985. Risk analysis and risk management: an historical perspective. *Risk Analysis* 5 (2): 103-120.

Cox, K.R. 1981. Bourgeois thought and the behavioral geography debate. In *Behavioral problems in geography revisited*, K.R. Cox and R.G. Golledge (eds), 256-279. New York: Methuen.

Craik, K.H. 1970. Environmental psychology. *New Directions in Psychology* 4: 1-121.

Crow, B. 1981. Why are the Ganges and Brahmaputra underdeveloped? politics and stagnation in the rivers of South Asia. *Bulletin of Concerned Asian Scholars* 13 (4): 35-63.

Crow, B., A. Lindquist and D. Wilson 1995. *Sharing the Ganges: the politics and technology of river development*. Dhaka: University Press Ltd.
Crozier, M.J. 1981. The character of natural hazards in different physical and social settings. *Proceedings of the conference on geography and development policy*, 106-109. The 11th New Zealand Geographical Conference, Auckland.
Cunningham, A. 1924. *Ancient geography of India*. S.N. Majumdar (ed.). Calcutta: Bengal Secretariat Book Depot.
Cuny, F.C. 1983. *Disasters and development*. New York: Oxford University Press.
Curray, J.R. and Moore, D.G. 1971. Growth of the Bengal deepsea fan and denudation in the Himalayas. *Geological Society of America Bulletin* 82: 563-572.
Currey, B. 1979. Mapping areas liable to famine in Bangladesh. Doctoral dissertation. Honolulu: University of Hawaii.
Cutter, S.L. 1993. *Living with risk*. London: Edward Arnold.
Cutter, S.L. 1994. Isn't any place safe anymore? In *Environmental risks and hazards*, S.L. Cutter (ed.), xi-3. Englewood Cliffs, New Jersey: Prentice-Hall.
Dacy, D.C. and H. Kunreuther 1969. *The economics of natural disasters*. New York: The Free Press.
Daly, H.E. and J.B. Cobb 1989. *For the common good: redirecting the economy toward community, the environment and a sustainable future*. Boston: Neacon Press.
Dasgupta, P. 1993. *An inquiry into well-being and destitute*. Oxford: Calrendon Press.
Dassman, R.F., J.P. Milton, and P.H. Freeman 1973. *Ecological principles for economic development*. New York: John Wiley & Sons.
Davis, K. 1951. *The population of India and Pakistan*. Princeton: Princeton University Press.
Davis, K. and J. Blake 1956. Social structure and fertility: an analytical framework. *Economic Development and Cultural Change* 4 (3): 211-235.
Degg, M. 1992. Natural disasters: recent trends and prospects. *Geography* 77: 198-209.
Degg, M. 1993. The 1992 'Cairo earthquake': cause, effect and response. *Disasters* 17 (3): 226-238.
Desai, A.R. 1959. *Social background of Indian nationalism*. Bombay: Popular Book Depot.
Desai, H.J. 1968. The Brahmaputra. In *Mountains and rivers of India*, B.C. Law (ed.), 341-437. Calcutta: National Committee for Geography.
Desai, M. 1974. *Marxian economic theory*. London: Grey-Mills.
Descola, P. and G. Palsson 1996. *Nature and society: anthropological perspectives*. London: Routledge.
de Vries, J. 1985. Analysis of historical climate-society interaction. In *Climate impact assessment*, R.W. Kates *et al.* (eds), SCOPE 27, 273-292. New York: John Wiley & Sons.
de Vylder, S. 1982. *Agriculture in chains*. London: Zed Press.
Dixon, C. and B. Leach 1978. *Sampling methods for geographical research*. Norwich: Geo Abstract.

Downing, T.E. 1991. *Climate change and vulnerable places*. Environmental Change Unit, Research Report No. 1. Oxford: University of Oxford.

Doyal, L. and Gough, I. 1991. *A theory of human need*. London: Macmillan and Guilford Press.

Drabek, T.E. 1986. *Human system response to disaster: an inventory of sociological findings*. New York: Springer-Verlag.

Dube, S.K., M.C. Sinha and V. Balasubramanyam 1982. The effect of coastal geometry on the location of peak surge. *Mausam* 33: 445-450.

Du Bois, V.D. 1974. The drought in Niger, Part 3: the flight of the Malian Tuareg. *American Universities Fieldstaff Reports: West African Series* 16 (6): 1-13.

Duncan, D.S. and D. Ley 1982. Structural Marxism and human geography: a critical assessment. *Annals of the Association of American Geographers* 72: 30-59.

Dupuy, A. and P.V. Fitzgerald 1977. A contribution to the critique of the world system perspectives. *The Insurgent Sociologist* 7 (2): 113-128.

Dworkin, J. 1974. *Global trends in natural disasters 1947-73*. Boulder, Colorado: Institute of Behavioral Science, University of Colorado.

Dynes, R.R. 1970. *Organized behavior in disaster*. Lexington, Mass.: D.C. Heath & Co.

Eckstein, S. 1977. *The poverty of revolution: the state and the urban poor in Mexico*. Princeton, New Jersey: Princeton University Press.

Eichenbaum, J. 1975. A matrix of human movement. *International Migration* 13 (1 & 2): 21-41.

Elahi, K.M. 1972. Urbanization in Bangladesh: a geodemographic study. *Oriental Geographer* 16 (1).

Elahi, K.M. 1974. Aspects of age-sex structure of the rural population of Bangladesh. *Rural Demography* 1 (1): 35-47.

Elahi, K.M. and L.T. Ruzika 1981. Trends and differentials in mortality. In *Population of Bangladesh*, Country Monograph No. 8, 43-63. New York: United Nations.

El-Hinnawi, E. 1985. *Environmental refugees*. Nairobi: United Nations Environmental Programme.

Ellickson, J. 1974. A believer among believers: the religious beliefs, practices and meanings in a village in Bangladesh. Doctoral dissertation. East Lansing, Michigan: Michigan State University.

Emmanuel, A. 1972. *Unequal exchange: a study of imperialism of trade*. New York: Monthly Review Press.

Engineering Consultants Inc. 1970. *Design report on bank protection of Serajganj from erosion by the River Jamuna*. Dacca: EPWAPDA.

English, P.W. and R.C. Mayfield 1972. The cultural landscape. In *Man, space, and environments*. P.W. English and R.C. Mayfield (eds), 3-9. New York: Oxford University Press.

EPWAPDA (East Pakistan Water and Power Development Authority) 1968. *Report on completion of Brahmaputra flood embankment project: Kaunia to Hurasagar River*, IDA Credit 39-Pak. Dacca.

Erikson, K.T. 1976. *Everything in its path*. New York: Simon and Schuster.

ESCAP (Economic and Social Commission for Asia and the Pacific) 1981. Size, growth and distribution of population. In *Population of Bangladesh*, Country Monograph Series No. 8, 10-29. New York: United Nations.

Eyles, J. and Lee, R. 1982. Human geography in explanation. *Transactions of the Institute of British Geographers* 7 (New Series): 117-122.

Fergusson, J. 1863. On recent changes in the delta of the Ganges. *Quarterly Journal of the Geological Society of London* 19: 321-354.

Festinger, L. 1957. *Theory of cognitive dissonance*. Evaston: Row Peterson.

Foster, H. 1980. *Disaster planning: the preservation of life and property*. New York: Springer-Verlag.

FPCO (Flood Plan Coordination Organization) 1993. *Bank protection and river training (AFPM) pilot project, FAP 21/22*. Vol. I-B. Dhaka: Ministry of Water Resources.

FPCO (Flood Plan Coordination Organization) 1995. *Bangladesh water and flood management strategy*. Dhaka: Ministry of Water Resources.

Frank, A.G. 1966. The development of underdevelopment. *Monthly Review* 18(4).

Frazier, K. 1979. *The violent face of nature*. New York: William Morrow.

Freeman, T.W. 1961. *A hundred years of geography*. London: Gerald Duckworth.

Fried, M. 1963. Grieving for a lost home. In *The urban condition: people and policy in the Metropolis*, L.J. Duhl (ed.). New York: Basic Books.

Fried, M. 1982. Residential attachment: sources of residential and community satisfaction. *Journal of Social Issues* 38 (3): 107-119.

Gait, E.A. 1902. *Census of India, 1901*. Vol. 6, The lower provinces of Bengal and their feudatories. Calcutta: Bengal Secretariat Press.

Galay, V.J. 1980. River channel shifting on large rivers in Bangladesh. In *Proceedings of the international symposium on river sedimentation*. Beijing, March.

Garcia, E. 1992. *A future without refugees*. Manila: Ecumenical Commission for Displaced Families and Communities.

Geddes, A. 1960. The alluvial morphology of the Indo-Gangetic plain: its mapping and geographical significance. *Transactions of the Institute of British Geographers* 28: 253-276.

Gergen, K.J. 1982. *Toward transformation in social knowledge*. New York: Springger-Verlag.

Gergen, K.J. and K.E. Davis 1985. *The social construction of the person*. New York: Springer-Verlag.

Gerrity, E.T. and P. Steinglass 1994. Relocation stress following natural disasters. In *Individual and community responses to trauma and disaster: the structure of human chaos*, R.J. Ursano *et al.* (eds), 220-247. Cambridge: Cambridge University Press.

Gill, A.M. 1982. Residents' images of northern Canadian resource communities. Doctoral dissertation. Winnipeg, Manitoba: University of Manitoba.

Glacken, C.J. 1967. *Traces on the Rhodian shore: nature and culture in Western thought from ancient times to the end of eighteenth century*. Berkeley: University of California Press.

Glass, D.V. 1954. *Social mobility in Britain*. London: Routledge and Kegan Paul.

Gleser, G.C., B.L. Green and C. Winget 1981. *Prolonged psychological effects of disaster: a study of Buffalo Creek*. New York: Academic Press.

Glittenberg, J.K. 1982. Reconstruction in four urban post disaster settlements. In *Recovery, change and development: a longitudinal study of the 1976 Guatemalan earthquake*, F.L. Bates (ed.), Vol. 2, 634-707. Athens, Georgia: The University of Georgia.

Golant, S. and I. Burton 1976. A semantic differential experiment in the interpretation and grouping of environmental hazards. In *Environmental knowing*, G.T. Moore and R.G. Golledge (eds), 364-374. Stroudsburg: Dowden, Hutchinson and Ross.

Gold, J.R. 1980. *An introduction to behavioral geography*. New York: Oxford University Press.

Goodwin-Gill, G. 1988. Refugees: the expanding mandate of the office of the United Nations High Commissioner for Refugees. Unpublished paper. Geneva: UNHCR.

Gordon, R.B. 1972. *Physics of the earth*. New York: Holt, Riehart and Winstone.

Goswami, D.C. 1985. Brahmaputra River, Assam, India: physiography, basin denudation, and channel aggradation. *Water Resources Research* 21 (7): 959-978.

Goulter, I.C. and D. Dubois 1991. Methods of presenting trends in channel capacity movement as an indicator of bank erosion in the Jamuna River. In *Riverbank erosion, flood and population displacement in Bangladesh*, K.M. Elahi *et al.* (eds), 46-60. Dhaka: Riverbank Erosion Impact Study, Jahangirnagar University.

Government of East Pakistan 1963. *Report of the land administration enquiry committee, East Pakistan 1962-63*. Dacca: East Pakistan Government Press.

Government of Pakistan 1960. *Pakistan census of agriculture*, Vol. 1. Karachi: Agricultural Census Organization.

Graves, T.D. 1967. Psychological acculturation in a tri-ethnic community. *Southeastern Journal of Anthropology* 23: 337-350.

Greenberg, C. 1986. The adaptation of riverbank erosion displacees in an urban environment: a case study of squatters in Serajganj, Bangladesh. Master's thesis. Winnipeg, Manitoba: University of Manitoba.

Gregory, D. 1981. Epistemology. In *The dictionary of human geography*, R.J. Johnston *et al.* (eds), 108. New York: The Free Press.

Grossman, L. 1977. Man-environment relationships in anthropology and geography. *Annals of the Association of American Geographers* 67: 126-144.

Grove, J.M. 1972. The incidence of landslides, avalanches and floods in Western Norway during the Little Ice Age. *Arctic and Alpine Research* 4: 131-138.

Guelke, L. 1971. Problems of scientific explanation in geography. *The Canadian Geographer* 15: 38-53.

Guha, R. 1963. *A rule of property for Bengal*. Paris: Monton.

Gupta, R.M.N. 1940. *Land system of Bengal*. Calcutta: University of Calcutta Press.

Habib, I. 1963. *The agrarian system of Mughal India (1550-1707)*. Bombay: Asia Publishing House.

Hagarstrand, T. 1976. Geography and the study of interaction between nature and society. *Geoforum* 7: 329-334.

Hamilton, L.S. 1987. What are the impacts of Himalayan deforestation on the Ganges-Brahmaputra lowlands and delta? Assumptions and facts. *Mountain Research and Development* 7: 256-263.

Hammond, R. and P. McCullagh 1991. *Quantitative techniques in geography*, Second edition. Oxford: Clarendon Press.

Haque, C.E. 1983a. Population versus economic activity: perspectives on temporal changes and regional disparities in Bangladesh. *Asian Profile* 11 (2): 175-194.

Haque, C.E. 1983b. Human adjustments to riverbank slumping: the case of the lower Brahmaputra-Ganges floodplain. In *South Asian horizons*, V.S. Pendakur (ed.), 70-86. Ottawa: Canadian Asian Studies Association.

Haque, C.E. 1983c. Patterns and changes in regional employment structure in Bangladesh. *Singapore Journal of Tropical Geography* 4 (2): 119-130.

Haque, C.E. 1984. Human adjustments to riverine environment: some theoretical and conceptual issues. Paper presented at the international seminar on Man and Environment, University of Dhaka, Dhaka, April 19-21.

Haque, C.E. 1986a. Characteristics of human strategies of coping with hazards in the Jamuna floodplain of Bangladesh. Paper presented at the 20th Bengal Studies Conference, Lake Geneva, Wisconsin, May 16-18.

Haque, C.E. 1986b. Impacts of riverbank erosion on population displacement in the lower Brahmaputra (Jamuna) floodplain. In *South Asian horizons*, V.S. Pendakur and O.P. Dwivedi (eds.), 34-46. Montreal: Canadian Asian Studies Association.

Haque, C.E. 1987a. Introduction: the need for a focus on the political economy of Bangladesh. In *Bangladesh: politics, economy and society*, C.E. Haque (ed.), 1-14. Winnipeg, Manitoba: Bangladesh Studies Assemblage, University of Manitoba.

Haque, C.E. 1987b. Underdevelopment, urbanization and industrialization in Bangladesh: a comparative perspective. In *Bangladesh: politics, economy and society*, C.E. Haque (ed.), 63-91. Winnipeg, Manitoba: Bangladesh Studies Assemblage, University of Manitoba.

Haque, C.E. 1987c. The changing face of North Bengal: a study of human occupance in the Brahmaputra-Jamuna floodplain. Paper presented at the 21st Bengal Studies Conference, University of Wisconsin, Oshkosh, May 8-10.

Haque, C.E. 1988a. Human adjustments to riverbank erosion hazard in the Jamuna floodplain, Bangladesh. *Human Ecology* 16 (4): 421-437.

Haque, C.E. 1988b. Perspectives on the integration of physical and human geography. In *The Moose Mountain papers*, A.H. Paul and R.W. Widdis (eds.), 124-137. Regina, Saskatchewan: University of Regina.

Haque, C.E. 1988c. Impact of riverbank erosion hazard in the Brahmaputra-Jamuna floodplain: a study of population displacement and response strategy. Doctoral dissertation. Winnipeg, Manitoba: University of Manitoba.

Haque, C.E. 1991. Human responses to riverbank erosion hazard in Bangladesh: some lessons from indigenous adjustment strategies. In *Riverbank erosion, flood and population displacement in Bangladesh*, K.M. Elahi *et al.* (eds), 191-217. Dhaka: Riverbank Erosion Impact Study, Jahangirnagar University.

Haque, C.E. 1993. Flood prevention and mitigation actions in Bangladesh: the "sustainable floodplain development" approach. *Impact Assessment* 11 (4): 367-390.

Haque, C.E. 1994. Flood prevention and mitigation in Bangladesh: the need for sustainable floodplain development. In *Environmental assessment and development*, R. Goodland and V. Edmundson (eds), 101-113. Washington, D.C.: The World Bank.

Haque, C.E. 1995a. The climatic hazards warning process in Bangladesh: experience of, and lessons from the 1991 April cyclone. *Environmental Management* 19 (5): 719-734.

Haque, C.E. 1995b. Water resource management and the problem of riverbank erosion: the context of Vikrampur (in Bengali). *Monthly Vikrampur* 18 (2 & 3): 8-11.

Haque, C.E. 1996a. The dilemma of 'nationhood' and religion: a survey and critique of studies on population displacement resulting from the partition of the Indian subcontinent. *Journal of Refugee Studies* 8 (2): 189-209.

Haque, C.E. 1996b. The integration of regional economic impact assessment with social impact assessment. *Impact Assessment* 14 (4): 343-370.

Haque, C.E. and D. Blair 1992. Vulnerability to tropical cyclones: evidence from the April 1991 cyclone in coastal Bangladesh. *Disasters* 16 (3): 217-229.

Haque, C.E. and M.Z. Hossain 1988. Riverbank erosion in Bangladesh. *Geographical Review* 78 (1):20-31.

Haque, C.E. and M. Parvin 1986. Impact of riverbank erosion on poverty and starvation. Paper presented at the annual meeting of the Canadian Association of Geographers, University of Calgary, Calgary, June 18-24.

Haque, C.E. and M. Rahman 1994. Water resources management. In *Policy issues in Bangladesh*, H. Zafarullah *et al.* (eds), 216-237. New Delhi: South Asian Publishers Pvt. Ltd.

Haque, C.E. and M.Q. Zaman 1989. Coping with riverbank erosion hazard and displacement in Bangladesh: survival strategies and adjustments. *Disasters* 13 (4): 300-314.

Haque, C.E. and M.Q. Zaman 1993. Human responses to riverine hazards in Bangladesh: a proposal for sustainable floodplain development. *World Development* 21 (1): 93-107.

Haque, C.E. and M.Q. Zaman 1994. Vulnerability and responses to riverine hazards in Bangladesh: a critique of flood control and mitigation approaches. In *Disasters, development and environment*, A. Varley (ed.), 65-79. London: John Wiley & Sons.

Haque, C.E. *et al.* 1984. Riverbank erosion hazard in Bangladesh: perspectives on population displacement. Paper presented at the international seminar on Man and Environment, University of Dhaka, Dhaka, April 19-21.

Haque, C.E. *et al.* 1985a. Impacts of erosion on the land and people in Kazipur: an application of Landsat images. Paper presented at the workshop on the Impact of Riverbank Erosion and Flood Hazards in Bangladesh, Jahangirnagar University, Savar, March 5-7.

Haque, C.E. *et al.* 1985b. Survey methods and sampling design for the survey on impacts of riverbank erosion in rural Bangladesh. Paper presented at the workshop on the Impact of Riverbank Erosion and Flood Hazards in Bangladesh, Jahangirnagar University, Savar, March 5-7.

Halli, S.S. 1988. Impact of riverbank erosion on economic activities in Kazipur Upazila, Bangladesh. Paper presented at the International symposium on the Impact of Riverbank Erosion, Flood Hazard and the Problem of Population Displacement, Dhaka, April 11-14.

Hansen, A. and A. Oliver-Smith (ed.) 1982. *Involuntary migration and resettlement.* Boulder, Colorado: Westview Press.

Hanson, S. *et al.* 1979. Natural disaster: long-range impact on human response to future disaster threats. *Environment and Planning* 11: 268-284.

Harbert, D.T. 1979. Introduction: geographical perspectives and urban problems. In *Social problems and the city*, D.T. Herbert and D.M. Smith (eds.), 1-9. Oxford: Oxford University Press.

Hare, F.K. 1985. Climatic variability and change. In *Climate impact assessment*, R.W. Kates *et al.* (eds), SCOPE 27, 37-68. New York: John Wiley & Sons.

Hare, F.K. 1995. Contemporary climate change: the problem of uncertainty. In *Resource and environmental management in Canada: addressing conflict and uncertainty*, Second edition, B. Mitchell (ed.), 10-28. Toronto: Oxford University Press.

Harvey, D. 1969. *Explanation in geography*. London: Arnold.

Hathaway, J. 1991. Reconceiving refugee law as human rights protection. *Journal of Refugee Studies* 4 (2): 113-131.

Hayden, H.H. and E.H. Pascoe 1919. *Report on the Hoogley River and its headwaters.* Calcutta: The Bengal Secretariat Book Depot.

Heathcote, R.L. 1969. Drought in Australia: a problem of perception. *Geographical Review* 59 (2): 175-194.

Heathcote, R.L. 1974. Drought in South Australia. In *Natural hazards - local, national, global*, G.F. White (ed.), 128-136. New York: Oxford University Press.

Heller, T. 1982. The effects of involuntary residential relocation: a review. *American Journal of Community Psychology* 10 (4): 471-492.

Henderson, F.M. 1963. Stability of alluvial channels. *Transactions of the American Society of Civil Engineers* 128 (Part 1): 657-686.

Hewitt, K. 1983. The idea of calamity in a technological age. In *Interpretations of calamity*, K. Hewitt (ed.), 3-32. Boston: Allen & Unwin Inc.

Hewitt, K. 1997. *Regions of risk: a geographical introduction to disasters.* Harlow, Essex: Addison Wesley Longman Limited.

Hewitt, K. and I. Burton 1971. *The hazardousness of a place: a regional ecology of damaging events.* Department of Geography Research Paper No. 6. Toronto: University of Toronto Press.

Hewitt, K. and F.K. Hare 1973. *Man and environment: conceptual framework.* Commission on College Geography Resource Paper No. 20. Washington, D.C.: Association of American Geographers.

Hindness, B. and P. Hirst 1977. *Modes of production and social formation.* London: Macmillan.

Hirst, F.C. 1916. *Report on the Nadia Rivers.* Calcutta: Bengal Secretariat Book Depot.

Hoey, T.B. 1992. Temporal variations in bedload transport rates and sediment storage in gavel-bed rivers. *Progress in Physical Geography* 16: 319-338.

Holeman, J.M. 1968. The sediment yield of major rivers of the world. *Water Resources Research* 4: 737-747.

Holt-Jensen, A. 1980. *Geography - its history and concepts.* B. Fullerton (tr.). Totowa, New Jersey: Berns and Noble Books.

Hong, L.B. and T.R.H. Davies 1979. A study of stream braiding: summary. *Geological Society of America Bulletin* 90: 1094-1095.

Hoogvelt, A.M.M. 1978. *The sociology of developing countries.* London: Macmillan.

Hooke, J.M. 1979. An analysis of the processes of river bank erosion. *Journal of Hydrology* 42: 39-42.

Hooke, J.M. and Kain, R.J.P. 1982. *Historical changes in the physical environment.* London: Butterworth Scientific.

Hoque, N. 1983. Application of remote sensing technology for detecting fluvial changes of the Ganges and Brahmaputra-Jamuna River course in Bangladesh. In *Landsat image processing applied to natural resource development in Bangladesh*, M. Fink (ed.). Albuquerque, New Mexico: Technology Application Center, University of New Mexico.

Hornborg, A. 1996. Ecology as semiotics: outlines of a contextualist paradigm for human ecology. In *Nature and society: anthropological perspectives*, P. Descola and G. Palsson (eds), 45-62. London: Routledge.

Hossain, M. 1980. *Foodgrain production in Bangladesh: performance, potential and constraints*. Dhaka: Bangladesh Institute of Development Studies.

Hossain, M. 1985. Effect of 1984 flood in Serajganj district with special reference to population displacement due to erosion in River Jamuna. Paper presented at the workshop on the Impact of Riverbank Erosion and Flood Hazards in Bangladesh, Jahangirnagar University, Savar, March 5-7.

Hossain, M.Z. 1984. Riverbank erosion and population displacement: a case of Kazipur in Pabna. Master's thesis. Savar, Dhaka: Jahangirnagar University.

Hossain, M.Z. 1988. Riverbank erosion and population displacement: a study of the Serajganj urban squatters. Master's thesis. Winnipeg, Manitoba: University of Manitoba.

Hossain, M.Z. and C. Greenberg 1985. Population displacement due to riverbank erosion and urban squatters: a case of Serajganj. Paper presented at the workshop on the Impact of Riverbank Erosion and Flood Hazards in Bangladesh, Jahangirnagar University, Savar, March 5-7.

Hossain, M.Z. and J.L. Roopnarine 1992. On the fringes: urban living among squatters of Serajganj town in Bangladesh. *Urban Anthropology* 21 (1): 45-65.

Howarth, R.B. and Norgaard, R.B. 1990. Intergenerational resource rights, efficiency, and social optimality. *Land Economics* 66: 1-11.

Hudson, R. 1980. Personal construction theory, the repertory grid method and human geography. *Progress in Human Geography* 4: 346-359.

Hunker, H.L. 1964. *Introduction to world resources*. New York: Harper and Row.

Hunter, W.W. 1875. *A statistical account of Bengal*, Vol. 1: districts of 24 Parganas and Sundarbans. London: Trubner & Co. (reprint by D.K. Publishing, Delhi, 1973).

Hunter, W.W. 1877. *A statistical account of Bengal*, Vol. 5: districts of Dacca, Bakarganj, Faridpur and Mymensingh. London: Trubner & Co. (reprint by D.K. Publishing, Delhi, 1973).

Huq, M. 1974. Geological framework of Bangladesh. In *Studies in Bangladesh*, A.F.M. Kamaluddin (ed.), 1-21. Dhaka: Bangladesh National Geographical Association.

Huq, M. 1980. *Note on flood policy and national planning in Bangladesh*. Dhaka: Bangladesh Institute of Development Studies.

Hviding, E. 1996. Nature, culture, magic, science: on meta-languages for comparison in cultural ecology. In *Nature and society: anthropological perspectives*, P. Descola and G. Palsson (eds), 165-184. London: Routledge.

IBRD (International Bank for Reconstruction) and World Bank 1990. *Flood control in Bangladesh: a plan for action*. World Bank Technical Paper No. 119. Washington, D.C.: The World Bank.

ICIHI (Independent Commission on International Humanitarian Issues) 1986. *Refugees: dynamics of displacement*. London: Zed Books Ltd.

IECO (International Engineering Company) 1964. *East Pakistan water and power development authority master plan*, Vol. II. San Francisco, California.

IECO (International Engineering Company) 1977. *Special studies*. San Francisco, California.

IFRCRCS (International Federation of Red Cross and Red Crescent Societies) 1995. *World disasters report 1995*. N. Cater (ed.). Geneva.

Ingold, T. 1996. The optimal forager and economic man. In *Nature and society: anthropological perspectives*, P. Descola and G. Palsson (eds), 25-44. London: Routledge.

IOM (International Organization for Migration) 1996. *Statement of principles: international symposium on environmentally-induced population displacements and environmental impacts resulting from mass migrations*. Geneva, April 21-24.

IOM (International Organization for Migration) and RPG (Refugee Policy Group) 1992. *Migration and the environment*. Geneva: IOM.

IPCC (Intergovernmental Panel on Climate Change) 1990. *Climate change: the IPCC scientific assessment*. New York: Cambridge University Press.

Irish, J.L. and B. Falconer 1979. Reaction to flood warning. In *Natural hazards in Australia*. R.L. Heathcote and B.G. Thom (eds), 313-329. Canberra: Australia Academy of Sciences.

Islam, M. Ecological catastrophes and refugees in Bangladesh. Paper presented at the conference on Worldwide Refugee Movements, Development and Politics and Human Rights, New York, November 8-9.

Islam, M. 1992. Natural calamities and environmental refugees in Bangladesh. *Refuge* 12 (1): 5-10.

Islam, M.A. 1968. Changes in cropping patterns in East Pakistan. *The Oriental Geographer* 12 (1):1-23.

Islam, M.A. 1974. Tropical cyclones: coastal Bangladesh. In *Natural hazards - local, national, and global*, G.F. White (ed.), 19-24. New York: Oxford University Press.

Islam, M.A. 1980. Agricultural adjustments to flooding in Bangladesh: a preliminary report. *National Geographical Journal of India* 26: 50-59.

Islam, M.A. 1989. Floods in Bangladesh: causes, consequences and adjustments. In *Proceedings of the international conference on Bangladesh floods: what are the solutions?* 54-68. Montreal: International Society of Bangladesh.

Islam, M.A. 1993. Flood control and environment in Bangladesh: experience and alternatives. In *Proceedings of the international conference on floods in Bangladesh: an interdisciplinary analysis of alternative solution strategies*, 175-194. Urbana-Champaign: College of Commerce and Business Administration: University of Urbana-Champaign.

Islam, M.S. and A.M.Z. Islam 1985. A brief account of bank erosion, model studies and bank protective works in Bangladesh. *REIS Newsletter* 2: 11-13.

Islam, N. 1976. *The urban poor in Bangladesh.* Dacca: Center for Urban Studies, University of Dacca.

Islam, N. 1979. *Squatters in Bangladesh cities.* Dhaka: Centre for Urban Studies, University of Dhaka.

ISPAN (Irrigation Support Project for Asia and the Near-East) 1993. *ISPAN Char Study - Jamuna inventory.* Dhaka: Flood Action Plan.

Ives, J.D. 1989. Deforestation in the Himalayas: the cause of increased flooding in Bangladesh and Northern India. In *Proceedings of the international conference on Bangladesh floods: what are the solutions?* 23-27. Montreal: International Society of Bangladesh.

Ives, J.D. and B. Messerli 1989. *The Himalayan dilemma: reconciling development and conservation.* London: Routeledge.

Jack, J.C. 1916. *The economic life of a Bengal district: a study.* Oxford: Clarendon Press.

Jackson, E.L. and T. Mukerjee 1974. Human adjustment to the earthquake hazard of San Francisco. In *Natural hazards - local, national, and global*, G.F. White (ed.), 160-166. New York: Oxford University Press.

Jackson, J.C. 1979. Urban squatters in Southeast Asia. *Geography* 59: 24-30.

Jacobson, J.L. 1988. *Environmental refugees: a yardstick of habitability.* Washington, D.C.: Worldwatch Institute.

Jacobson, J.L. 1991. The root causes of migration/refugees: the environmental dimensions. *Reformed World* 41 (7 & 8): 237-243.

Jacobson, K. 1994. *The impact of refugees on the environment: a review of the evidence.* Washington, D.C.: Refugee Policy Group.

Jahan, R. 1980. Members of parliament in Bangladesh. In *Bangladesh politics: problems and issues.* Dacca: University Press Ltd.

Jahangir, B.K. 1981. *Rural society, power structure and class practice*, DERAP Working Papers. Bergen: The Mitchelson Institute.

Jannuzi, F.T. and J.T. Peach 1980. *The agrarian structure of Bangladesh: an impediment to development.* Boulder, Colorado: Westview Press.

Jensen-Butler, C. 1981. *A critique of behaviourial geography: an epistemological analysis of cognitive mapping and of Hagarstrands's time-space model.* Aarhus: Geographical Institute, University of Aarhus.

Johnson, B.L.C. 1982. *Bangladesh*, Second edition. London: Heinemann Educational Books.

Johnston, R.J. 1976. On regression coefficients of comparative studies of the 'friction of distance'. *Tijdschrift voor Economische en Sociale Geografie* 67: 15-18.

Johnstone, R.J. 1982. *Geography and the state: an introductory essay.* London: Macmillan.

Johnston, R.J. 1983a. *Geography and geographers: Anglo-American human geography since 1945*, Second edition. London: Edward Arnold.

Johnston, R.J. 1983b. *Philosophy and human geography: an introduction to contemporary approaches*. London: Edward Arnold.

Johnston, R.J. 1983c. Resource analysis, resource management and the integration of physical and human geography. *Progress in Physical Geography* 7 (1): 127-147.

Jones, E. 1980. Social geography. In *Geography, yesterday and tomorrow*, E.H. Brown (ed.), 251-262. Oxford: Oxford University Press.

Jones, S. 1979. An evaluation of rural development programme in Bangladesh. *Journal of Social Studies* 6: 19-31.

Kabir, L. 1969. *Land laws in East Pakistan*. Dacca: Law House.

Kabir, M. 1977. Fertility trends in Bangladesh since 1951. *Genus* 33 (3/4): 163-173.

Kamaluddin, A.F.M. 1973. Changing river courses in Bangladesh. *Journal of the Bangladesh National geographical Association* 1 (2): 1-9.

Kamaluddin, A.F.M. 1985. *Feasibility study of IRDP in Serajganj*. Savar, Dhaka: Jahangirnagar University.

Kates, R.W. 1962. *Hazard and choice perception in flood plain management*. Department of Geography Research Report No. 78. Chicago: University of Chicago.

Kates, R.W. 1971. Natural hazards in human ecological perspective: hypotheses and models. *Economic Geography* 47: 438-451.

Kates, R.W. 1975. Natural disasters and development. Paper presented at Wingspread conference, Recine, Wisconsin, October.

Kates, R.W. and I. Burton (eds) 1986a. *Geography, resources and environment*, Vol I. selected writings of Gilbert F. White. Chicago: The University of Chicago Press.

Kates, R.W. and I. Burton (eds) 1986b. *Geography, resources and environment*, Vol II. themes from the work of Gilbert F. White. Chicago: The University of Chicago Press.

Kates, R.W. *et al.* 1973. Human impact of the Managua earthquake: transitional societies are peculiarly vulnerable to natural disasters. *Science* 182 (4416): 981-989.

Kayastha, S.L. and R.P. Yadava 1985. Flood induced population migration in India: a case study of Ghaghara zone. In *Population redistribution and development in South Asia*, L.A. Kosinski and K.M. Elahi (eds), 79-88. Dordrecht: D. Reidel Publishing Co.

Kazemi, F. 1980. *Poverty and revolution in Iran*. New York: New York University Press.

Keat, R. and J. Urry 1975. *Social theory as science*. London: Routeledge and Kegan Paul.

Keesing, F. 1952. The Papuan Orokalva vs. Mt. Lemington: cultural shock and its aftermath. *Human Organization* 11 (1).

Khalil, G.M. 1990. Floods in Bangladesh: a question of disciplining the rivers. *Natural Hazards* 3: 379-401.

Khan, A.A.M. 1982. Rural-urban migration and urbanization in Bangladesh. *The Geographical Review* 72 (4): 379-394.

Khan, A.R. 1977. Poverty and inequality in rural Bangladesh. In *Poverty and landlessness in rural Asia*, K. Griffin and A.R. Khan (eds), 137-160. Geneva: International Labour Office.

Khan, A.R. and A. Rashid 1985. Morphological processes and erosion phenomenon of Brahmaputra right bank and Bhola area of lower Meghna estuary. Paper presented at the workshop on the Impact of Riverbank Erosion and Flood Hazards in Bangladesh, Jahangirnagar University, Savar, March 5-7.

Khan, A.R. and L.T. Ruzicka 1981. Trends and patterns of fertility. In *Population of Bangladesh*, Country Monograph Series No. 8, 64-74. New York: United Nations.

Khan, M.A.O.A. 1988. Statistical analysis of bank erosion of the Brahmaputra River in Bangladesh. Master's thesis. Winnipeg, Manitoba: University of Manitoba.

Khan, M.A.O.A. 1995. Predicting bank movement of the Brahmaputra River in Bangladesh. Doctoral dissertation. Winnipeg, Manitoba: University of Manitoba.

Khan, M.A.O.A. and C. Booy 1991. Statistical analysis of bank erosion of the Jamuna River. In *Riverbank erosion, flood and population displacement in Bangladesh*, K.M. Elahi *et al.* (eds), 30-45. Dhaka: Riverbank Erosion Impact Study, Jahangirnagar University.

Khan, Z.R. 1983. *Leadership in the least developed nation: Bangladesh*. Syracuse, New York: Maxwell School of Citizenship and Public Affairs.

Kibreab, G. 1991. The system of asylum in African economies under pressures. In *Hundred flowers bloom: essays in honor of Bo Gustafsson*, K. Ullenhag (ed.). Stockholm: Almqivist and Wiksell International.

Kibreab, G. 1994. Migration, environment and refugeehood. In *Environment and population change*, B. Zaba and J. Clarke (eds.), 115-129. Leige: Derouaux Ordina.

Kirk, W. 1963. Problems in geography. *Geography* 48: 357-371.

Kirkby, A.V. 1973. The use of land and water resources in the past and present valley of Oaxaca, Mexico. *Memoirs of the Museum of Anthropology* 5. Ann Arbor: University of Michigan.

Kirkby, A.V. 1974. Individual and community response to rainfall variability in Oaxaca, Mexico. In *Natural hazards - local, national, and global*, G.F. White (ed.), 119-128. New York: Oxford University Press.

Kish, L. 1965. *Survey sampling*. New York: Wiley.

Klimek, K. 1974. The retreat of alluvial river banks in the Wisloka valley (South Poland). *Geographica Polanica* 28: 59-75.

Kling, B.B. 1960. The Bengal indigo disturbances, 1859-62: a case study in the origins of political activity in modern Bengal. Doctoral dissertation. Pennsylvania, Philadelphia: University of Pennsylvania.

Knighton, A.D. 1984. *Fluvial forms and processes*. London: Edward Arnold.

Knighton, A.D. and G.C. Nanson 1993. Anastomosis and the continuum of channel pattern. *Earth, Surface Processes and Landforms* 18: 613-625.

Knorr-Cetina, K. 1981. The micro-sociological challenge of macro-sociology: towards a reconstruction of social theory and methodology. In *Advances in social theory and methodology: towards an integration of micro- and macro sociologies*, K. Knoor-Cetina and A.V. Circourel (eds), 1-47. Boston: Routeledge and Kegan Paul.

Kolenic, P. 1974. Characteristics and patterns of movement. Master's thesis. Columbus, Ohio: Ohio University.

Kollmorgen, W.M. 1953. Settlement control beats flood control. *Economic Geography* 29: 208-215.

Kosinski, L.A. and K.M. Elahi. 1985. Introduction. In *Population redistribution and development in South Asia*, L.A. Kosinski and K.M. Elahi (eds), 3-14. Dordrecht: D. Reidel Publishing Co.

Kovach, R.L. 1995. *Earth's fury: an introduction to natural hazards and disasters*. Englewood Cliffs, New Jersey: Prentice-Hall.

Krausse, G. 1977. Economic adjustment of migrants in the city: the Jakarta experience. *International Migration Review* 13 (1): 46-70.

Kreimer, A. and M. Munasinghe (eds) 1992. *Environmental management and urban vulnerability*. World Bank Discussion Paper No. 168. Washington, D.C.: The World Bank.

Kreimer, A., M. Munasinghe and M. Preece 1992. Reducing environmental vulnerability and managing disasters in urban areas. In *Environmental management and urban vulnerability*, Kreimer, A. and M. Munasinghe (eds), 21-50. World Bank Discussion Paper No. 168. Washington, D.C.: The World Bank.

Krigstorm, A. 1962. Geomorphological studies of Sandur plains and their braided rivers in Iceland. *Geographiska Annaler* 44: 328-346.

Kritz, M.M. (ed.) 1981. *Global trends in migration: theory and research on international population movements*. New York: Center for Migration Studies.

Lane, E.W. 1957. *A study of the shape of channels formed by natural streams flowing in erodible material*. MRD Sediment Series 9, US Army Engineer Division, Missouri River Corps Engineers, Omaha, Nebraska.

Lassailly-Jacob, V. and M. Zmolek 1992. Environmental refugees. *Refuge* 12 (1): 1-4.

Latif, A. 1963. Water resources development in East Pakistan. *Proceedings of the fifth regional conference on water resources development in Asia and the Far East*. United Nations Water Resource Series No. 23, 101-110. New York: United Nations.

Latif, A. 1969. Investigation of Brahmaputra River. *Journal of the Hydraulics Division, American Society of Civil Engineers* 95 (HY5): 1685-1698.

La Touche, T.H.D. 1910. Relics of the great Ice Age of the plains of Northern India. *Geological Magazine* 67: 21-22.

Law, B.C. 1954. *Hydrological geography of ancient India*. Paris: Societe Asiatique de Paris.

Lazaras, D.S. 1990. Environmental refugees: new strangers at the door. *Our Planet* 2 (3): 19-90.

Leach, B. 1978. Geography, behavior and Marxist philosophy. *Antipode* 10 (2): 33-37.

Leiderman, S.M. and J. Hall 1995. A beginner's guide to environmental refugees. *Bulletin for April 30 - May 6, 1995*. Durham: Environmental Response/Fourth World Project.

Leigh, C.H. and L.K. Sim 1983. Attitudes and adjustments to the flood hazards in a mixed ethnic community in Malacca town, peninsular Malaysia. *Singapore Journal of Tropical Geography* 4 (1): 40-52.

Lele, S.M. 1991. Sustainable development: a critical review. *World Development* 19 (6): 607-621.

Leopold, L.B. and T. Maddock 1953. The hydraulic geometry of stream channels and some physiographic implications. *United States Geological Survey Professional Paper* 252.

Leopold, L.B. and M.G. Wolman 1957. River channel patterns - braided, meandering and straight. *United States Geological Survey Professional Paper* 282-B.

Leopold, L.B. et al. 1964. *Fluvial processes in geomorphology*. San Francisco, California: W.H. Freeman.

Levin, H.M. 1986. A benefit-cost analysis of nutritional programs for anemia reduction. *World Bank Research Observer* 1.

Lewin, J. and B.J. Brindle 1977. Confined meanders. In *River channel changes*, K.J. Gregory (ed.), 221-233. Chichester: Wiley.

Lewis, J. 1975. *Disaster management with special reference to pre-disaster planning*. DRU Research Paper No. 2. Bradford, York: Disaster Research Unit, University of Bradford.

Lewis, O. 1978. The culture of poverty. *Scientific American* 215 (4): 19-28.

Ley, D. 1978. Social geography and social action. In *Humanistic geography: problems and prospects*, D. Ley and M.S. Samuels (eds), 41-57. London: Croom Helm.

Libbee. M.J. and D.E. Sopher 1972. Marriage migration in rural India. Paper presented at the conference of the IGU Commission on Population Geography, Edmonton, Alberta, July 28, Aug. 3.

Lillesand, T.M. and R.W. Kiefer 1979. *Remote sensing and image interpretation*. New York: John Wiley & Sons.

Lipset, S.M. 1959. Social mobility and urbanization. In *Cities and societies*, P. Hatt and A. Reiss (eds). Glencoe, Illinois: Free Press.

Lloyd, P.C. 1979. *Slums of hope? shanty towns of the Third World*. New York: St. Martin's Press.

Logue, J.N., M.E. Melick and H. Hansen 1981. Research issues and direction in the epidemiology of health effects of disasters. *Epidemiology Review* 3: 140-162.

Lomnitz, L.A. 1977. *Networks and marginality: life in a Mexican shantytown*. New York: Academic Press.

Lonergan, S. 1995. Population movements and the environment. *Refugee Participation Network* 18: 4-6.

McCormick, J. 1989. *Reclaiming paradise: the global environmental movement*. Bloomington, Indiana: Indiana University Press.

McDonald, N.S. 1979. Hazard perception in northern South Wales. In *Natural hazards in Australia*, R.L. Heathcote and B.G. Thom (eds), 260-269. Canberra: Australian Academy of Sciences.

McGee, T.G. 1971. *The urbanization process in the Third World*. London: G. Bell & Sons.

McGregor, J. 1993. Refugees and the environment. In *Geography and refugees: patterns and processes of change*, R. Black and V. Robinson (eds), 157-170. London: Belhaven Press.

McGregor, J. 1995. Are there environmental refugees? *Refugee Participation Network* 18: 7-8.

Mackin, J.H. 1956. Cause of braiding by a graded river. *Bulletin of the Geological Society of America* 67: 1717-1718.

Maddock. T. 1969. The behavior of straight open channels with moveable beds. *United States Geological Survey Professional Paper* 622-A.

Mahmud, N. 1994. Live through this: environmental refugees in the Phillipines. Master's thesis. Halifax, Nova Scotia: Saint Mary's University.

Majumdar, R.C. 1943. *History of Bengal*. Dacca: University of Dacca.

Majumdar, R.C. 1971a. *History of ancient Bengal*. Calcutta: G. Bhardwaj & Co.

Majumdar, R.C. 1972b. *Ancient India*. Sixth edition. Delhi: Motilal Banarsidass.

Majumdar, R.C. 1973. *History of medieval Bengal*. Calcutta: G. Bhardwaj & Co.

Majumdar, S.C. 1942. *Rivers of the Bengal delta*. Calcutta: Calcutta University Press.

Malik, S. 1983. Land reclamation. *Bangladesh Today* 1 (1): 25-28.

Maloney, C. 1990-91. Environmental and project displacement of population in South Asia, part II: land and water. *American Universities Field Staff Reports, Asia* 19: 1-12.

Mamun, M. 1975. *Chirosthay bondobantha o Bangali samaj* (in Bengali). Dhaka: Maola Bros.

Mamun, M.Z. 1996. Awareness, preparedness and adjustment measures of riverbank erosion-prone people: a case study. *Disasters* 20 (1): 68-74.

Mangalam, J.J. 1968. *Human migration: a guide to migration literature in English, 1955-1962*. Lexington, Kentucky: University of Kentucky Press.

Mangin, W. 1967. Squatter settlements. *Scientific American* 217 (5): 21-29.

Maniruzzaman, T. 1975. Bangladesh in 1974: economic crisis and political polarization. *Asian Survey* 15 (2): 117-128.

Manning, E.W. 1990. Conservation strategies: providing the vision for sustainable development. *Alternatives* 16: 24-29.

Martorell, R. 1985. Child growth retardation: a discussion of its causes and relationship to health. In *Nutritional adaptation in man*, K. Blaxter and J.C. Waterlow (eds). London: John Libbey.

Martorell, R. 1990. Body size, adaptation and function. *Human Organization* 48.

Maslow, A. 1970. *Motivation and personality*. Second edition. New York: Harper & Row.

Matin, M.A. and M.A. Husain 1989. Hydrological aspects of the 1988 flood. In *Flood in Bangladesh*, M. Ahmad (ed.), 61-76. Dhaka: Community Development Library.

Maybury, R.H. (ed.) 1986. *Violent forces of nature*. Mt. Airy, Maryland: Lomond Publications and UNESCO.

Melick, M.E. 1978. Life change and illness: illness behavior of males in the recovery period of a natural disaster. *Journal of Health and Social Behavior* 19: 335-342.

Merrington, J. 1975. Town and countryside in the transition to capitalism. *New Left Review* 93: 71-92.

Merryman, J.L. 1982. Pastoral nomad settlement in response to drought: the case of Kenya Somali. In *Involuntary migration and resettlement: the problems and responses of dislocated people*, A. Hansen and A. Oliver-Smith (eds), 105-120. Boulder, Colorado: Westview Press.

Mey, W. 1984. *Genocide in the Chittagong Hill Tracts, Bangladesh*. Copenhagen: IWGIA.

Miah, M.M. 1988. *Flood in Bangladesh: a hydromorphological study of the 1987 flood*. Dhaka: Academic Publishers.

Miall, A.D. 1977. A review of the braided-river depositional environment. *Earth Science Reviews* 13: 1-62.

Micklin, M. 1973. *Population, environment and social organization*. Hinsdale, Illinois: Dryden Press.

Mileti, D.S. 1980. Human adjustment to the risk of environmental extremes. *Sociology and Social Research* 64 (3): 327-347.

Mileti, D.S. *et al.* 1975. *Human systems in extreme environment: a sociological perspective*. Boulder, Colorado: Institute of Behavioral Science, University of Colorado.

Miller, J.P. and F. Wendorf 1958. Alluvial chronology of the Tesuque Valley, New Mexico. *Journal of Geology* 66: 177-194.

Milliman, J.D. and R.H. Meade 1983. World-wide delivery of river sediments to the oceans. *Journal of Geology* 91 (1): 1-21.

Mills, C.W. 1959. *The sociological imagination*. Oxford: Oxford University Press.

Milne, G. 1977. Cyclone Tracy: I some consequences of the evacuation for adult victims. *Australian Psychologist* 12: 39-54.

Mitchell, B. 1979. *Geography and resource analysis*. New York: Longman.

Mitchell, J.K. 1990. Human dimensions of environmental hazards: complexity, disparity, and the search for guidance. In *Nothing to fear: risks and hazards in American society*, A. Kirby (ed.), 131-175. Tucson, Arizona: University of Arizona Press.

Mitra, S.N. 1979. Infant and childhood mortality in Bangladesh: levels and differentials. Master's thesis. Canberra: Australian National University.

Molner, P. 1986. The structure of Mountain ranges. *Scientific American* 225.

Mookerjee, R. 1926. *Fundamental unity of India*. Calcutta: New Nation.

Morgan, J.P. and W.G. McIntire 1959. Quaternary geology of the Bengal basin, East Pakistan and India. *Bulletin of the Geological Society of America* 70: 319-342.

Morisawa, M. 1985. *Rivers: form and processes*. K.M. Clayton (ed.). London: Longman.

Morren, Jr. G.E.B. 1983. A general approach to identification of hazards and responses. In *Interpretations of calamity*, K. Hewitt (ed.), 284-297. Boston: Allen & Unwin Inc.

Morris, D.M. 1968. Towards a re-interpretation of the 19th century Indian economic history. *Indian Economic and Social History Review* 5: 10-11.

Moser, C.A. and Kalton, G. 1977. *Survey method and social investigation*. Second edition. London: Heinemann.

Mosley, M.P. 1975. Channel changes on the river Bollin, Cheshire, 1872-1973. *East Midland Geographer* 6: 185-199.

MPO (Master Plan Organization) 1985. *National water plan project, draft final report*, Vol. 1 & 2. Dhaka: Ministry of Irrigation, Water Development and Flood Control.

Mueller, W.H. and E. Pollitt 1982. The Becon Chow study: effects of nutrition supplementation on sibling - sibling anthropometric correlations. *Human Biology* 54.

Mujeeb, M. 1972. *Islamic influence on Indian society*. Delhi: Meenakshi Prakashan.

Mukerjee, R. 1938. *The changing face of Bengal: a study in riverine economy*. Calcutta: University of Calcutta.

Muqtada, M. 1981. Poverty and famines in Bangladesh. *The Bangladesh Development Studies* 9 (1): 1-34.

Murphy, L.M. and A.B. Moriarty 1976. *Vulnerability, coping and growth from infancy to adolescence*. New Haven, Connecticut: Yale University Press.

Murthy, K.V.S. and L. Sastri 1981. Tectonic influence on the course of Brahmaputra. *Proceedings of the seminar on fluvial processes and geomorphology of the Brahmaputra basin*, Geological Survey of India, Misc. Publication No. 46, 129-132. New Delhi.

Myers, N. 1993. Environmental refugees in a globally warmed world. *BioScience* 43 (11): 752-761.

Myers, N. and J. Kent 1995. *Environmental exodus: an emergent crisis in the global arena*. Washington, D.C.: The Climate Institute.

Naess, A. 1983. Philosophical aspects of the deep ecological movement. Unpublished manuscript, Oslo.

Nagpaul, H. 1984. Sociology of disaster in Indian society: a case study of floods from recovery to rehabilities strategies in rural communities of North India. *Indian Journal of Social Research* 25 (1): 68-78.

Nandy, S. Water management organizations in Bangladesh: a brief overview. In *Proceedings of the international conference on floods in Bangladesh: an interdisciplinary analysis of alternative solution strategies*, 216-226. Urbana-Champaign: College of Commerce and Business Administration: University of Urbana-Champaign.

Nelson, J.M. 1969. *Migrants, urban poverty, and instability in developing nations*. Occasional Papers in International Affairs No. 22. Cambridge, Mass.: Center for International Affairs, Harvard University.

Nevins, T.H.F. 1969. *River training - the single thread channel*. New Zealand Engineering December Issue: 367-373.

Newland, K. 1981. *Refugees: the new international politics of displacement*. Paper No. 43. Washington, D.C. Worldwatch Institute.

Nicholas, R. 1962. Villages of the Bengal delta: a study of ecology and peasant society. Doctoral dissertation. Chicago: University of Chicago.

Notestain, F.W. 1960. Introduction. In *On population: three essays*, F.W. Notestain (ed.), i-x. New York: The New American Library.

Obaidullah, M. n.d. *Setting for demographic analysis in East Pakistan*. Dacca: Institute of Statistical Research and Training.

Oberai, A.S. and H.K.M. Singh 1984. Migration, employment and the urban labour market: a study in the Indian Punjab. *International Labour Review* 123 (4): 507-523.

Oberlander, T.M. and Muller, R.A. 1987. *Essentials of physical geography today*, Second Edition. New York: Random House.

O'Donnell, C.J. 1893. *Census of India, 1891*, Vol. 3: the lower provinces of Bengal and their feudatories. Calcutta: Bengal Secretariat Press.

O'Keefe, J. 1975. *African drought: a review*. Occasional Paper No. 8. Bradford, York: Disaster Research Unit, University of Bradford.

O'Keefe, P., K. Westgate and B. Wisner 1976. Taking the naturalness out of natural disasters. *Nature* 260 (15): 566-567.

O'Keefe, P., K. Westgate and B. Wisner 1994. Taking the naturalness out of natural disasters. In *Environmental risks and hazards*, S.L. Cutter (ed.), 94-96. Englewood Cliffs, New Jersey: Prentice Hall.

Okrent, D. 1980. Comment on social risk. *Science* 208: 372-375,

Oliver, J. 1975. The significance of natural hazards in a developing area: a case study from North Queensland. *Geography* 60: 99-110.

Oliver-Smith, A. 1982. Here there is a life: the social and cultural dynamics of successful resistance to resettlement in post-disaster Peru. In *Involuntary migration and resettlement: the problems and responses of dislocated people*, A. Hansen and A. Oliver-Smith (eds), 85-104. Boulder, Colorado: Westview Press.

Oliver-Smith, A. 1986. Disaster context and causation: an overview of changing perspectives in disaster research. In *Natural disasters and cultural responses*, A. Oliver-Smith (ed.), Studies in Third World Societies, 1-34. Williamsburg, Virginia: College of William and Mary.

Oliver-Smith, A. 1991. Successes and failures in post-disaster resettlement. *Disasters* 15 (1): 12-23.

Oliver-Smith, A. 1994. Peru's five hundred year earthquake: vulnerability in historical context. In *Disasters, development and environment*, A. Varley (ed.), 31-48. London: John Wiley & Sons.

Oliver-Smith, A. and A. Hansen 1982. Introduction: involuntary migration and resettlement - causes and contexts. In *Involuntary migration and resettlement: the problems and responses of dislocated people*, A. Hansen and A. Oliver-Smith (eds), 1-12. Boulder, Colorado: Westview Press.

Olson, M.E. 1979. Refugees as a special case of population distribution. In *Population redistribution: patterns, policies and prospects*, L.A.P. Gosling and L.Y.C. Lim (eds), 130-151. New York: United Nations Fund for Population Activities.

Olsson, G. 1975. Distance and human interaction: a migration study. *Geographiska Annaler* 47B: 3-43.

Orovwuje, R.P. 1989. Refugee migration: theory and practice. In *Refugee resettlement and wellbeing*, M. Abbott (ed). Auckland: Auckland Mental Health Foundation.

Osgood, C.E. et al. 1957. *Measurement of meaning*. Urbana: University of Illinois Press.

Ossowski, S. 1969. Old notions and new problems: interpretations of social structure in modern society. In *Social inequality*, A. Beteille (ed.), 79-90. London: Penguin Books.

Osterkamp, W.R. 1978. Gradient, discharge, and particle size relations of alluvial channels in Kansas with observations on braiding. *American Journal of Science* 278: 1253-1268.

Palacio, J.O. 1982. Post hurricane resettlement in Belize. In *Involuntary migration and resettlement: the problems and responses of dislocated people*, A. Hansen and A. Oliver-Smith (eds), 121-135. Boulder, Colorado: Westview Press.

Palm, R. 1981. Public response to earthquake hazard information. *Annals of the Association of American Geographers* 71: 388-399.

Palm, R. 1990. *Natural hazards: an integrative framework for research and planning*. Baltimore, Maryland: The John Hopkins University Press.

Panandikar, S.G. 1926. *The wealth and welfare of the Bengal delta*. Calcutta: Calcutta University Press.

Papagaroufali, F. 1996. Xenotransplantation and transgenesis: im-moral stories about human-animal relations in the West. In *Nature and society: anthropological perspectives*, P. Descola and G. Palsson (eds), 240-255. London: Routledge.

Pargiter, F.F. 1885. *A revenue history of the Sundarbans from 1765 to 1870*. Calcutta: Bengal Secretariat Press.

Parker, D.J. and D.M. Harding 1979. Natural hazard perception: evaluation and adjustment. *Geography* 64: 307-316.

Parker, G. 1975. Psychological disturbances in Darwin evacuees following Cyclone Tracy. *Medical Journal of Australia* 1: 650-652.

Parker, G. 1976. On the cause and characteristics scale of meandering and braiding in rivers. *Journal of Fluid Mechanics* 76: 459-480.

Parker, R.S. 1992. Issues and trends in emergency preparedness for technological disaster: moving beyond the 'natural and technological levels'. In *Environmental management and urban vulnerability*, A. Kreimer and M. Munasinghe (eds), 237-259. World Bank Discussion Paper No. 168. Washington, D.C.: The World Bank.

Parvin, M. 1987. Agrarian structure and entitlement to food in Bangladesh. Paper presented at the 21st Bengal Studies Conference, University of Wisconsin, Oshkosh, may 8-10.

Paul, B.K. 1982. Non-tribal settlement in a tribal area: a case of the Chittagong Hill Tracts. Paper presented at the 91st meeting of the Ohio Academy of Sciences, Ohio State University, Columbus, Ohio, April 23-25.

Paul, B.K. 1984. Perception of and agricultural adjustments to floods in Jamuna floodplain, Bangladesh. *Human Ecology* 12: 3-19.

Paul, B.K. 1995. Farmers' responses to the Flood Action Plan (FAP) of Bangladesh: an empirical study. *World Development* 23 (2): 299-309.

Paul, B.K. 1997. Women's responses to the Flood Action Plan (FAP) of Bangladesh. Unpublished manuscript. Manhattan, Kansas.

Pavlov, V.I. 1978. *Historical premises for India's transition to capitalism*. Moscow: Nauka Publishing, Central Department of Oriental Literature.

Pearce, D.W. 1986. *The MIT dictionary of modern economics*. Cambridge, Mass.: The MIT Press.

Pearce, F. and G. Tickell 1993. West sinks Bangladesh flood plan. *New Scientist* 139: 4.

Peet, R.J. and J.V. Lyons 1981. Marxism's dialectical materialism, social formation and the geographic relations. In *Themes in geographic thought*, M.E. Harvey and B.P. Holly (eds), 187-221. London: Croom Helm.

Peil, M. 1976. African squatter settlements: a comparative study. *Urban Studies* 13: 155-166.

Penning-Roswell, E.C. and L. Underwood 1972. *Flood hazard and flood plain management: survey of existing studies*. Flood Hazard Research Project, Progress Report No. 2. Middlesex: Middlesex Polytechnic.

Penning-Roswell, E.C. 1972. *Problems and methods of flood damage assessment*. London: Middlesex Polytechnic Flood Hazard Research Project.

Perlman, J.E. 1976. *The myth of marginality: urban poverty and politics in Rio de Jeneiro*. Berkeley, California: University of California Press.

Petersen, W. 1958. A general typology of migration. *American Sociological Review* 23: 256-265.

Petersen, W. 1975. *Population*, Third edition. New York: Macmillan.

Peterson, W. 1984. *Income, employment and economic growth*, Fifth edition. New York: W.W. Norton & Co.

Poddar, M.C. 1952. Preliminary report of the Assam earthquake, 15th August, 1950. *Bulletin of the Geological Society of India* 2: 11-13.

Porteous, J.D. 1977. *Environmental behavior*. Reading, Mass.: Addison-Wesley.

Porter, A.E. 1933. *Census of India, 1931*, Vol. 5: Bengal and Sikkim, Part 1. Calcutta: Central Publication Branch.

Porter, G. and J.W. Brown 1996. *Global environmental politics*, Second edition. Boulder, Colorado: Westview Press.

Powell, J.W. and R. Raynor 1952. *Progress notes: disaster investigation July 1, 1951 - June 30, 1952*. Army Chemical Center, Chemical Corps Medical Laboratories, Englewood, Maryland.

Prakash, B. 1983. *The urban dead-end? patterns of employment among slum-dwellers*. Bombay: Somaiya Publications Pvt Ltd.

PRB (Population Reference Bureau) 1985. *World population data sheet 1985*. Washington, D.C.

Pred, A. 1967. *Behavior and location: frontier for a geographic and dynamic location theory*. Part I. Lund: C.W.K. Gleerup.

Qadir, S.A. 1975. *Modernization of an agrarian society*. Mymensingh: Bangladesh Agricultural University.

Quarantelli, E.L. 1960. A note on the protective function of the family in disasters. *Marriage and Family Living* 22: 263-264.

Quarantelli, E.L. 1970. Emergent accommodation groups: beyond current collective behavior typologies. In *Human nature and collective behavior: papers in honor of Herbert Blumer*, T. Shibutani (ed.), 111-113. Englewood Cliffs, New Jersey: Prentice-Hall.

Quarantelli, E.L. 1977. *Disasters: theory and research*. Beverly Hills, California: Sage Publications.

Quarishi, F.A. 1987. Geopolitical realities in the northeastern region of Indo-Bangladesh subcontinent. In *Bangladesh: politics, economy and society*, C.E. Haque (ed.), 4-19. Winnipeg, Manitoba: Bangladesh Studies Assemblage, University of Manitoba.

Rahman, A. 1981. *Rural power structure: a study of local level leaders in Bangladesh*. Dhaka: Bangladesh Books International.

Rahman, A.H.H. 1969. Floodplain morphology of the Meghna. *The Oriental Geographer* 13: 127-138.

Rahman, H. 1985. From our correspondent: urbanization and the problem of slums in Bangladesh. *Community Development Journal* 20 (1): 52-57.

Rahman, H.Z. 1988. Monitoring rural poverty: conceptual framework and organizational options. Mimeograph. Dhaka: Bangladesh Institute of Development Studies.

Rahman, M.A. 1985. *Situation and future trends of energy and water resources development in Bangladesh*. United Nations Water Resources Series No. 60. New York: United Nations.

Rahman, M.M. 1992. Peasants' adjustment to natural hazards in Bangladesh: a case study of two upazilas in the Brahmaputra floodplain. Doctoral dissertation. Winnipeg, Manitoba: University of Manitoba.

Rahman, S. 1984. The Brahmaputra right bank flood protection embankment: its problems and probable solutions. Paper presented at the seminar on Brahmaputra Right Bank Flood Protection Embankment, Bangladesh Rural Development Academy, Bogra, December 18.

Ralph, K.A. 1975. Perception and adjustment to flood in the Meghna flood plain. Master's thesis. Honolulu, Hawaii: University of Hawaii.

Rashid, H.E. 1977. *Geography of Bangladesh*. Dhaka: University Press Ltd.

Rasid, H. 1966. Morphology of the Jamuna floodplain. *The Oriental Geographer* 10 (2): 57-71.

Rasid, H. 1989. Mini projects for a mega problem: a pragmatic solution for the flood problem in Bangladesh. In *Proceedings of the international conference on Bangladesh floods: what are the solutions?* 39-53. Montreal: International Society of Bangladesh.

Rasid, H. 1993. Preventing flooding or regulating flood levels? case studies on perception of flood alleviation in Bangladesh. *Natural Hazards* 8: 39-57.

Rasid, H. and A. Mallik 1993. Poldering vs compartmentalization: the choice of flood control techniques in Bangladesh. *Environmental Management* 17 (1): 59-71.

Rasid, H. and A. Mallik 1995. Flood adaptations in Bangladesh: is the compartmentalization scheme compatible with indigenous adjustments of rice cropping to flood regimes? *Applied Geography* 15 (1): 3-7.

Rasid, H. and A.U. Mallik 1996. Living on the edge of stagnant water: an assessment of environmental impacts of construction-phase drainage congestion along Dhaka City flood control embankment, Bangladesh. *Environmental Management* 20 (1): 89-98.

Rasid, H. and B.K. Paul 1987. Flood problems in Bangladesh: is there an indigenous solution? *Environmental Management* 11 (2): 155-173.

Ray, R. 1979. *Change in Bengal agrarian society 1760-1850*. Delhi: Monohar Publishers.

Read, M.S. 1977. Malnutrition and human performance. In *Malnutrition, behaviour and social organization*, L.S. Greene (eds.). New York: Academic Press.

Redclift, M. 1987. *Sustainable development: exploring the contradictions*. London: Routeledge.

Redclift, M.R. 1973. Squatter settlements in Latin American cities: the response from government. *The Journal of Development Studies* 10 (1): 92-109.

Redfield, R., R. Linton and M.J. Herskovits 1936. Memorandum on the study of acculturation. *American Anthropologist* 38: 149-152.

Rennell, J.J. 1781. An account of the Ganges and the Brahmapooter Rivers. *Philosophical Transactions of the Royal Society of London* 71: 91-103.

Rennell, J.J. 1793. *An account of the Ganges and Brahmapooter Rivers*, Appendix to Memoir of a map of Hindoostan or the Moghul empire. London: W. Bulmer & Co.

Richards, K. 1982. *Rivers: form and processes in alluvial channels*. London: Methuen.

Rizvi, A.I.H. 1955. Comparative physiography of the lower Ganges and lower Mississippi valleys. Doctoral dissertation. Baton Rouge, Louisiana: Louisiana State University.

Robinson, W.C. 1967. Recent mortality trends in Pakistan. In *Studies in the demography of Pakistan*, W.C. Robinson (ed), 2028. Karachi: Pakistan Institute of Development Economics.

Roder, W. 1961. Attitudes and knowledge on the Topeka flood plain. In *Papers on flood problems*, G.F. White (ed.), 62-83, Department of Geography Research Paper No. 70. Chicago: University of Chicago.

Rogers, P., P. Lydon and D. Seckler 1989a. *Eastern water study: strategies to manage flood and drought in the ganges-Brahmaputra basin*. Arlington, Virginia: ISPAN.

Rogers, P. *et al.* 1989b. Notes on time series and kringing of rainfall records in the Ganges-Brahmaputra basin. *Eastern Waters Study Computational Notes*. Cambridge, Mass.: Harvard University.

Rogge, J.R. 1977. Refugee migrations in Africa: some problems of definition. Paper presented at the meeting of the Canadian Association of Geographers, University of Regina, Regina, Saskatchewan.

Rogge, J.R. 1985. *Too many, too long: Sudan's twenty-year refugee dilemma*. Totowa, New Jersey: Rowman & Allanheld.

Rogge, J.R. (ed.) 1987. *Refugees: a Third World dilemma*. Totowa, New Jersey: Rowman & Allenheld.

Rossi, P.H. *et al.* 1983. *Victims of the environment: loss from natural hazards in the United States, 1970-1980*. New York: Plenum Press.

Rostow, W.W. 1960. *The stage of economic growth: a non-communist menifesto*. Cambridge, Mass.: Cambridge University Press.

Rotter, J.B. 1966. Generalized expectancies for internal versus external control of reinforcement. *Psychological Monographs* 80 (1): 1-28.

Rowntree, R.A. 1974. Coastal erosion: the meaning of a natural hazard in a cultural and ecological context. In *Natural hazards - local, national, global*, G.F. White (ed.), 70-79. New York: Oxford University Press.

RPG (Refugee Policy Group) 1992. Migration and environment. Paper presented at the conference on Migration and the Environment, Nyon, January 19-22.

Rudra, K. 1981. Identification of the ancient mouths of the Ganga as described by Ptolemy. *Geographical Review of India* 43 (2): 97-104.

Rutton, V.W. 1986. Assistance to expand agricultural production. *World Development* 14 (1).

Saarinen, T.F. 1966. *Perception of drought hazard on the Great Plains*. Department of Geography Research Paper No. 106. Chicago: University of Chicago.

Saarinen, T.F. 1974. Problems in the use of a standardized questionnaire for cross-cultural research on perception of natural hazards. In *Natural hazards - local, national, global*, G.F. White (ed.), 180-184. New York: Oxford University Press.

Saarinen, T.F. 1976. *Environmental planning: perception and behavior*. Boston: Houghton Mifflin.

Saha, R. 1923. *Pabna jelar itihas* (in Bengali). Pabna: Sarwaswati Library.

Saha, S.K. 1980. A process of underdevelopment: the socioeconomic dynamics of Bangladesh. A paper presented at the workshop on Creativity and the Integration of Traditions and Modern Attitudes, University of Chittagong, Chittagong, December 18-19.

Satter, S.M.A. 1980. *Protection of Serajganj town against severe bank erosion by the mighty River Jamuna*. Serajganj: BWDB, Water Development Division.

Sauer, C.O. 1925. *The morphology of landscape*. University of California Publications in Geography 2: 19-54.

Sauer, C.O. 1959. The education of a geographer. *Annals of the Association of Geographers* 46: 287-299.

Sayer, R.A. 1979. Epistemology and conceptions of people and nature in geography. *Geoforum* 10: 19-44.

Sayer, R.A. 1982. Explanation in economic geography: abstraction versus generalization. *Progress in Human Geography* 6: 68-88.

Schaefer, F. 1953. Exceptionalism in geography. *Annals of the Association of American Geographers* 43: 226-249.

Schiff, M.R. 1977. Hazard adjustment, locus of control, and sensation seeking: some null findings. *Environment and Behavior* 9 (2): 233-254.

Schumm, S.A. 1963. A tentative classification of alluvial river channels. *United States Geological Survey Circular* 477.

Schumm, S.A. 1972. Fluvial geomorphology: the historical perspective. In *River morphology*, S.A. Schumm (ed.), Benchmark papers in geology, 365-395. Stroudsburg, Pennsylvania: Dowden, Hutchinson and Ross Inc.

Schumm, S.A. 1980. Some applications of the concept of geomorphic threshold. In *Threshold in geomorphology*, D.R. Coates and J.D. Vitek (eds), 473-485. Boston: George Allen & Urwin.

Schumm, S.A. and H.R. Khan 1972. Environmental study of channel patterns. *Bulletin of the Geological Society of America* 83: 1755-1770.

Schumm, S.A. and R.W. Lichty 1963. Channel widening and flood-plain construction along Cimarron River in South Western Kansas. *United States Geological Survey Professional Paper* 352-D.

Schwartzberg, J.E. 1978. *A historical atlas of South Asia*. Chicago: University of Chicago Press.

Scrimshaw, N.C., C.E. Taylor and J.E. Gordon 1968. *Interactions of nutrition and infections*. Geneva: World Health Organization.

Scudder, T. 1968. Social anthropology, man-made lakes and population relocation in Africa. *Anthropological Quarterly* 41 (1): 168-176.

Scudder, T. 1973a. Summary: resettlement. In *Man-made lakes: their problems and environmental facts*, W.C. Ackerman *et al.* (eds), 707-719, Geographical Monograph No. 17. Washington, D.C.: American Geographical Union.

Scudder, T. 1973b. The human ecology of big projects: river basin development and resettlement. In *Annual review of anthropology*, B. Seigel (ed.), 27-62. Palo Alto, California: Annual Reviews.

Scudder, T. 1991. The need for justification for maintaining transboundary flood regimes: the African case. *Natural Resources Journal* 31 (1): 75-107.

Scudder, T. and E. Colson 1982. From welfare to development: a conceptual framework for analysis of dislocated people. In *Involuntary migration and resettlement: the problems and responses of dislocated people*, A. Hansen and A. Oliver-Smith (eds), 267-288. Boulder, Colorado: Westview Press.

Seehan, L. and K. Hewitt 1969. *A pilot survey of global natural disaster of the past twenty years*. Natural Hazards Research Working Paper No. 11. Toronto: University of Toronto.

Sen, A.K. 1980 Famines. *World Development* 8: 613-621.

Sen, A.K. 1981a. *Poverty and famines: an essay of entitlement and deprivation*. Oxford: Clarendon Press.

Sen, A.K. 1981b. Ingredients of famine analysis: availability and entitlement. *Quarterly Journal of Economics* 96 (3): 433-464.

Sen, A.K. 1982. The food problem: theory and policy. *Third World Quarterly* 4 (3): 447-459.

Sen, B. 1962. *Evaluation of agrarian relations in India*. New Delhi: Peoples' Publishing.

Sen, S. 1968. Major changes in river courses in recent history. In *Mountains and rivers in India*, B.C. Law (ed.), 211-220. Calcutta: National Committee for Geography.

Sewell, W.R.D. 1965. *Water management and floods in the Fraser river basin.* Department of Geography Research Paper No. 100. Chicago: University of Chicago.

Sewell, W.R.D. 1969. Human response to floods. In *Water, earth and man*, R.J. Chorley (ed.), 431-454. London: Methuen.

Shacknov, A.E. 1985. Who is a refugee? *Ethics* 95 (2): 274-284.

Shaw, R. Living with floods in Bangladesh. *Anthropology Today* 5 (1): 11-13.

Shryock, H.S. *et al.* 1979. *The methods and materials of demography.* New York: Academic Press Inc.

Shukla, J. 1987. Interannual variability of monsoons. In *Monsoons*, J.S. Fein and P.L. Stephens (eds). New York: John Wiley & Sons.

Shukla, J. 1989. *Consultant's report: interannual variability of summer monsoon rainfall over Bangladesh.* Eastern Waters Study. College Park, Maryland: University of Maryland.

Shumaker, S.A. and G.J. Conti 1985. Understanding mobility in America: conflicts between stability and change. In *Home environments*, I. Altman and C.M. Werner (eds). New York: Plenum Press.

Shumaker, S.A. and R.B. Taylor 1983. Toward a clarification of people-place relationships. In *Environmental psychology: directions and perspectives*, N.R. Reimer and E.S. Geller (eds.). New York: Praeger Publishers.

Simmons, I.G. 1993. *Environmental history: a concise introduction.* Oxford: Blackwell.

Simon, H.A. 1959. Theories of decision-making in economic and behavioral science. *American Economic Review* 49: 253-283.

Simon, J.L. 1981. *The ultimate resource.* Princeton, New Jersey: Princeton University Press.

Simon, D.B. *et al.* 1979. *Report on Connecticut River streambank erosion study.* Report CSU-213. New Hampshire: Colorado State University.

Simpson-Housley, P. and P. Bradshaw 1978. Personality and the perception of earthquake hazard. *Australian Geographical Studies* 16: 65-72.

Slovic, P. *et al.* 1974. Decision-process, rationality and adjustment to natural hazards. In *Natural hazards - local, national, and global*, G.F. White (ed.), 187-205. New York: Oxford University Press.

Smith, D.I. and J.W. Handmaker (eds). 1986. *Flood warning in Australia. Centre for Resource and Environmental Studies.* Canberra: Australian National University.

Smith, D.M. 1981. Marxian economics. In *The dictionary of human geography*, R.J. Johnston et al. (eds.), 203-209. Oxford: Basil Blackwell.

Smith, G.C. 1976. Response of residents and policy-makers to urban environmental hazards. *Institute of Urban Geographers* 8 (4): 279-283.

Smith, K. 1992. *Environmental hazards: assessing risk and reducing disaster.* New York: Routledge.

Smith, K. and G.A. Tobin 1979. *Human adjustment to flood hazard.* London: Longman.

Smith, N. and P. O'Keefe 1980. Geography, Marx and the concept of nature. *Antipode* 12 (12): 30-39.

Smock, D.R. 1969. The role of anthropology in a Western Nigerian resettlement project. In *The anthropology of development in Sub-Saharan Africa*, D. Brokensha and M. Pearsall (eds), 40-47. Lexington, Kentucky: The University Press of Kentucky.

Smyser, W.R. 1985-86. Refugees: a never-ending story. *Foreigh Affairs* 64: 154-168.

Spate, O.H.K. 1954. *India and Pakistan: a general and regional geography*. New York: E.P. Dutton & Co.

Spiegal, J.P. 1957. The English flood of 1953. *Human Organization* 16: 3-5.

Spurr, G.B. 1988. Marginal malnutrition in childhood: implications for adult work capacity and productivity. In *Capacity for work in the Tropics*, K.J. Collins and D.F. Roberts (eds). Cambridge: Cambridge University Press.

Stavenhagen, R. 1964. *Development and underdevelopment*. Berkeley, California: University of California Press.

Stene, L.P. 1985. Some geomorphic observation on river erosion along the Jamuna River, 1972-1982. Paper presented at the workshop on the Impact of Riverbank Erosion and Flood Hazards in Bangladesh, Jahangirnagar University, Savar, March 5-7.

Stene, L.P. 1988. Jamuna bankline behavior, 1972-1984. Paper presented at the International symposium on the Impact of Riverbank Erosion, Flood Hazard and the Problem of Population Displacement, Dhaka, April 11-14.

Stewart, K. 1988. Post-flood assessment and nutritional status of children in Matlab, Bangladesh. Paper presented at the seminar on Regional and Global Environmental Perspectives, Dhaka, March 4-6.

Stoddart, D.R. 1967. Organism and ecosystem as geographical models. In *Models in geography*, R.J. Chorley and P. Haggett (eds.), 511-548. London: Methuen.

Stoett, P. 1995. Redefining 'environmental refugees': Canada and the UNHCR. *Canadian Foreign Policy* 2(3): 29-42.

Stokes, C.J. 1970. A theory of slums. In *A geography of urban places*. R.H. Putnam *et al.* (eds), 411-422. Toronto: Methuen Publications.

Stokes, E. 1978. *The peasant and the Raj: studies in agrarian society and peasant revolution in colonial India*. Delhi: Vikas Publishing House.

Stokols, D. 1978. Environmental psychology. *Annual Review of Psychology* 29: 253-296.

Stokols, D., S.A. Shumaker and J. Martinez 1983. Residential mobility and personal wellbeing. *Journal of Environmental Psychology* 3: 5-19.

Stouffer, S.A. 1960. Intervening opportunities and competing migrants. *Journal of Regional Science* 2: 1-26.

Strathern, M. 1992. *Reproducing the future: anthropology, kinship and the new reproductive technologies*. Manchester: Manchester university Press.

Street, J.H. and D. James 1982. Institutionalism, structuralism, and development in Latin America. *Journal of Economic Issues* 16 (3): 673-689.

Strickland, C. 1940. *Deltaic formation with special reference to the hydrographic processes to the Ganges and Brahmaputra*. Calcutta: Longman and Green.

Suharyanto 1992. Analysing movement of centroid of conveyance of braided river: an attempt to identify the channel migration of the Brahmaputra River in Bangladesh. Master's thesis. Winnipeg, Manitoba: University of Manitoba.

Suhrke, A. 1991. Ecological crisis and population displacement. Paper presented at the conference on Worldwide Refugee Movements, New School for Social Research, New York, November 8-9.

Suhrke, A. and A. Visentin 1991. The environmental refugee: a new approach. *Ecodecision*, September: 73-74.

Susman, P., P. O'Keefe and B. Wisner 1983. Global disasters, a radical interpretation. In *Interpretations of calamity*, K. Hewitt (ed.), 263-283. Boston: Allen & Unwin.

Sweet, L.E. 1965. Camel raiding among the Bedouins of North Arabia: a mechanism of ecological adaptation. *American Anthropologist* 67.

Sylvan, R. 1985. A critique of deep ecology. *Radical Philosophy* 40.

Tarafdar, M.R. 1974. Rivers and flood problems in Bangladesh. In *Studies in Bangladesh*, A.F.M. Kamaluddin (ed.), 22-61. Dhaka: Bangladesh National Geographical Association.

Tarafdar, M.R. 1978. Development of water resources in the deltaic areas of Bangladesh. *Proceedings of the third regional symposium on the Development of Deltaic Areas*, United Nations Water Resources Series No. 50, 162-169. New York: United Nations.

Taylor, V. 1977. Good news about disaster. *Psychology Today* 11: 93.

Terrill, C. 1984. Symposium on the problems and consequences of refugee migrations in the developing world. *Disasters* 8 (3): 164-165.

Thompson, W.H. 1923. *Census of India, 1921*, Vol. 5. Bengal, Part 1. Calcutta: Bengal Secretariat Book Depot.

Thorne, C.R. 1982. Processes and mechanisms of river bank erosion. In *Gravel-bed rivers: fluvial processes, engineering and management*, R.D. Hey *et al.* (eds), 227-259. Chichester: Wiley.

Thorne, C.R. and N.K. Tovey 1982. Stability of composite river banks. *Earth, Surface Processes and Landforms* 6: 469-468.

Tickell, C. 1989a. Environmental refugees: the human impact of global climate change. Paper presented at the Natural Environment Research Council Annual Lecture, Royal Society, London.

Tickell, C. 1989b. Climate change could cause world refugee crisis. *British Overseas Development* 7: 16.

Tickell, C. 1990. Climate change and development. The 1989 Gilbert Murrey Memorial Lecture. Oxford: Oxfam.

Tobias, M. 1984. *Deep ecology*. San Diego, California: Avant Books.

Todaro, M.P. 1969. A model of labor migration and urban unemployment. *American Economic Review* 69: 183-193.

Todaro, M.P. 1985. *Economic development in the Third World*. New York: Longman.

Torry, W.I. 1979a. Anthropological studies in hazardous environments: past trends and new horizons. *Current Anthropology* 20 (3): 517-540.

Torry, W.I. 1979b. Hazards, hazes, and holes: a critique of the environment as hazard and general reflections on disaster research. *Canadian Geographer* 23: 368-383.

Torry, W.I. 1986. Economic development, drought and famines: some limitations of dependency explanations. *Geojournal* 12 (1): 5-18.

Tuan, Y-F. 1974. *Topophilia - a study of environmental perception, attitude and values*. Englewood Cliffs, New Jersey: Prentice Hall.

Twidale, C.R. 1964. Erosion of an alluvial bank at Bairwood, South Australia. *Zeitschrift fur Geomorphogie* 8: 189-211.

Ullack. R. 1978. The role of urban squatter settlements. *Annals of the Association of American Geographers* 68 (4): 535-550.

Ullman, E.L. 1953. Human geography and area research. *Annals of the Association of American Geographers* 43: 54-66.

Umar, B. 1974. *Chirosthaye bondobasthe Bangladesher krishok* (in Bengali): Dacca: Maola Bros.

UNDP (United Nations Development Program) 1988. *Report of the mission on 1987 flood recurrence analysis and recommended actions*. Dhaka.

UNDP (United Nations Development Program)/ GOB (Government of Bangladesh) 1989. *Bangladesh flood policy study: final report*. Dhaka: UNDP.

UNDRO (United Nations Disaster Relief Office) 1980. *Flood in Jamaica: case report*. Geneva.

UNESCO 1970. *Annual summary of information on natural disasters, 1966*. Paris.

UNESCO 1974. *Discharge of selected rivers of the world*, Vol. 3, Part II, Mean Monthly and Extreme Discharges (1969-1972). Paris: The Unesco Paris.

UNESCO 1979. *Discharge of selected rivers of the world*, Vol. 3, Part III, Mean Monthly and Extreme Discharges (1972-1975). Paris: The Unesco Paris.

UNESCO 1982. *Migration and development: major features of migratory movements in India*. Paris.

UNHCR 1991. The challenge of the 1990s. *Refugees* May: 16.

UNHCR 1995. Environmental issues: UNHCR's experience and response. Paper presented at the ECHO expert consultation on Policy Issues on the Environmental Impact of Displacement of Population During the Emergency Phase, Brussels, September 23-24.

UNHCR (United Nations High Commissioner for Refugees), IOM (International Organization for Migration), and RPG (Refugee Policy Group) 1996. *Environmentally-induced population displacements and environmental impacts resulting from mass migrations*. Geneva: IOM.

United Nations 1958. *Multilingual demographic dictionary*. Population Studies No. 29. ST/SOA/Ser. A/29. New York: Bureau of Social Affairs, Population Branch.

Varley, A. 1994. The exceptional and the everyday: vulnerability analysis in the International Decade for Natural Disaster Reduction. In *Disaster, development and environment*, A. Varley (ed.), 1-12. New York: John Wiley & Sons.

Visaria, L. and P. Visaria 1983. Population (1757-1947). In *The Cambridge economic history of India*, Vol. 2, D. Kumar (ed.), 463-532. Cambridge: Cambridge University Press.

Viz, G.K. and R.C. Shenoy 1968. Hydrology of Indian rivers. In *Mountains and rivers in India*, B.C. Law (ed.), 258-283. Calcutta: National Committee for Geography.

Von Weizsacker, C.F. 1949. *The history of nature*. Chicago: University of Chicago Press.

Waddell, E. 1977. The hazards of scienticism: a review article. *Human Ecology* 5: 69-76.

Wagner, P. and M.W. Mikesell 1962. General introduction: the themes of cultural geography. In *Readings in cultural geography*, P.L. Wagner and M.W. Mikesell (eds), 1-5. Chicago: University Chicago Press.

Wahed, A. *et al.* 1983. Land of hope or despair? images of erosion and emergence. *Bangladesh Today* 1 (1): 16-24.

Wallace, A.F.C. 1956a. *Tornado in Worcester: an explanatory study of individual and community behavior in an extreme situation*. NAS-NRC Disaster Study 3. Washington, D.C.: National Academy of Sciences-National Research Council.

Wallace, A.F.C. 1956b. Revitalization movements. *American Anthropologist* 58.

Wallace, A.F.C. 1957. Mazeway disintegration: the individual's perception of socio-cultural disintegration. *Human Organization* 16(2):23-27.

Walmsley, D.J. and G.J. Lewis 1984. *Human geography - brhavioural approaches*. New York: Longman.

Walters, T. 1905. *On Yuen Chwang's travels in India*, Vol. 2. London.

Warrick, R.A. 1983. Drought in the US Great Plains: shifting social consequences? In *Interpretations of calamity*, K. Hewitt (ed.), 67-82. Boston: Allen & Unwin.

Warwick, D.P. and C.A. Lininger 1975. *The sample survey: theory and practice*. New York: McGraw-Hill.

Watts, M. On the poverty of theory: natural hazards research in context. In *Interpretations of calamity*, K. Hewitt (ed.), 231-262. Boston: Allen & Unwin.

WCED (World Commission on Environment and Development) 1987. *Our common future*. New York: Oxford University Press.

Westgate, K. and P. O'Keefe 1976. *Some definitions of disaster*. Occasional Paper No. 4. Bradford, York: Disaster Research Unit, University of Bradford.

Westing, A.H. 1992a. Environmental refugees: a growing category of displaced persons. Paper presented at the conference on Migration and the Environment, Nyon, January 19-21.

Westing, A.H. 1992b. Environmental refugees: a growing category of displaced persons. *Environmental Conservation* 19(3):201-207.

White, G.F. 1945. *Human adjustment to floods: a geographical approach to the flood problem in the United States*. Department of Geography Research Paper No. 29. Chicago: University of Chicago.

White, G.F. *et al.* 1961. *Papers on flood problems*. Department of Geography Research Paper No. 70. Chicago: University of Chicago.

White, G.F. 1964. *Choice of adjustment to floods*. Department of Geography Research Paper No. 93. Chicago: University of Chicago.

White, G.F. 1973. Natural hazard research. In *Directions in geography*, R.J. Chorley (ed.), 193-216. London: Methuen.

White, G.F. 1974a. Natural hazards research: concepts, methods, and policy implications. In *Natural hazards - local, national, global*, G.F. White (ed.), 3-16. New York: Oxford University Press.

White, G.F. 1974b. *Natural hazards - local, national, global*. New York: Oxford University Press.

White, G.F. and J.E. Hass 1975. *Assessment of research on natural hazards*. Cambridge, Mass.: The MIT Press.

Whitehead, R.G. 1989. Famine. In *The fragile environment*, L. Friday and R. Laskey (eds). Cambridge: Cambridge University Press.

Whittow, J. 1980. *Disaster: the anatomy of environmental hazards*. Athens, Georgia: University of Georgia Press and Penguin.

Wiest, R.E. 1979. Anthropological perspective on return migration: a critical commentary. *Papers in anthropology* 20 (1): 167-187.

Wiest, R.E. 1991. Domestic group dynamics in the resettlement process related to riverbank erosion in Bangladesh. In *Riverbank erosion, flood and population displacement in Bangladesh*, K.M. Elahi *et al.* (eds), 246-272. Dhaka: Riverbank Erosion Impact Study, Jahangirnagar University.

Wijkman, A. and L. Timberlake 1984. *Natural disasters: acts of god or acts of man?* London: International Institute for Environment and Development.

Williams, S.W. 1981. Realism, Marxism and human geography. *Antipode* 13 (2): 31-38.

Wilson, K. 1993. Crisis and opportunity: challenges of the transition (1992-1994). Paper presented at the conference on Humanitarian Aid in Complex Emergencies. Oxford, University of Oxford, September 5-9.

Wisner, B. 1980. Nutritional consequences of the articulation of capitalist and non-capitalist modes of production in Eastern Kenya. *Rural Africana* 8/9: 99-132.

Wisner, B. 1988. *Power and need in Africa: basic human needs and development policies*. London: Earthscan and Africa World Press.

Wisner, B., K. Westgate and P. O'Keefe 1976. Poverty and disaster. *New Society* 9: 547-548.

Wisner, B., P. O'Keefe and K. Westgate 1977. Global system and local disasters: the untapped potential of people's science. *Disasters* 1 (1): 47-57.

Woehlcke, M. 1992. Environmental refugees. *Aussen Politik* (German Forign Affairs Review) 43 (3): 287-296.

Wolman, M.G. and J.P. Miller 1960. Magnitude and frequency of forces in geomorphic processes. *Journal of Geology* 68: 54-74.

Wolpert, J. 1964. The decision process in spatial context. *Annals of the Association of American Geographers* 54: 337-358.

Wood, G. 1976. Class differentiation and power in Bandakgram: the minifundist case. In *Exploitation and the rural poor*, A. Huq (ed.). Comilla: Bangladesh Academy for Rural development.

Wood, G.D. 1981. Rural class formation in Bangladesh, 1940-1980. *Bulletin of Concerned Asian Scholars* 13 (4): 2-15.

World Bank 1970. *Proposal for an action program: East Pakistan agriculture and water development*, Vol. 3. The Water program Report No. PS-2a. Washington D.C.: The World Bank.

World Bank 1972. *Bangladesh: land and water resources sector study*, Vol. 8. The Flood Problem Technical Report No. 24. Washington, D.C.: The World Bank.

World Bank 1987. *Bangladesh: second small scale flood control, drainage and irrigation project*. Report No. 6904_BD. Washington, D.C.: The World Bank.

World Bank 1990. *Performance audit report: Bangladesh drainage and flood control project*. Report No. 8805, Asia Region. Washington, D.C.: The World Bank.

Young, M. 1989. *The convention refugee determination process in Canada: its reform*. Ottawa: Library of Parliament Research Branch.

Zaman, M. 1989. The Ganges basin development: a long-term problem and some short-term options. In *River basin development*, M. Zaman *et al.* (eds.), 99-109. Dublin: Tycooly International Publishing Ltd.

Zaman, M.Q. 1982a. Change and continuity in the patron-client relations among peasants of Bangladesh. *The Eastern Anthropologist* 35 (4).

Zaman, M.Q. 1982b. Crisis in Chittagong Hill Tracts: ethnicity and integration. *Economic and Political Weekly* 17: 75-80.

Zaman, M.Q. 1984. Urban squatter studies in Bangladesh: a review and some theoretical considerations. *Spectra of Anthropological Progress* 6: 51-67.

Zaman, M.Q. 1987. Adjustment to riverbank erosion hazards and population resettlement in Bangladesh. Paper presented at the annual meeting of the Society for Applied Anthropology, Oaxaca, Mexico, April 8-12.

Zaman, M.Q. 1988. The socioeconomic and political dynamics of adjustment to riverbank erosion hazard and population resettlement in the Brahmaputra-Jamuna floodplain. Doctoral dissertation. Winnipeg, manitoba: University of Manitoba.

Zaman, M.Q. 1989. The social and political context of adjustment to riverbank erosion hazard and population resettlement in Bangladesh. *Human Organization* 48 (3): 196-205.

Zaman, M.Q. 1991a. Social structure and process in charland settlement in the Brahmaputra-Jamuna floodplain. *Man* 26 (4): 673-690.

Zaman, M.Q. 1991b. Erosion and accretion of land: a case for reform in charland policies and administration. In *Riverbank erosion, flood and population displacement in Bangladesh*, K.M. Elahi *et al.* (eds), 273-289. Dhaka: Riverbank Erosion Impact Study, Jahangirnagar University.

Zaman, M.Q. 1991c. The displaced poor and resettlement policies of Bangladesh. *Disasters* 15 (2): 117-125.

Zaman, M.Q. 1995. Population displacement and resettlement in Brahmaputra-Jamuna floodplain, Bangladesh. In *Development, displacement and resettlement: focus on Asian experience*, H.M. Mathur (ed.), 209-223. New Delhi: Vikas Publishing.

Zaman, M.Q. and C.E. Haque, C.E. 1991. Working together: reflections on collaborative research in Bangladesh. *Canadian Journal of Development Studies* 12 (2): 387-403.

Zaman, M.Q. and R.E. Wiest 1985. Local level socioeconomic and political dynamics of accretional and depositional land utilization. Paper presented at the workshop on the Impact of Riverbank Erosion and Flood Hazards in Bangladesh, Jhangirnagar University, Savar, March 5-7.

Zaman, M.Q. and R.E. Wiest 1991. Riverbank erosion and population resettlement in Bangladesh. *Practicing Anthropology* 13 (3): 29-33.

Zolberg, A.R. 1989. The next waves: migration theory for a changing world. *International Migration Review* 23 (3): 403-430.

Zolberg, A.R., A. Suhrke and S. Aguayo 1989. *Escape from violence: conflict and refugee crisis in the developing world*. New York: Oxford University Press.

INDEX

Advances in Natural and Technological Hazards Research

Series Editor: Prof. Dr. Mohammed I. El-Sabh, *Département d'Océanographie, Université du Québec à Rimouski, 310 Allée des Ursulines, Rimouski, Québec, Canada G5L 3A1*

Publications

1. S. Tinti (ed.): *Tsunamis in the World.* Fifteenth International Tsunami Symposium (1991). 1993 ISBN 0-7923-2316-5
2. J. Nemec, J.M. Nigg and F. Siccardi (eds.): *Prediction and Perception of Natural Hazards.* Symposium Perugia, Italy (1990). 1993 ISBN 0-7923-2355-6
3. M.I. El-Sabh, T.S. Murty, S. Venkatesh, F. Siccardi and K. Andah (eds.): *Recent Studies in Geophysical Hazards.* 1994 ISBN 0-7923-2972-4
4. Y. Tsuchiya and N. Shuto (eds.): *Tsunami: Progress in Prediction, Disaster Prevention and Warning.* 1995 ISBN 0-7923-3483-3
5. A. Carrara and F. Guzzetti (eds.): *Geographical Information Systems in Assessing Natural Hazards.* 1995 ISBN 0-7923-3502-3
6. V. Schenk (ed.): *Earthquake Hazard and Risk.* 1996 ISBN 0-7923-4008-6
7. M.I. El-Sabh, S. Venkatesh, H. Denis and T.S. Murty (eds.): *Land-based and Marine Hazards.* Scientific and Management Issues. 1996 ISBN 0-7923-4064-7
8. J.M. Gutteling and O. Wiegman: *Exploring Risk Communication.* 1996 ISBN 0-7923-4065-5
9. G. Hebenstreit (ed.): *Perspectives on Tsunami Hazard Reduction.* Observations, Theory and Planning. 1997 ISBN 0-7923-4811-7
10. C. Emdad Haque: *Hazards in a Fickle Environment: Bangladesh.* 1998 ISBN 0-7923-4869-9

Kluwer Academic Publishers – Dordrecht / Boston / London

Advances in Natural and Technological Hazards Research

GPSR Compliance

The European Union's (EU) General Product Safety Regulation (GPSR) is a set of rules that requires consumer products to be safe and our obligations to ensure this.

If you have any concerns about our products, you can contact us on ProductSafety@springernature.com

In case Publisher is established outside the EU, the EU authorized representative is:

Springer Nature Customer Service Center GmbH
Europaplatz 3
69115 Heidelberg, Germany

Batch number: 10755284

Printed by Printforce, the Netherlands